Introduction to Quantitative Methods in Business

Introduction to Quantitative Methods in Business

With Applications Using Microsoft® Office Excel®

Bharat Kolluri
Michael J. Panik
Rao Singamsetti
Barney School of Business
University of Hartford

Published by John Wiley & Sons, Inc., Hoboken, New Jersey
Published simultaneously in Canada

Library of Congress Cataloging-in-Publication Data

Names: Kolluri, Bharat, author. | Panik, Michael J., author. | Singamsetti, Rao, author.
Title: Introduction to quantitative methods in business : with applications using Microsoft Office Excel / Bharat Kolluri, Michael J. Panik, Rao Singamsetti.
Description: Hoboken, New Jersey : John Wiley & Sons, Inc., [2017] | Includes index.
Identifiers: LCCN 2016017312| ISBN 9781119220978 (cloth) | ISBN 9781119220992 (epub)
Subjects: LCSH: Business mathematics. | Management--Mathematical models. | Microsoft Excel (Computer file)
Classification: LCC HF5691 .K71235 2017 | DDC 650.0285/54--dc23 LC record available at https://lccn.loc.gov/2016017312

Printed in the United States of America

10 9 8 7 6 5 4 3 2 1

To our Wives
Vijaya, Paula, and Andal

Table of Contents

Preface

A proper understanding of Economics as well as Business disciplines, such as Finance, Marketing, and Operations, requires a basic knowledge of linear and nonlinear relationships and, in general, the fundamental techniques of quantitative methods. The following techniques are some of the examples:

1. *Basic optimization:* Businesses are, for the most part, typically trying to maximize or minimize something. And that "something" is usually depicted by an equation that is or is not subject to side conditions or constraints.
2. *Basic statistical methods:* Acquiring an understanding of descriptive statistical tools for presenting and analyzing data is crucial for making sound business decisions under uncertainty.

Indeed, a student's ability to model a business situation and apply these quantitative techniques via Microsoft® Office Excel® and other computer software has to be an essential component of any modern business curriculum. In this book, we have incorporated Excel as a tool for problem solving. We have included specific instructions for the usage of Excel Add-ins in selected chapter examples and exercises. For the most part, there are not many texts that adequately deal with these techniques at a beginning or foundational level. To rectify this deficiency, we start first with principles – we develop a solid foundation in arithmetic operations, functions and graphs, and elementary differentiation (rates of change) and integration. Once this foundation is built, we then develop both linear and nonlinear models of business activity and their solutions. And, at all times, the meaning of the solution is fully discussed.

The intended audience includes students enrolled in undergraduate courses in business programs proper, or even in pre-MBA programs, who need a review of, or first exposure to, the basics of mathematics and statistics so that they can move on to the more advanced courses in their curriculum. In this regard, our objective is to close the skills gap that many students have in the quantitative area.

This textbook has many strengths:

- Getting students to think in terms of modeling business activity.
- An applications orientation.
- Fully classroom tested.
- Starts with basic mathematics and offers more advanced concepts once a solid grounding in fundamentals is attained.

- Students do not get "bogged down" in tedious calculations since the accompanying computer routines perform the "heavy lifting."
- A supplemental Web site contains data sets, lecture slides, sample exams, and quizzes.

In fact, the aforementioned (Excel-based) computer software enable students to perform an assortment of tasks such as graphing, formulas usage, solving equations, and data analysis. More specifically, these are

- copy and paste,
- filling a range: the Fill handle,
- entering a formula,
- copying formulas,
- absolute versus relative cell addresses,
- common error messages, and
- decisions and optimization (solving simultaneous equations and linear programming problems) using the 'Solver' Add-in program.

In this text, each quantitative technique is illustrated with many examples that are fully solved and that are accompanied by a variety of practice problems emphasizing business applications and designed to gauge the student's understanding of the concepts involved. Given the nature of the subject matter, the text is organized into six chapters, with the first three chapters dealing with the use of mathematical modeling and quantitative techniques in business. The next three chapters deal with business applications of descriptive statistics, probability, and probability distributions.

In terms of topics covered, this text is not at all encyclopedic and, with respect to its size, is quite manageable and should not intimidate or overwhelm students. Our objective is to simply *do a few things well* and to provide a semester's worth of material that can readily be taught at a reasonable and unhurried pace.

Chapter 1 reviews the basic mathematical concepts and methods with business applications. Chapter 2 deals with linear and nonlinear applications of quantitative techniques in business. Optimization techniques, both unconstrained and constrained, are presented in Chapter 3. Chapter 4 covers descriptive statistics, including measures of location and variability. Chapter 5 includes probability, probability distribution, rules of probability, and contingency table applications. Finally, Chapter 6 introduces both discrete and continuous random variables, expected values, and decision making with payoff tables, and presents general types of distributions such as the binomial distribution and its applications. Throughout the text, special emphasis is placed upon providing many examples that deal with real data. Abbreviated solutions to odd-numbered exercises are presented at the end of the textbook.

Why such an emphasis on "tools?" For those who follow employment issues, it has been estimated that the typical worker/professional will hold an average of seven jobs during his/her working career – and at this point in time, at least half of those jobs have not been even invented yet. The moral of this point is that students need, in terms of their skill set, to be flexible and nimble. Clearly, a firm grounding in analytical tools makes one highly trainable and adaptable. To build any modern "skills edifice," one needs a solid foundation.

But this is only part of the story. In the latest issue of *Economic Letter* (Vol. 9, No. 5, May 2014) published by the Dallas Federal Reserve Bank, it was mentioned that "the number of people performing low-skill, low-pay manual labor tasks has grown with the number undertaking high-skill, high-pay, non-routine principally problem-solving jobs." Middle-skill jobs have been lost in this U.S. labor market polarization or hollowing-out. It is the cognitive nonroutine jobs that are high-skilled jobs that require performing abstract tasks such as problem solving, intuition, and persuasion – jobs that typically require a college degree and some analytical expertise. Let us face it, the high-skill, high-pay jobs will not be landed by the innumerate.

Our sincere thanks go to our colleagues, Profs. Jim Peta and Frank Dellolacono at the University of Hartford, for their support and encouragement. A special note of thanks goes to our office coordinator, Alice Schoenrock, and to our graduate assistant, Lavanya Raghavan, for their timely typing of the various sections and revisions of the text. We are also grateful to our previous graduate assistant, Mustafa Atalay, for helping with the preparation of examples and solutions.

We also acknowledge the support and encouragement of Susanne Steitz-Filler, Senior Editor, Mathematics and Statistics; Allison McGinniss, Project Editor; and Sari Friedman, Editorial Program Coordinator, at John Wiley & Sons, Inc.

Bharat Kolluri
Michael Panik
Rao Singamsetti

About the Companion Website

This book is accompanied by a companion website:

www.wiley.com/go/Kolluri/QuantitativeMethods

The Instructor's website includes:

- Instructor's Solutions Manual
- Sample exams and quizzes
- Lecture slides
- Excel data sets
- Supplementary exercises
- Chapter summaries

The Student's website includes:

- Chapter summaries
- Excel data sets
- Supplementary exercises
- Sample exams and quizzes with questions only

Chapter 1

The Mathematical Toolbox

1.1 INTRODUCTION

This chapter is essentially about applicable analytical tools. We cover items such as linear equations, simultaneous linear equations (involving two equations in two unknowns), the summation operator, sets (and set operations), general functions and graphs, and a brief exposure to differentiation and integration.

Our experience has indicated that some students might benefit from a review of the basic mathematical concepts typically studied in high school mathematics classes or in courses that might be offered in the first year of college. This is why we have included an optional appendix at the end of this chapter for those students in need of a refresher in topics such as exponents, factoring, fractions, and decimals and percents, among others. This review provides the student with a firm grounding in the fundamental mathematical tools needed for problem solving. Indeed, these basic foundation skills in elementary mathematics are necessary for developing and effecting business applications. For instance, derivatives of functions are not calculated for their own sake. One usually has to simplify a derivative in order for it to be useful in, say, determining the maximum of a profit function or the minimum of a cost function. While many students can readily apply the rules of differentiation, they often lack the algebraic skills required to simplify a derivative. (This is why some students describe the calculus as "too difficult.") The appendix to this chapter attempts to overcome this difficulty.

As to the approach followed throughout this text, and within this chapter proper, a few key notions have motivated our thinking. First, we want students to think in terms of the *modeling* of business activity. To this end, operating in terms of *functions* is critical. That is, any decision maker must identify those variables (called *independent* or *decision variables*) that are under his/her control. Once such variables are recognized and manipulated, other variables (called *response variables*) typically react in a deterministic and predictable fashion.

For instance, suppose a manager wants to increase the sales of some consumer product. What, if anything, can be done to facilitate an increase in sales revenue?

Introduction to Quantitative Methods in Business: With Applications Using Microsoft® Office Excel®, First Edition. Bharat Kolluri, Michael J. Panik, and Rao Singamsetti.

Companion website: www.wiley.com/go/Kolluri/QuantitativeMethods

What sort of decision variable can a manager manipulate? How about increasing the organizations advertising budget? Generally, firms advertise to increase sales (and/or to maintain market share). Thus, advertising expenditure is the decision or control variable and sales level is the response variable.

What might a revenue response look like? Possibly, the relationship between advertising (X) and sales revenue (Y) is one of direct proportionality (e.g., a dollar increase in advertising expenditure might always precipitate, say, 3 dollars in sales revenue), or maybe the advertising response increases at an increasing rate. Obviously, we are describing, in the first case, a linear relationship between advertising and revenue, and, in the second instance, a nonlinear one. Each is an example of a *functional relationship* between these two variables, which is of the general form $Y = f(X)$. For the linear case, structurally,

$$Y = f(X) = b_0 + b_1X.$$

And for the nonlinear (quadratic) case,

$$Y = f(X) = b_0 + b_1X + b_2X^2.$$

To reiterate, decision makers must think in terms of functions—in terms of relationships between decision variables and response variables.

Second, one might benefit from thinking in terms of *sets* of items. Here the use of set operations enables us to "create order" by organizing information in a very systematic fashion. This process will prove particularly useful when we introduce *uncertainty* into the decision-making picture and work with probability calculations.

To consider an additional decision-making theme/approach, let us return to the nonlinear advertising expenditure versus sales revenue relationship posited earlier. Specifically, how does one know if, at a particular advertising level, sales revenue is actually increasing at an increasing rate? To answer this question, we can employ a very useful and versatile analytical device, namely the requisite derivatives of a function at a given point—but more on this calculus approach later on.

1.2 LINEAR FUNCTIONS

An equation like $Y = b_0 + b_1X$ is known as a *linear function*. Here, b_1 is called the *slope* of the linear function; it measures the change in the value of the *dependent variable* Y as a result of a one unit change in the value of the *independent variable* X. It is also obtained as the ratio of the change in Y to the change in X. That is, $b_1 = \frac{\Delta Y}{\Delta X}$. b_1 is also called the coefficient of X. b_0 is known as the *Y-intercept*. It measures the value of Y when $X = 0$. Both b_0 and b_1 are constants.

Note the following:

- A horizontal line has zero slope.
- A vertical line has no slope or its slope is undefined.
- A line rising from left to right has positive slope.
- A line falling from left to right has negative slope.

1.3 SOLVING A SIMPLE LINEAR EQUATION FOR ONE UNKNOWN VARIABLE

Finding the particular value of the unknown variable in a linear equation is a useful algebraic exercise. When solving for the unknown variable in a single linear equation, all other values are assumed to be known. For example, let us solve the equation $6X + 42 = 0$. If we subtract 42 from both sides, the equation will be $6X = -42$. Dividing both sides by 6, X will be-7. It is a useful practice to substitute the answer into the original equation so as to check the solution. In our case, $6(-7) + 42 = 0$ is correct, so $X = -7$ is the solution. Let us say we have the equation, $4abX - 24 = 4$. To solve for X, all other values are assumed to be known. So, $4abX = 28$ and thus $X = 28/4ab = 7/ab$.

EXAMPLE 1.1

a. Solve the following equation:

$$4X - 24 = 0.$$

SOLUTION: Rewrite the given equation as

$$\begin{aligned} 4X &= 24, \\ X &= \frac{24}{4} = 6. \end{aligned}$$

b. Solve the following equation:

$$17X + 60 = 9.$$

SOLUTION: Rewrite the given equation as

$$\begin{aligned} 17X &= 9 - 60, \\ 17X &= -51, \\ X &= -\frac{51}{17} = -3. \end{aligned}$$

c. Solve the following equation:

$$\frac{1}{2}X - 10 = 2.$$

SOLUTION: Rewrite the given equation as

$$\frac{1}{2}X = 10 + 2 = 12,$$

$$X = 24.$$

d. Solve the following equation:

$$\frac{X}{2} + \frac{3}{4} = 5.$$

SOLUTION: Rewrite the given equation as

$$\frac{X}{2} = 5 - \frac{3}{4} = \frac{20-3}{4} = \frac{17}{4},$$

$$X = \frac{17}{4} \times 2 = \frac{17}{2} = 8.5.$$

e. Solve the following equation:

$$0.2X + 10.4 = 6.2.$$

SOLUTION: Rewrite the given equation as

$$0.2X = -10.4 + 6.2 = -4.2,$$

$$X = \frac{-4.2}{0.2} = \frac{-42}{2} = -21.$$

■

1.3.1 Solving Two Simultaneous Linear Equations for Two Unknown Variables

Solving simultaneous linear equation sets involves finding the values of two unknown variables that satisfy each equation at the same time. Graphically, the solution values of both the unknown variables indicate the point of intersection of two lines. This is useful in break-even analysis, market equilibrium analysis, economic order quantity (EOQ) calculations, and finding corner points in linear programming applications. For example, let us solve the following set of equations for X and Y:

$$\begin{aligned} 2X + 3Y &= 13, \\ X - 2Y &= 17. \end{aligned}$$

Although there is more than one method of solving for two unknowns X and Y, we adopt the *method of elimination* as follows:

Denote the equation $2X + 3Y = 13$ as (1) and $X - 2Y = 17$ as (2).

The first step in this procedure is to make the coefficients on one of the variables, say, X, in both equations the same by appropriate arithmetic operations. Thus, multiplying equation (2) by 2 we get $2X - 4Y = 34$. Call this equation (2′). Subtract equation (2′) from equation (1), that is,

$$2X + 3Y = 13 \quad (1)$$

$$\underline{2X - 4Y = 34} \quad (2')$$

$$0 + 7Y = -21 \quad (3)$$

Then, solving equation (3) renders $Y = (-21)/7 = -3$. We have “eliminated” X.

The second step is to substitute $Y = -3$ into equation (1).
Thus, $2X + 3(-3) = 13$, and so

$$2X - 9 = 13,$$
$$2X = 13 + 9 = 22,$$

and thus $X = 22/2 = 11$. Thus, the simultaneous solution is $X = 11$ and $Y = -3$. Equation systems that have at least one solution are termed *consistent*.

A simple *test for consistency* is the following: For the system

$$aX + bY = e,$$
$$cX + dY = f,$$

if $ad - cb \neq 0$, then this equation system is consistent.

EXAMPLE 1.2

a. Solve the following equation set:

$$X_1 + X_2 = 0,$$
$$2X_1 + (1/3)X_2 = 5.$$

SOLUTION: Designate the first equation as (1) and the second equation as (2). Multiply equation (2) by 3 to make the coefficient on X_2 the same in both equations. Then subtract equation (1) from this new equation. Thus,

3 × equation (2)	$6X_1 + X_2 = 15$
subtract equation (1)	$X_1 + X_2 = 0$
	$5X_1 + 0 = 15$

(Here variable X_2 has been eliminated.) Solving this resulting equation for X_1 we get

$$X_1 = 15/5 = 3.$$

Then substituting $X_1 = 3$ into equation (1) yields $3 + X_2 = 0$ or $X_2 = -3$.

b. Solve the following equation set:

$$0.4P + 0.2Q = 7,$$
$$7P - 0.2Q = -2.$$

SOLUTION: Designating the first equation as (1) and the second equation as (2) and adding the two equations, we get

$7.4P = 5$. Solving this single equation in one unknown gives

$$P = \frac{5}{7.4} = 0.6757.$$

Substituting P = 0.6757 into equation (1) we get

$$0.4(0.6757) + 0.2Q = 7,$$

$$0.2703 + 0.2Q = 7,$$

$$0.2Q - 7 - 0.2703 = 6.7297,$$

and thus $Q = \frac{6.7297}{0.2} = 33.6485$.

Thus, the simultaneous solution is P = 0.6757 and Q = 33.6485.

c. Solve the following equation set:

$$3X_1 - 6X_2 = 1. \quad (1)$$

$$6X_1 - 12X_2 = 2. \quad (2)$$

Multiplying equation (1) by 2 and subtracting equation (2) from 2 × equation (1) yields "0 = 0." What happened? This example reveals that an equation system may not possess a simultaneous solution? Here, the two equations are parallel lines—they do not intersect. Such equations are known as *dependent* equations. (Note that via the consistency test, 3 (−12) − 6 (−6) = 0.)

In sum, to solve two simultaneous linear equations in two unknowns:

Step 1: Perform the above test for *consistency*. If the system is *consistent* or *independent*, go to step 2.

Step 2: Use the method of elimination to solve for the unknowns. ■

1.4 SUMMATION NOTATION

Since the operation of addition occurs frequently in statistics, the special Greek symbol, $\sum$(pronounced "sigma"), is used to denote a sum. For example, if we have a set of n values for a variable X, the expression, $\sum_{i=1}^{n} X_i$ means that these n values, running from i = 1, 2, 3, ...,n, are to be added together. Thus,

$$\sum_{i=1}^{n} X_i = X_1 + X_2 + X_3 + \cdots + X_n.$$

Mathematically, "Σ" is an *operator*—It operates only on those terms with an i index. And the operation itself is addition.

The use of summation notation is illustrated in the following examples.

EXAMPLE 1.3

Assume that we have five values of a variable X, as given below:

$$X_1 = 12,$$

$$X_2 = 0,$$

$$X_3 = -1,$$

$$X_4 = -5,$$

$$X_5 = 3.$$

Then, $\sum_{i=1}^{5} X_i = X_1 + X_2 + X_3 + X_4 + X_5 = 12 + 0 + (-1) + (-5) + 3 = 9.$

In statistics, we frequently deal with summing the squared values of a variable. Thus, in our example,

$$\sum_{i=1}^{5} X_i^2 = X_1^2 + X_2^2 + X_3^2 + X_4^2 + X_5^2 = 12^2 + 0^2 + (-1)^2 + (-5)^2 + 3^2$$
$$= 144 + 0 + 1 + 25 + 9 = 179.$$

We should realize that $\sum_{i=1}^{n} X_i^2$, the *summation of the squares*, is not the same as $\left(\sum_{i=1}^{n} X_i\right)^2$, the *square of the sum*, that is,

$$\sum_{i=1}^{n} X_i^2 \neq \left(\sum_{i=1}^{n} X_i\right)^2.$$

In our example, the summation of squares is equal to 179. This is not equal to the square of the sum, which is $9^2 = 81$.

Another frequently used operation involves the summation of the product of sets of values. That is, suppose that we have two variables, X and Y, each having n observations. Then,

$$\sum_{i=1}^{n} X_i Y_i = X_1 Y_1 + X_2 Y_2 + X_3 Y_3 + \cdots + X_n Y_n.$$

■

EXAMPLE 1.4

Continuing with our previous example involving five values, suppose that there is also a second variable Y whose five values are

$$Y_1 = 1,$$
$$Y_2 = 3,$$
$$Y_3 = -2,$$
$$Y_4 = 4,$$
$$Y_5 = 6.$$

Then, $\sum_{i=1}^{5} X_i Y_i = X_1 Y_1 + X_2 Y_2 + X_3 Y_3 + X_4 Y_4 + X_5 Y_5 = 12(1) + (0)(3) + (-1)(-2)$

$$+(-5)(4) + 3(6) = 12 + 0 + 2 - 20 + 18 = 12.$$

In computing $\sum_{i=1}^{n} X_i Y_i$, we must realize that the first value of X is multiplied by the first value of Y, the second value of X is multiplied by second value of Y, and so on. These cross-products are then summed in order to obtain the desired result. However, we should note here that the summation of cross-products is not equal to the product of the individual sums, that is,

$$\sum_{i=1}^{n} X_i Y_i \neq \left(\sum_{i=1}^{n} X_i\right)\left(\sum_{i=1}^{n} Y_i\right).$$

Table 1.1a Summation Values

Observation	X_i	X_i^2	Y_i	Y_i^2	X_iY_i
1	12	144	1	1	12
2	0	0	3	9	0
3	−1	1	−2	4	2
4	−5	25	4	16	−20
5	3	9	6	36	18
Total	$\sum_{i=1}^{5} X_i = 9$	$\sum_{i=1}^{5} X_i^2 = 179$	$\sum_{i=1}^{5} Y_i = 12$	$\sum_{i=1}^{5} Y_i^2 = 66$	$\sum_{i=1}^{5} X_iY_i = 12$

In our example, $\sum_{i=1}^{5} X_i = 9$ and $\sum_{i=1}^{5} Y_i = 1 + 3 + (-2) + 4 + 6 = 12$, so that

$$\left(\sum_{i=1}^{5} X_i\right)\left(\sum_{i=1}^{5} Y_i\right) = 9 \times 12 = 108.$$

This is not the same as $\sum_{i=1}^{5} X_iY_i$, which equals 12.

Before studying the four basic rules of performing operations with summation notation, it would be helpful to present the values for each of the five observations of X and Y in a tabular format (see Table 1.1a).

Rule 1: The sum of the set of sums of two variables is equal to the sum of their individual sums. That is,

$$\sum_{i=1}^{n} (X_i + Y_i) = \left(\sum_{i=1}^{n} X_i\right) + \left(\sum_{i=1}^{n} Y_i\right).$$

Thus, in our example,

$$\sum_{i=1}^{5} (X_i + Y_i) = (12 + 1) + (0 + 3) + (-1 + (-2)) + (-5 + 4) + (3 + 6) = 21,$$

$$\text{which is equal to} \left(\sum_{i=1}^{n} X_i\right) + \left(\sum_{i=1}^{n} Y_i\right) = 9 + 12 = 21.$$

Rule 2: The sum of the set of differences between two variables is equal to the difference between their individual sums, or

$$\sum_{i=1}^{n} (X_i - Y_i) = \left(\sum_{i=1}^{n} X_i\right) - \left(\sum_{i=1}^{n} Y_i\right).$$

Thus, in our example,

$$\sum_{i=1}^{5} (X_i - Y_i) = (12 - 1) + (0 - 3) + (-1 - (-2)) + (-5 - 4) + (3 - 6) = -3,$$

$$\text{which is equal to} \left(\sum_{i=1}^{n} X_i\right) - \left(\sum_{i=1}^{n} Y_i\right) = 9 - 12 = -3.$$

Rule 3: The sum of a constant times a variable equals that constant times the sum.

$$\sum_{i=1}^{n} cX_i = c\sum_{i=1}^{n} X_i, \quad \text{where c is the constant.}$$

Thus, in our example, if c = 3,

$$\sum_{i=1}^{5} cX_i = 3(12) + 3(0) + 3(-1) + 3(-5) + 3(3) = 36 + 0 - 3 - 15 + 9 = 27,$$

$$\text{which is equal to } 3\left(\sum_{i=1}^{5} X_i\right) = 3(9) = 27.$$

Rule 4: The sum of a constant taken n times equals n times the constant, that is,

$$\sum_{i=1}^{n} c = nc, \quad \text{where c is the constant.}$$

Thus, if the constant c = 3 is summed five times, we would have

$$\sum_{i=1}^{5} c = 3 + 3 + 3 + 3 + 3 = 5(3) = 15.$$

To illustrate how these summation rules are used, we may demonstrate one of the mathematical properties pertaining to the average or arithmetic mean of a sample of X values. To this end, if the sample mean is $\overline{X} = \sum_{i=1}^{n} X_i/n$, then

$$\sum_{i=1}^{n} (X_i - \overline{X}) = 0.$$

This property states that the summation of the differences between each observation and the arithmetic mean is zero.

This can be proven mathematically by the following steps:

Step 1: As just indicated, the arithmetic mean of a sample of X values can be defined as

$$\overline{X} = \left(\sum_{i=1}^{n} X_i\right)/n.$$

Using summation rule 2, we have

$$\sum_{i=1}^{n} (X_i - \overline{X}) = \sum_{i=1}^{n} X_i - \sum_{i=1}^{n} \overline{X}.$$

Step 2: Since, for any fixed set of data, $\overline{X}$ can be considered a constant, then, and from summation rule 4, we have

$$\sum_{i=1}^{n} \overline{X} = n\overline{X}.$$

$$\text{Therefore, } \sum_{i=1}^{n} (X_i - \overline{X}) = \sum_{i=1}^{n} X_i - n\overline{X}.$$

Step 3: However, from step 1, since

$$\overline{X} = \left(\sum_{i=1}^{n} X_i\right)/n \quad \text{and} \quad n\overline{X} = \sum_{i=1}^{n} X_i,$$

we consequently obtain

$$\sum_{i=1}^{n} (X_i - \overline{X}) = \sum_{i=1}^{n} X_i - \sum_{i=1}^{n} X_i = 0.$$

We have thus shown that $\sum_{i=1}^{n}(X_i - \overline{X}) = 0$, and this is true for any sample data set. Why? Because $\overline{X}$ is the "center of gravity" of the X values—It is where the X values would "balance."

■

EXAMPLE 1.5

Using the data in Table 1.1a, namely, $X_1 = 12$, $X_2 = 0$, $X_3 = -1$, $X_4 = -5$, $X_5 = 3$ and $Y_1 = 1$, $Y_2 = 3$, $Y_3 = -2$, $Y_4 = 4$, $Y_5 = 6$,

(a) Evaluate the Pearson correlation coefficient

$$r = \sum_{i=1}^{5} \frac{(X_i - \overline{X})(Y_i - \overline{Y})}{\sqrt{\sum_{i=1}^{5}(X_i - \overline{X})^2 \sum_{i=1}^{5}(Y_i - \overline{Y})^2}},$$

where $\overline{X} = \frac{\sum X_i}{5}$ and $\overline{Y} = \frac{\sum Y_i}{5}$.

(b) Also, reevaluate the correlation coefficient using the following "short formula":

$$r = \frac{\sum X_i Y_i - \frac{\sum X_i \sum Y_i}{n}}{\left(\left(\sum X_i^2 - \frac{(\sum X_i)^2}{n}\right)\left(\sum Y_i^2 - \frac{(\sum Y_i)^2}{n}\right)\right)^{1/2}}.$$

SOLUTION:

(a)

$$r = \sum_{i=1}^{5} \frac{(X_i - \overline{X})(Y_i - \overline{Y})}{\sqrt{\sum_{i=1}^{5}(X_i - \overline{X})^2 \sum_{i=1}^{5}(Y_i - \overline{Y})^2}}$$

$$= \frac{-9.6}{[(162.8)(37.2)]^{1/2}}$$

$$= -0.1104 = -0.1104.$$

From the short formula,

(b)

$$r = \frac{12 - \frac{(9)(12)}{5}}{\left[\left(179 - \frac{9^2}{5}\right)\left(66 - \frac{12^2}{5}\right)\right]^{1/2}}$$

$$= \frac{-9.6}{[(162.8)(37.2)]^{1/2}}$$

$$= -0.1104.$$

■

EXAMPLE 1.6

Using the data and the results in Example 1.5, compute $\frac{r\times\sqrt{3}}{\sqrt{1-r^2}}$.

SOLUTION:

$$\begin{aligned}\frac{r\times\sqrt{3}}{\sqrt{1-r^2}} &= \frac{-0.1104\times 1.4422}{\sqrt{1-(-0.1104)^2}}\\ &= \frac{-0.1592}{\sqrt{1-0.0122}}\\ &= \frac{-0.1592}{\sqrt{0.9878}}\\ &= -0.1602.\end{aligned}$$

■

EXAMPLE 1.7

Using the data and results in Example 1.5:

(a) Compute the slope value, $b=\frac{\sum_{i=1}^{5}(X_i-\overline{X})(Y_i-\overline{Y})}{\sum_{i=1}^{5}(X_i-\overline{X})^2}$, and the intercept value, $a=\overline{Y}-b\overline{X}$, of the linear equation $Y=a+bX$. Using the resulting equation, predict the value of Y when $X=14$.

(b) Also, reevaluate the slope value b using the following "short formula":

$$b=\frac{\sum X_iY_i-\frac{\sum X_iY_i}{n}}{\left(\sum X_i^2-\frac{\left(\sum X_i\right)^2}{n}\right)}.$$

SOLUTION:

(a) Referring to Table 1.1b,

$$\begin{aligned}b &= \frac{\sum_{i=1}^{5}(X_i-\overline{X})(Y_i-\overline{Y})}{\sum_{i=1}^{5}(X_i-\overline{X})^2}\\ &= \frac{-9.6}{162.8}\\ &= -0.0589.\end{aligned}$$

$$\begin{aligned}a &= \overline{Y}-b\overline{X}\\ &= 2.4-(-0.0589)(1.8)\\ &= 2.5060.\end{aligned}$$

$$\begin{aligned}Y &= a+bX\\ &= 2.5060+(-0.0589)(14)\\ &= 1.6814.\end{aligned}$$

Table 1.1b Summation Values

X_i	$X_i - \overline{X}$	$(X_i - \overline{X})^2$	Y_i	$Y_i - \overline{Y}$	$(Y_i - \overline{Y})^2$	$(X_i - \overline{X})(Y_i - \overline{Y})$
12	10.2	104.04	1	−1.4	1.96	−14.28
0	−1.8	3.24	3	0.6	0.36	−1.08
−1	−2.8	7.84	−2	−4.4	19.36	12.32
−5	−6.8	46.24	4	1.6	2.56	−10.88
3	1.2	1.44	6	3.6	12.96	4.32
$\sum_{i=1}^{5} X_i = 9$	0	$\sum (X_i - \overline{X})^2 = 162.8$	$\sum_{i=1}^{5} Y_i = 12$		$\sum (Y_i - \overline{Y})^2 = 37.2$	$\sum (X_i - \overline{X})(Y_i - \overline{Y}) = -9.6$

Table 1.2 Summation Values

X_i	X_i^2	Y_i	Y_i^2	$X_i Y_i$
12	144	1	1	12
0	0	3	9	0
−1	1	−2	4	2
−5	25	4	16	−20
3	9	6	36	18
$\sum_{i=1}^{5} X_i = 9$	$\sum_{i=1}^{5} X_i^2 = 179$	$\sum_{i=1}^{5} Y_i = 12$	$\sum_{i=1}^{5} Y_i^2 = 66$	$\sum_{i=1}^{5} X_i Y_i = 12$

(b) Referring to Table 1.2,

$$b = \frac{\sum X_i Y_i - \frac{\sum X_i \sum Y_i}{n}}{\left[\sum X_i^2 - \frac{(\sum X_i)^2}{n}\right]}$$

$$= \frac{12 - \frac{(9)(12)}{5}}{179 - \frac{9^2}{5}}$$

$$= \frac{-9.6}{162.8}$$

$$= -0.0589.$$

■

1.5 SETS

A *set* is a collection or grouping of items (called *elements*) without regard to structure or order. The elements may be people, sheep, desks, cars, files in a cabinet, or even numbers. We may define our sets as the collection of all sheep in a given pasture, all people in a room, all cars in a given parking lot at a given time, all values between 0 and 1, or all integers. The number of elements in a set may be infinite. For example, the numbers in the set of integers are called *countably infinite*, while the numbers between any two points on the real line, such as 0 and 1, are *uncountably infinite*.

A type of notation used to define a set is the *roster method*. In the roster method, we simply list the names of the members separated by commas. Then we enclose the names with braces. The names can be written in any order. For example, if we want to construct the set A consisting of the first five nonnegative integers using the roster method, we simply write

$$A = \{0,1,2,3,4\}.$$

For some infinite (unending) sets, we can write names (or numbers) for some of the members and then write dots to indicate the rest of the numbers. For example, if we want to write a set B of nonnegative even integers using the roster method, we form

$$B = \{0,2,4,6,8,\ldots\}.$$

When we solve an equation, we look for the set of all solutions. We call such a set the *solution set*. For example, the solution set of the equation $2x + 7 = 15$ is $\{4\}$. In some contexts, one might just list the solution as 4. Similarly, the solutions of $x(x - 2) = 0$ are 0 and 2. One might use the roster method and say that the solution set is $\{0, 2\}$. The symbol "$\in$" is used to denote membership in a set or *element inclusion*. For example, the expression 3 $\in$ "the set of natural numbers" is true because 3 is a natural number. But the expression, Hence $2/3 \notin \{0, 3, 7, 9\}$ is false, because $2/3$ is not a member of the given set.

1.5.1 Subset, Empty Set, Universal Set, and Complement of A Set

When all the members of one set are also members of another, we say that the first set is a *subset* of the second. For example, $A \subset B$ means that "A is a subset of B." The symbol $\subset$ is used to indicate *set inclusion*. It can also be written as $B \supset A$, which again means that "A is a subset of B."

The set without any members is called the *empty set*, and is denoted by "$\varnothing$" (also called the *null set*). The empty set is a subset of every set.

A *universal set* is the set containing everything in a given context. We denote the universal set by U. For example, all integers comprise a universal set. All registered voters in a given election form a universal set. As another example, the set of all whole numbers, both positive and negative, form a universal set.

Given a set A, we may define its *complement* as the set containing all the elements in the universal set U that are not members of set A (Figure 1.1a). We denote the complement of A by $\overline{A}$ or $\tilde{A}$, often called "not A." With regard to die tossing, all odd integers form the complement of the set of all even integers. In a deck of playing cards, the suit "spades" forms the complement of the three remaining suits (hearts, diamonds, and clubs).

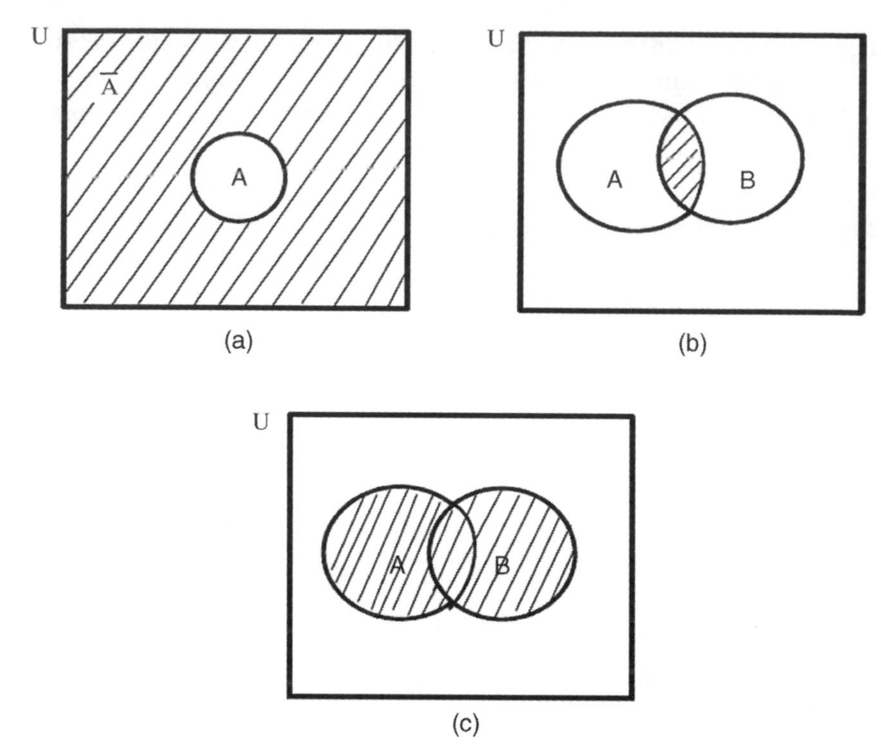

Figure 1.1 Set operations. (a) $\overline{A}$is the complement of A. (b) Intersection of A and B. (c) Union of A and B.

EXAMPLE 1.8

Determine whether the following is true or false:

$$\{1,2,3,4\} \subset \{0,1,2,3,4,5\}.$$

This is true because every member of the first set is also a member of the second. ■

EXAMPLE 1.9

List all the subsets of $\{1,2,3\}$.

$\{1,2,3\}$	Every set is a subset of itself.
$\{1,2\},\{1,3\},\{2,3\}$	There are three subsets with two members each.
$\{1\},\{2\},\{3\}$	There are three subsets with one member each.
$\varnothing$	The empty set is a subset of any set.

■

1.5.2 Intersection and Union

The *intersection* of two sets A and B is the set of all members that are common to A and B or reside in both A and B (Figure 1.1b). We denote the intersection of sets A and B as A∩B, where "∩" stands for "and." For example, the intersection of the two sets A = {1,2,3,4,5} and B = {−2, − 1,0,1,2,3} is

$$A \cap B = \{1,2,3,4,5\} \cap \{-2, -1,0,1,2,3\} = \{1,2,3\}.$$

The numbers 1, 2, and 3 are common to both sets, so the intersection is {1, 2, 3}.

The *union* of two sets A and B is formed by pooling their elements and is defined as those that are in A, in B, or in both A and B (Figure 1.1c). We denote the union of sets A and B as A∪B, where the symbol "∪" stands for "or"/"both." For example, the union of the two sets A = {1, 2, 3} and B = {3, 4, 5} is

$$A \cup B = \{1,2,3\} \cup \{3,4,5\} = \{1,2,3,4,5\}.$$

The numbers in either or both sets are 1,2,3,4, and 5, so the union is {1,2,3,4,5}.

Finally, four specialized sets that will be useful in subsequent sections are as follows:

Open interval from a to b

$$(a,b) = \{a < X < b\}.$$

Closed interval from a to b

$$[a,b] = \{a \le X \le b\}.$$

Left open, right closed from a to b

$$(a,b] = \{a < X \le b\}.$$

Left closed, right open from a to b

$$[a,b) = \{a \le X < b\}.$$

1.6 FUNCTIONS AND GRAPHS

Given variables x and y, let f be a "rule" or "law of correspondence" that associates with each value of x a unique value of y. Then, y is said to be a *function* of x and written $y = f(x)$. Here, y is termed the *image* of x under rule f. The expression f(x) does not mean "f times x." Instead, f(x) is the symbol for the indicated correspondence.

For the function $y = f(x)$, the set of x values is called the *domain of the function*, and the set corresponding to the y values, which are the image of at least one x, is known as the *range of the function*. Here x and y are, respectively, known as the *independent* and *dependent variables* in the context of the equation $y = f(x)$.

For the function $y = 3 + 2x$, what is the "rule" that tells us how to get a y from x? For each x value, there is one and only one y value. However, a y value may be the image of more than one x. For example, let $y = x^2$ and let $x = \pm 2$ (Figure 1.2). But if $x_1 \neq x_2$ implies $f(x_1) \neq f(x_2)$, then f is said to be *one-to-one*.

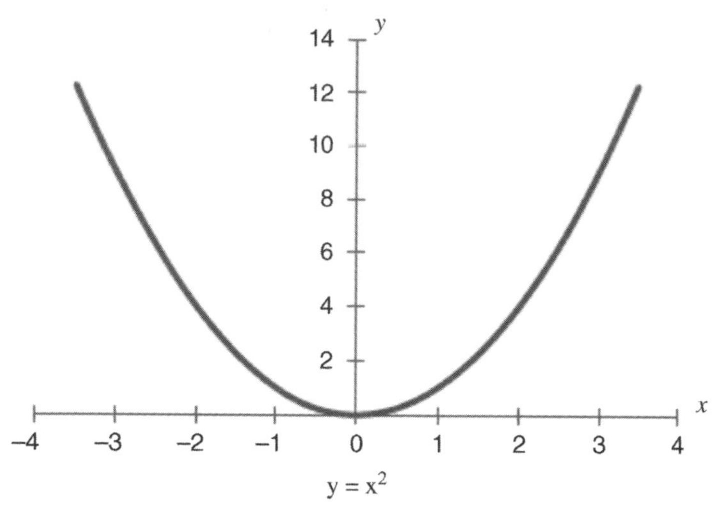

Figure 1.2 Graph of $y = x^2$.

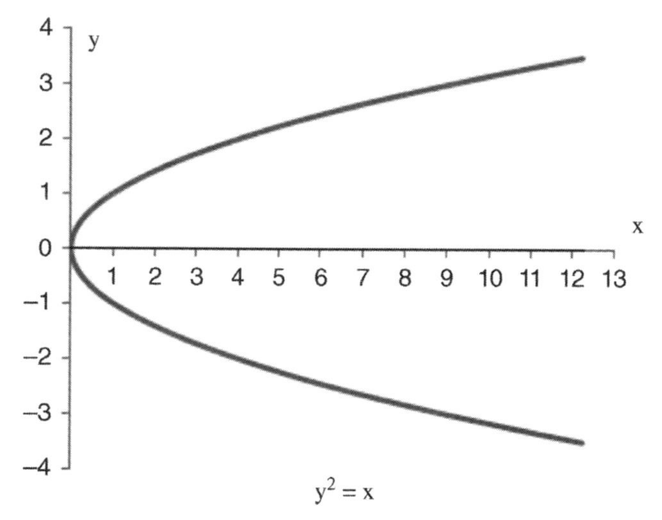

Figure 1.3 Graph of $y = \pm\sqrt{x}$.

It should be evident that $y^2 = x$ is not a function. This is because, if we solve the equation for y, we obtain $y = \pm\sqrt{x}$. For a given value of x, y can be both positive and negative. So, there is not one and only one y value for a given x. The graph is shown in Figure 1.3.

1.6.1 Vertical Line Test

To determine whether or not a particular graph represents a function, we can use the *vertical line test*. For a given value of x, find y by drawing a vertical line at x. If the vertical line cuts the graph at more than one value of y, then the graph does not represent a function.

For instance, if in Figure 1.2 we draw a vertical line at $x = 2$, the said line cuts the graph at $y = 4$ only. Thus, $y = x^2$ is a function. In a similar way, we can easily find that in Figure 1.3, $y^2 = x$ is not a function.

Why are functions important? What is their place in decision making? As we shall see later on, when we undertake the process of model building and we deal with a concept such as "cost," this notion will be represented by a "cost function"; and if we desire to maximize "profit," then to do so we shall employ a "profit function," and so on. Hence, a function is an analytical device useful for representing an abstract concept.

EXAMPLE 1.10

Show that

$x^2 + y^2 = k^2$, where k, a constant, is not a function (it actually represents a circle). From $y^2 = k^2 - x^2$, we obtain

$y = \pm\sqrt{(k^2 - x^2)}$. This shows that for a given value of x, there is more than one value of y. ■

1.7 WORKING WITH FUNCTIONS

Evaluating functions given by a formula can involve algebraic simplification, as the following examples show. Similarly, solving for the input or independent variable involves solving an equation algebraically.

1.7.1 Evaluating Functions

A formula like $f(x) = \left(x^2 + 2/6 + x\right)$ is a rule that tells us what the function f does with its input value x. In the formula, the letter x is a "placeholder" for the input value. Thus, to evaluate f(x), we replace each occurrence of x in the formula with the value of the input.

EXAMPLE 1.11

Let $g(x) = \frac{2x^2+3}{4+x}$. Evaluate the following expressions. Some of your answers will contain c, a constant.

a. $g(3)$

b. $g(-3)$

c. $g(c)$

d. $g(c - 2)$

e. $g(c) - 4$

f. $g(c) - g(3)$

SOLUTIONS:

a. To evaluate g(3), replace every x in the formula with 3:

$$g(3) = \frac{2(3^2)+3}{4+3} = \frac{21}{7} = 3.$$

b. To evaluate g(−3), replace every x in the formula with (−3):

$$g(-3) = \frac{2(-3)^2+3}{4-3} = 21.$$

c. To evaluate g(c), replace every x in the formula with c:

$$g(c) = \frac{2c^2+3}{4+c}.$$

d. To evaluate g(c − 2), replace every x in the formula with c − 2:

$$g(c-2) = \frac{2(c-2)^2+3}{4+(c-2)} = \frac{2(c^2-4c+4)+3}{2+c} = \frac{2c^2-8c+11}{2+c}.$$

e. To evaluate g(c) −4, first evaluate g(c) and then subtract 4:

$$g(c)-4 = \frac{2c^2+3}{4+c} - 4 = \frac{2c^2+3}{4+c} - 4\left(\frac{4+c}{4+c}\right) = \frac{2c^2+3}{4+c} - \left(\frac{16+4c}{4+c}\right)$$
$$= \frac{2c^2+3-16-4c}{4+c} = \frac{2c^2-4c-13}{4+c}.$$

f. To evaluate g(c) −g(3), subtract g(3) from g(c):

$$g(3) = 3 \quad \text{from part (a).}$$

$$\text{From part c,} \quad g(c) = \frac{2c^2+3}{4+c}.$$

Thus,

$$g(c)-g(3) = \frac{2c^2+3}{4+c} - 3 = \frac{(2c^2+3)-3(4+c)}{(4+c)} = \frac{2c^2-3c-9}{(4+c)}.$$

■

1.7.2 Graphing Functions

For the sake of clarity, a function can also be described in graphical form. Generally, the independent variable is shown on the x-axis (horizontal axis), the dependent variable is shown on they-axis (vertical axis), and the points are plotted and connected. Such a diagram is called a *graph*. Graphs may include lines and curves that consist of sets of points. For each x value, there is a value for f(x), and each ordered pair (x,f(x)) constitutes a point on the function's graph. To draw a linear function, the easiest way to proceed is to find the points where the line cuts the x- and y-axes. To find those values, we have to solve the function for x when y = 0 and for y when

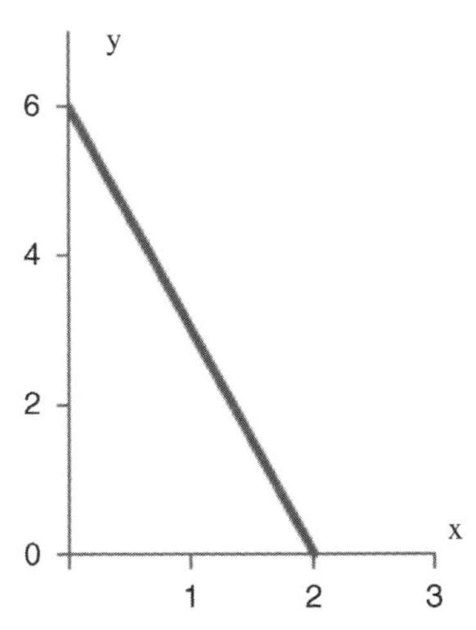

Figure 1.4 Graph of the straight line $3x + y = 6$ using *basic points*.

$x = 0$. As those values are, respectively, *horizontal* and *vertical intercepts*, this method is called the *intercept method*. For example, to draw the function $3x + y = 6$, we proceed as follows:

$$\text{If } x = 0, \text{ then } y = 6 \text{ and so } 6 \text{ is the y intercept.}$$
$$\text{If } y = 0, \text{ then } 3x = 6 \text{ and so } 2 \text{ is the x intercept.}$$

The straight-line graph is obtained by joining the two points $(0, 6)$ and $(2, 0)$ (called *basic points*). See Figure 1.4.

***Note*:** Two points are enough to draw a unique straight line.

Another way to graph a function is to connect more than two points. This method is called the *point method* and can be used for both linear and nonlinear functions. The resulting graph is smoother and more accurate when dealing with nonlinear functions. To illustrate the point method, let us draw the same function, $3x + y = 6$.

First, find y values for different x values (Table 1.3). Here the x values are selected arbitrarily around the origin, and selecting such x values depends on the nature of the given function.

Next, plot the points and connect them as shown in Figure 1.5.

Table 1.3 Values of y for Selected x Values

x	y
2	0
1	3
0	6
−1	9
−2	12

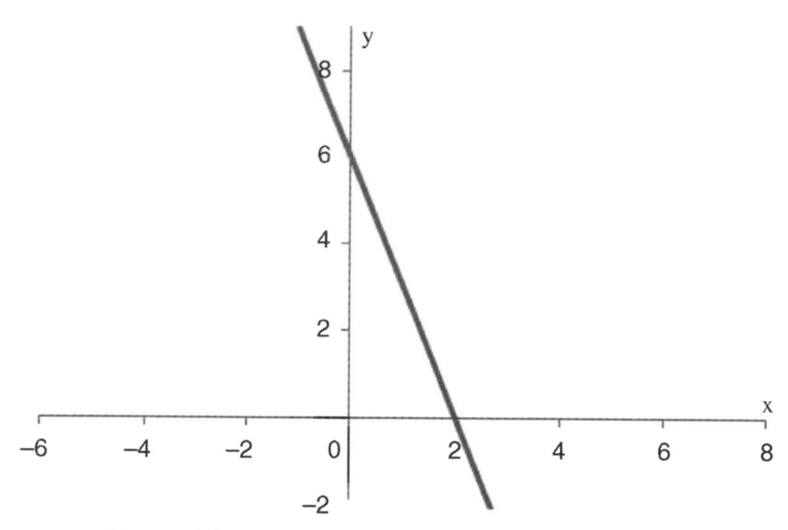

Figure 1.5 Graphing $3x + y = 6$ using the *point method.*

EXAMPLE 1.12

Graph $y = |x|$ (the absolute value of x). Here, $|x| = x$ when $x > 0$; $|x| = -x$ when $x < 0$. *Note*: $|7| = 7$; $|-7| = 7$.

First, select a set of x values and then construct Table 1.4.

Then, draw the graph corresponding to the points in this table (Figure 1.6).

Table 1.4 x and y (= |x|) Values

x	y
3	3
2	2
1	1
0	0
−1	1
−2	2
−3	3

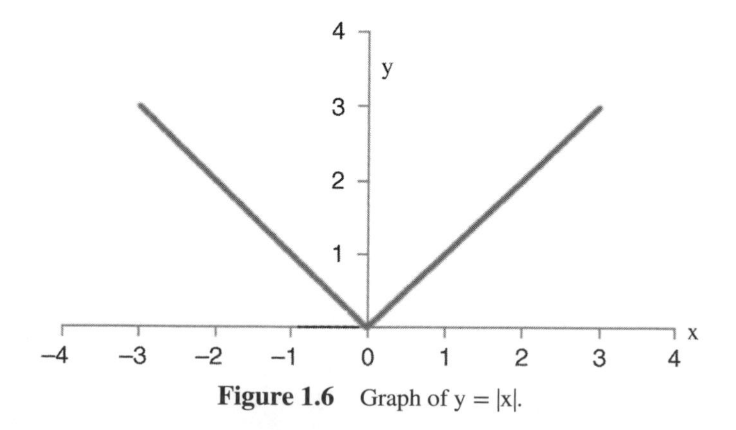

Figure 1.6 Graph of $y = |x|$.

■

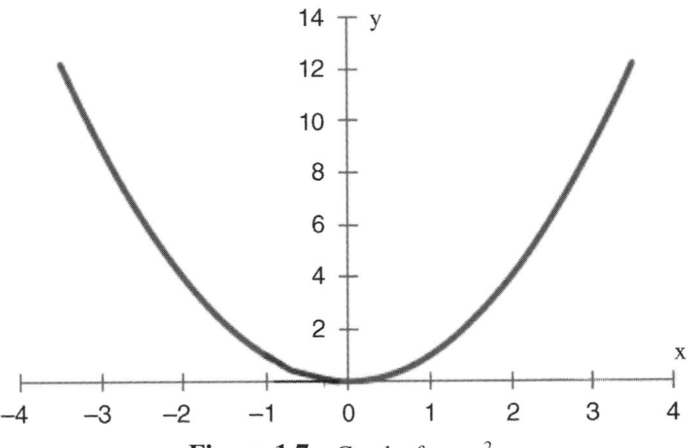

Figure 1.7 Graph of $y = x^2$.

Table 1.5 x and y ($=x^2$) Values

x	y
3	9
2	4
1	1
0	0
−1	1
−2	4
−3	9

EXAMPLE 1.13

Graph $y = x^2$ (Figure 1.7, Table 1.5). ■

EXAMPLE 1.14

Graph the exponential function $y = 2^x$ (Figure 1.8, Table 1.6). ■

Table 1.6 x and y ($= 2^x$) Values

x	y
3	8
2	4
1	2
0	1
−1	1/2
−2	1/4
−3	1/8

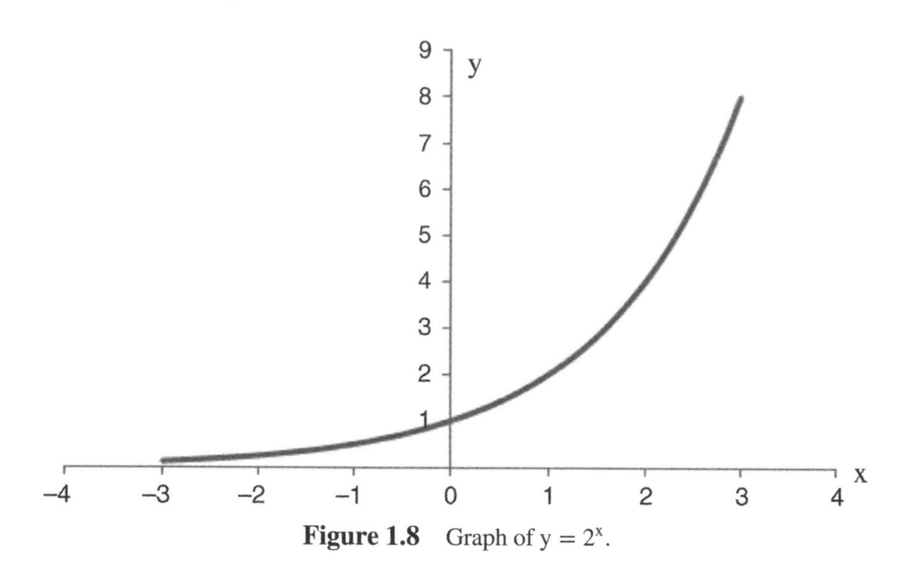

Figure 1.8 Graph of $y = 2^x$.

1.8 DIFFERENTIATION AND INTEGRATION

In many business applications, we encounter a variety of instances where we have to deal with *rates of change*. For example, we may want to know the effect of a change in interest rates on business investments or we may want to determine the effect of change in the disposable income of families on their consumption. Rates of change such as these can be described by the concept of a "derivative." Likewise, "integration" is used to compute a sum. For example, integration can be used to find the sum of accumulated net returns of an investment over a given period of time.

1.8.1 Derivative

The function $y = f(x)$ is said to approach the *limit* L as x tends to a if "f(x) is near L" whenever "x is near a." This statement is symbolized as

$$\lim_{x \to a} f(x) = L.$$

Now, if it is also the case that $L = f(a)$, then f is said to be *continuous* at $x = a$ and is written as

$$\lim_{x \to a} f(x) = f(a).$$

And if f is continuous at each point of its domain, then f is called a *continuous function*. Intuitively, think of a function as being continuous if its graph is without breaks.

Next, if $x = a$ is any point of the interval (x_1, x_2), and if a function $y = f(x)$ is defined over (x_1, x_2), then

$$\lim_{\Delta x \to 0} \frac{f(a + \Delta x) - f(a)}{\Delta x}$$

is called the *derivative of f(x) at x = a* and denoted f′(a). In general, *the derivative of f with respect to x* is defined as

$$\lim_{\Delta x \to 0} \frac{\Delta y}{\Delta x} = \lim_{\Delta x \to 0} \frac{f(x + \Delta x) - f(x)}{\Delta x} = f'(x) = dy/dx.$$

Here dy/dx is the instantaneous rate of change in y per unit change in x as Δx gets smaller and smaller.

For instance, let us find the derivative of $y = f(x) = x^2$ at x = 2. We first find

$$\frac{\Delta y}{\Delta x} = \frac{f(2 + \Delta x) - f(2)}{\Delta x} = \frac{(2 + \Delta x)^2 - f(2)}{\Delta x}$$
$$= \frac{4 + 4\Delta x + (\Delta x)^2 - 4}{\Delta x} = \frac{4\Delta x + (\Delta x)^2}{\Delta x} = 4 + \Delta x.$$

Then,

$$\lim_{\Delta x \to 0} \frac{\Delta y}{\Delta x} = \lim_{\Delta x \to 0} (4 + \Delta x) = 4 = dy/dx.$$

So, when Δx is near zero, 4 + Δx is near 4—which is the derivative of $y = f(x) = x^2$ at x = 2.

Note that if f is differentiable (it has a derivative) at x = a, then it is also continuous at x = a. However, the converse does not hold.

Geometrically, the derivative f′(a) represents the *slope* of the function y = f(x) at x = a (it is the slope of the line tangent to f(x) at x = a). In various (business) applications, dy/dx depicts the *marginal change in y with respect to x.*

Since the direct calculation of the derivative of a function via the limit process can be quite tedious, various rules of differentiation will now be offered as a convenience. These rules will expedite the calculation of derivatives.

Some Rules of Differentiation

1. *Power rule:* If $y = f(x) = x^n$, then

$$\frac{dy}{dx} = nx^{n-1}.$$

EXAMPLE 1.15

Find $\frac{dy}{dx}$ for the following:

a. $y = x^3$.

b. $y = x$.

c. $y = \frac{1}{x} = x^{-1}, \quad x \neq 0$.

d. $y = \frac{1}{\sqrt{x}} = x^{-(1/2)}, \quad x \neq 0$.

SOLUTIONS:

a. $\frac{dy}{dx} = 3x^{3-1} = 3x^2$.

b. $\frac{dy}{dx} = 1$.

c. $\frac{dy}{dx} = -1x^{-1-1} = -1x^{-2} = \frac{-1}{x^2}$.

d. $\frac{dy}{dx} = -\frac{1}{2}x^{\{-(1/2)-1\}} = -\frac{1}{2}x^{\{-(3/2)\}} = -\frac{1}{2}\left(\frac{1}{\sqrt{x^3}}\right)$. ■

2. The derivative of a constant is zero. If $y = f(x) = k$, where k is a constant, then $dy/dx = 0$.

EXAMPLE 1.16

If $y = 10$, then $dy/dx = 0$. It is obvious that the graph of $y = 10$ is a straight line parallel to x-axis, and its slope dy/dx is 0. This is due to the fact that even when x changes, y does not change ■

3. *Coefficient rule:* If $y = f(x) = kg(x)$, where k is a constant, then

$$\frac{dy}{dx} = kg'(x) = k\frac{dg}{dx}.$$

EXAMPLE 1.17

Find $\frac{dy}{dx}$ for the following:

a. $y = 5x^4$.

b. $y = \left(\frac{1}{2}\right)x^2$.

c. $y = \left(\frac{1}{2x^2}\right), \quad x \neq 0$.

SOLUTION:

a. $\frac{dy}{dx} = 5\left(\frac{d(x^4)}{dx}\right) = 5 \times 4x^3 = 20x^3$.

b. $\frac{dy}{dx} = \left(\frac{1}{2}\right)\left(\frac{d(x^2)}{dx}\right) = \left(\frac{1}{2}\right) \times 2x = x$.

c. $\frac{dy}{dx} = \left(\frac{1}{2}\right)\left(\frac{d(x^{-2})}{dx}\right) = \left(\frac{1}{2}\right) \times (-2)x^{-2-1} = -x^{-3} = -\left(\frac{1}{x^3}\right)$. ■

4. *Sum (difference) rule:* If $y = f(x) = g(x) \pm h(x)$, then

$$\frac{dy}{dx} = g'(x) \pm h'(x) = \frac{dg}{dx} \pm \frac{dh}{dx}.$$

EXAMPLE 1.18

Find the derivatives of the following:

a. $y = 5x^3 + 7x^4$.

b. $y = 2x^2 - \left(\frac{3}{x^2}\right), \quad x \neq 0$.

c. $y = 3x + 10$.

SOLUTION:

a. $\frac{dy}{dx} = 5 \times 3x^{3-1} + 7 \times 4x^{4-1} = 15x^2 + 28x^3$.

b. $\frac{dy}{dx} = 2 \times 2x^{2-1} - 3(-2)x^{-2-1} = 4x + 6x^{-3} = 4x + \left(\frac{6}{x^3}\right)$.

c. $\frac{dy}{dx} = 3 + 0 = 3$. ■

5. *Product rule:* If $y = f(x) \times g(x)$, then

$$\frac{dy}{dx} = f'(x) \times g(x) + f(x) \times g'(x) = \frac{df}{dx} \times g(x) + f(x) \times \frac{dg}{dx}.$$

EXAMPLE 1.19

Find the derivative of $y = f(x) \times g(x)$, where (a) $f(x) = 7x^2$ and $g(x) = 3x^3$, and (b) $f(x) = 9x^3$ and $g(x) = 2x^5$.

SOLUTION:

$\frac{dy}{dx} = f'(x) \times g(x) + f(x) \times g'(x)$.

(a) $\frac{dy}{dx} = (14x)(3x^3) + (7x^2)(9x^2) = 42x^4 + 63x^4 = 105x^4$.

(b) $\frac{dy}{dx} = 27x^2(2x^5) + 9x^3(10x^4) = 54x^7 + 90x^7 = 144x^7$. ■

6. *Quotient rule:* If $y = f(x)/g(x)$, $g(x) \neq 0$, then

$$\frac{dy}{dx} = \frac{g(x)f'(x) - f(x)g'(x)}{(g(x))^2}.$$

EXAMPLE 1.20

Find the derivative of $y = \frac{f(x)}{g(x)}$, where

(a) $f(x) = 3x^2 + 2$ and $g(x) = 5x^3 + 7$.

(b) $f(x) = 7x^3 + 3$ and $g(x) = 6x^2 - 5x$.

SOLUTION:

(a)

$$\frac{dy}{dx} = \frac{g(x)f'(x) - f(x)g'(x)}{(g(x))^2},$$

$$\frac{dy}{dx} = \frac{(5x^3 + 7)(6x) - (3x^2 + 2)(15x^2)}{(5x^3 + 7)^2},$$

$$\frac{dy}{dx} = \frac{(30x^4 + 42x) - (45x^4 + 30x^2)}{25x^6 + 70x^3 + 49},$$

$$\frac{dy}{dx} = \frac{-15x^4 - 30x^2 + 42x}{25x^6 + 70x^3 + 49}.$$

(b) $$\frac{dy}{dx} = \frac{(6x^2 - 5x)(21x^2) - (7x^3 + 3)(12x - 5)}{(6x^2 - 5x)^2}$$

$$= \frac{126x^4 - 105x^3 - 84x^4 - 36x + 35x^3 + 15}{36x^4 - 60x^3 + 25x^2} = \frac{42x^4 - 70x^3 - 36x + 15}{36x^4 - 60x^3 + 25x^2}.$$

■

1.8.2 Derivatives of Logarithmic and Exponential Functions

If $y = \log_a x$, then

$$\frac{dy}{dx} = \frac{1}{x} \times \log_a e = \frac{1}{x} \times \frac{\ln e}{\ln a} = \frac{1}{x \ln a}.$$

This result follows from the rule

$$\log_a b = \frac{\ln b}{\ln a} \quad \text{and} \quad \ln e = 1.$$

$$\text{If } y = \ln x, \quad \text{then} \quad \frac{dy}{dx} = \frac{1}{x}.$$

$$\text{If } y = e^x, \quad \text{then} \quad \frac{dy}{dx} = e^x.$$

***Note*:** For proofs of the above rules of differentiation (including other rules of differentiation), see any standard calculus textbook.

1.8.3 Higher Order Derivatives

Given $y = f(x)$, we previously denoted the first derivative of f with respect to x as dy/dx or $f'(x)$. For instance, if $y = f(x) = 2x^3 + 4x^2 - 2x + 10$, then

$$\frac{dy}{dx} = f'(x) = 6x^2 + 8x - 2.$$

Since dy/dx is also a function of x, we can find its derivative with respect to x. That is, we desire to find the derivative of the first derivative, which we shall term the *second derivative* of f and denote as

$$\frac{d}{dx}\left(\frac{dy}{dx}\right) = \frac{d^2y}{dx^2} \quad \text{or} \quad f''(x).$$

So given dy/dx above,

$$\frac{d^2y}{dx^2} = f''(x) = 12x + 8.$$

Note that the same basic rules of differentiation still apply—only the notation has changed.

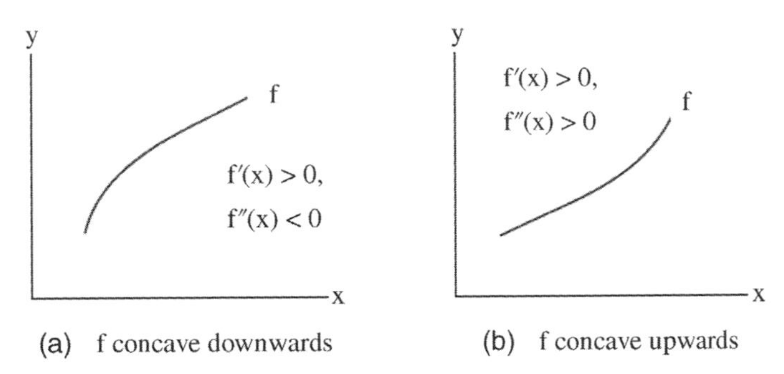

Figure 1.9 Behavior of the second derivative of f. (a) f concave downward. (b) f concave upward.

Geometrically, the second derivative of f is the rate of change of the slope of f at a particular point on the graph of f. For example, in Figure 1.9a, as we move along the curve from left to right, $f' > 0$ with $f'' < 0$, that is, the slope of f is decreasing as x increases. In Figure 1.9b, again $f' > 0$, but now $f'' > 0$, that is, the slope of f is increasing as x increases. In the former case we say that f is *concave downward*; in the latter case f is termed *concave upward.* (As an exercise, the reader should draw comparable graphs for $f' < 0$. Also, describe the behavior of the slope of f in terms of f'' and the concavity of f for each case.)

Furthermore, since d^2y/dx^2 also depends on x, we can find its derivative with respect to x, which we shall call the *third derivative* of f, by determining the derivative of the second derivative. Our notation for the third derivative of f is

$$\frac{d}{dx}\left(\frac{d^2y}{dx^2}\right) = \frac{d^3y}{dx^3} \quad \text{or} \quad f'''(x).$$

From d^2y/dx^2 above,

$$\frac{d^3y}{dx^3} = f'''(x) = 12.$$

In general, the *nth order derivative* of f with respect to x is denoted as

$$\frac{d^ny}{dx^n} = \frac{d}{dx}\left(\frac{d^{n-1}y}{dx^{n-1}}\right) \quad \text{or} \quad f^n(x).$$

For our purposes, we shall mostly rely on the first and second derivatives of a function in our applications to profit (and to revenue) maximization or to cost minimization.

To gain some practice with the calculation of higher order derivatives, let us find f', f'', and f''' when $y = f(x) = x^6 - x^{-1} + 2x - 6$, $x \neq 0$. It is straightforward to show that

$$\begin{aligned} f' &= 6x^5 + x^{-2} + 2, \\ f'' &= 30x^4 - 2x^{-3}, \\ f''' &= 120x^3 + 6x^{-4}. \end{aligned}$$

EXAMPLE 1.21

Find f′, f″ when

(a) $f(x) = 3x^4 + 9x^{-1}, \quad x \neq 0.$

(b) $f(x) = x^3 + 5x^{-2} + 3x^2.$

SOLUTION:

(a)

$$\begin{aligned} f(x) &= 3x^4 + 9x^{-1}, \\ f'(x) &= 12x^3 - \frac{9}{x^2}, \\ f''(x) &= 36x^2 + 18x^{-3} \\ &= 36x^2 + \frac{18}{x^3}. \end{aligned}$$

(b)

$$\begin{aligned} f(x) &= x^3 + 5x^{-2} + 3x^2, \\ f'(x) &= 3x^2 - 10x^{-3} + 6x, \\ f''(x) &= 6x + 30x^{-4} + 6, \\ &= 6x + \frac{30}{x^4} + 6. \end{aligned}$$

■

1.8.4 Integration

There are two distinct ways to define the concept of integration. First, an *integral* is the limit of a specific summation or addition process. Geometrically, it corresponds to the area under a given curve. This type of integral is termed a *definite integral.*

Second, an integral can be viewed as the result of reversing the process of differentiation. That is, if the derivative of $y = F(x)$ exists, then this derivative is also a function of x and can be denoted as f(x). The reverse problem then amounts to finding, from the given function f(x), a new function F(x) (or functions that have f(x) as their derivative). If F(x) can be found, then F(x) is termed the *indefinite integral* of f(x).

Specifically, suppose we are given the derivative $dy/dx = f(x)$, $a < x < b$, and asked to find $y = F(x)$. The function $y = F(x)$ is a solution to $dy/dx = f(x)$ if, over the domain $a < x < b$, F(x) is differentiable and

$$\frac{dF(x)}{dx} = f(x).$$

Hence, F(x) is termed the *integral of f(x)* with respect to x. In fact, if $y = F(x)$ is a particular solution of $dy/dx = f(x)$, then "all" solutions are subsumed within the expression

$$y = F(x) + c, \quad \text{where c is an arbitrary constant.}$$

This is indicated by writing

$$\int f(x)dx = F(x) + c,$$

where the symbol "$\int$" is termed the "integral sign."

Let us take a closer look at these two types of integrals.

1.8.5 The Definite Integral

Suppose we wish to find the area under the curve $y = f(x)$ and above the x-axis between $x = a$ and $x = b$. For example, we may be interested in determining the area under $y = f(x) = 10 - x$ and above the x-axis between $x = 2$ and $x = 7$ (Figure 1.10). A first approximation to this total area can be obtained by calculating the areas of a finite set of rectangles and then summing these separate areas. That is, we first consider the rectangle whose base (Δx) is from $x = 2$ to $x = 3$. The exact area of this rectangle is height x base $= f(2) \times \Delta x = 8 \times 1 = 8$. Clearly, the area of this rectangle is only an approximation to the actual area under f(x) between $x = 2$ and $x = 3$ since an error component (an overestimate) corresponding to the shaded triangular portion of the rectangle occurs.

Next, consider the area of the approximating rectangle between $x = 3$ and $x = 4$. It amounts to $f(3) \times \Delta x = 7 \times 1 = 7$ and also admits an error (it is shaded) or overestimate of the true area under f(x) between $x = 3$ and $x = 4$. Continuing this process three more times (we determine rectangular areas from $x = 4$ to $x = 5$, from $x = 5$ to $x = 6$, and from $x = 6$ to $x = 7$) enables us to approximate the entire desired area as the sum of the areas of the five approximating rectangles as

$$\sum_{x=2}^{7} f(x)(\Delta x) = f(2) \times 1 + f(3) \times 1 + f(4) \times 1 + f(5) \times 1 + f(6) \times 1$$
$$= 8 + 7 + 6 + 5 + 4 = 30.$$

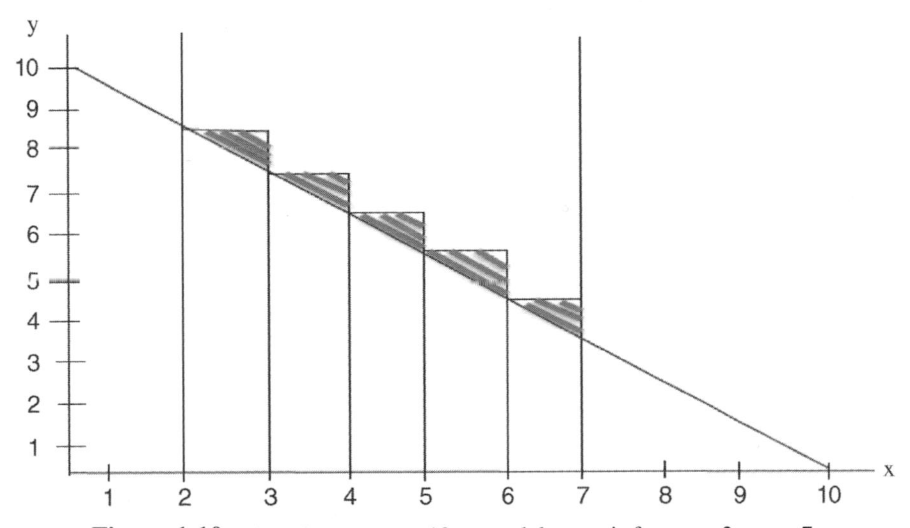

Figure 1.10 Area between $y = 10 - x$ and the x-axis from $x = 2$ to $x = 7$.

Thus, 30 is our approximation to the total area under $y = f(x) = 10 - x$ and above the x-axis between $x = 2$ and $x = 7$.

Let us now generalize this procedure. Given a function $y = f(x)$, we seek to find the approximate area under f(x) and above the x-axis between $x = a$ and $x = b$ or over [a,b]. First, determine the base Δx of each approximating rectangle by dividing [a,b] into n equal parts or

$$\Delta x = \frac{b - a}{n}.$$

Thus, the interval is subdivided according to

$$\begin{aligned} x_0 &= a, \\ x_1 &= a + \Delta x, \\ x_2 &= a + 2(\Delta x), \\ \vdots &\quad \vdots \\ x_n &= a + n(\Delta x) = b. \end{aligned}$$

Next, the sum of the areas of the n approximating rectangles is given by

$$\sum_{i=1}^{n} f(x_i)(\Delta x) = f(x_1)(\Delta x) + f(x_2)(\Delta x) + \cdots + f(x_n)(\Delta x).$$

A glance back at Figure 1.10 reveals that if n increases or Δx decreases in size, the error incurred in using the sum of the rectangular areas to estimate the true area also decreases. And if, "in the limit," we let $\Delta x \to 0$, then the sum of the errors also tends to zero. Hence, an exact measure of the area under $y = f(x)$ and above the x-axis over [a,b] is given by

$$\begin{aligned} \lim_{n\to\infty} \sum_{i=1}^{n} f(x_i)(\Delta x) &= \lim_{\Delta x\to 0} \sum_{i=1}^{n} f(x_i)(\Delta x) \\ &= \int_a^b f(x)dx, \end{aligned}$$

the *definite integral* of f(x). Note that "$\int$" is an "operator" and the operation is summation. In sum, the definite integral is the limiting sum of a set of rectangles as the number of rectangles increases without bound; it amounts to the area under a curve over a closed interval. Note also that the above limit exists provided f is continuous.

How is a definite integral computed? To answer this question, we need to only look to the *Fundamental Theorem of Integral Calculus*: If f(x) is a given continuous function on [a, b] and F is any differentiable function such that $F'(x) = f(x), x \in [a,b]$, then

$$\int_a^b f(x)dx = F(x)]_a^b = F(b) - F(a).$$

To facilitate the implementation of this theorem, we need to first look to how indefinite integrals are evaluated. Why? Because the fundamental theorem utilizes

F(x) and F(x) is obtained via the indefinite integral. As was the case with differentiation, we shall rely upon a set of rules for integration.

1.8.6 Some Rules of Integration

1. Power rule:

$$\int x^n dx = \frac{x^{n+1}}{n+1} + c.$$

Here, c is the constant and $n \neq -1$. If $n = -1$, then

$$\int x^{-1} dx = \int \frac{1}{x} dx = \ln x + c.$$

EXAMPLE 1.22

$$\int x^5 dx = \frac{x^6}{6} + c.$$

■

2. The integral of zero is a constant, including zero:

$$\int 0\, dx = c.$$

3. The integral of a constant times a function is the constant times the integral of the function:

$$\int kg(x)dx = k\int g(x)dx, \quad \text{k is a constant.}$$

EXAMPLE 1.23

a. $\int 6x^3 dx = 6\frac{x^4}{4} + c = \frac{3x^4}{2} + c.$

b. $\int \frac{\sqrt{3}}{x} dx = \sqrt{3}\int \frac{1}{x} dx = \sqrt{3}\ln x + c.$

c. $\int 3x^2(y-1)^2 dx = 3(y-1)^2 \int x^2 dx = 3(y-1)^2 \frac{x^3}{3} + c = x^3(y-1)^2 + c.$ ■

4. Integral of a sum or difference rule:

$$\int (f(x) \pm g(x))dx = \int f(x)dx \pm \int g(x)dx.$$

EXAMPLE 1.24

a. Integrate $\int (f(x) + g(x))dx$, where $f(x) = 10ax$ and $g(x) = 4bx^3$.

$$\begin{aligned}\int (f(x) + g(x))dx &= \int f(x)dx + \int g(x)dx = \int 10axdx + \int 4bx^3 dx \\ &= 10a\frac{x^2}{2} + 4b\frac{x^4}{4} + c = 5ax^2 + bx^4 + c.\end{aligned}$$

b.
$$\int (y-1)^2 dy = \int (y^2 - 2y + 1)dy = \int y^2 dy - \int 2y dy + \int 1 dy$$
$$= \frac{y^3}{3} - \frac{2y^2}{2} + y + c = \frac{1}{3}y^3 - y^2 + y + c. \quad ■$$

5. Integration by parts:

Recall from the product rule for differentiation that

$$\frac{d}{dx}(f(x)g(x)) = f'(x)g(x) + f(x)g'(x).$$

Integrating both sides, we obtain

$$f(x)g(x) = \int f'(x)g(x) + \int f(x)g'(x),$$

or

$$\int f(x)g'(x) = f(x)g(x) - \int f'(x)g(x).$$

Generally, most textbooks present this rule in an abbreviated form. That is, let $u = f(x)$ and $v = g(x)$. Then,

$$\int u\,dv = uv - \int v\,du.$$

EXAMPLE 1.25

Solve the following using integration by parts.

$$\int x(x-2)^2 dx.$$

SOLUTION: Let $u = x$ and $dv = (x-2)^2 dx$,

$$du = 1$$
$$v = \int (x-2)^2 dx = \frac{(x-2)^3}{3}.$$

Using the formula

$$\int udv = uv - \int vdu, \text{ we obtain}$$
$$\int x(x-2)^2 dx = x\left[\frac{(x-2)^3}{3}\right] - \int \frac{(x-2)^3}{3} dx$$
$$= \frac{x(x-2)^3}{3} - \frac{(x-2)^4}{3(4)} + c$$

$$= \left[\frac{(x-2)^3}{3}\right]\left(x - \frac{x-2}{4}\right) + c$$

$$= \left[\frac{(x-2)^3}{3}\right]\left(\frac{3x+2}{4}\right) + c.$$

■

6. Integrals of logarithmic and exponential functions:

$$\int \ln x\, dx = x(\ln x - 1) + c,$$

$$\int e^x dx = e^x + c,$$

$$\int e^{kx} dx = \frac{1}{k}e^{kx} + c.$$

EXAMPLE 1.26a

$$\int (3e^x - 2x + e)dx = 3\int e^x dx - 2\int x dx + e\int dx = 3e^x - 2\frac{x^2}{2} + ex + c$$

$$= 3e^x - x^2 + ex + c.$$

■

EXAMPLE 1.26b

$$\int e^{2x} dx = \frac{1}{2}e^{2x} + c.$$

■

Armed with the essentials of indefinite integration, we are now ready to explore the computation of definite integrals.

EXAMPLE 1.27

Let us evaluate the area under $y = f(x) = 10 - x$ and above the x-axis between $x = 2$ and $x = 7$ (see the preceding introduction to definite integrals). To this end, we have

$$\int_2^7 (10 - x)dx = \left(10x - \frac{1}{2}x^2\right)\Big]_2^7 = F(7) - F(2)$$

$$= \left[70 - \frac{1}{2}(49)\right] - \left[20 - \frac{1}{2}(4)\right]$$

$$= \frac{91}{2} - 18 = 55/2.$$

■

EXAMPLE 1.28

Find

$$\int_1^2 (2x + 5)dx = \int_1^2 2xdx + \int_1^2 5dx$$

$$= x^2\Big]_1^2 + 5x\Big]_1^2 = (4 - 1) + (10 - 5) = 8.$$

■

EXAMPLE 1.29

Evaluate

$$\int_0^1 \left(x^2 - 2x + 3\right)dx \int_0^1 x^2dx - \int_0^1 2xdx + \int_0^1 3dx$$

$$= \frac{1}{3}x^3\Big]_0^1 - x^2\Big]_0^1 + 3x\Big]_0^1 = \left(\frac{1}{3} - 0\right) - (1 - 0) + (3 - 0)$$

$$= \frac{7}{3}.$$

■

1.9 EXCEL APPLICATIONS

EXAMPLE 1.30

For the values of X and Y given in Example 1.5, evaluate r using the long and short formulas with EXCEL.

SOLUTION: Using Excel and the long formula, r is evaluated as follows:

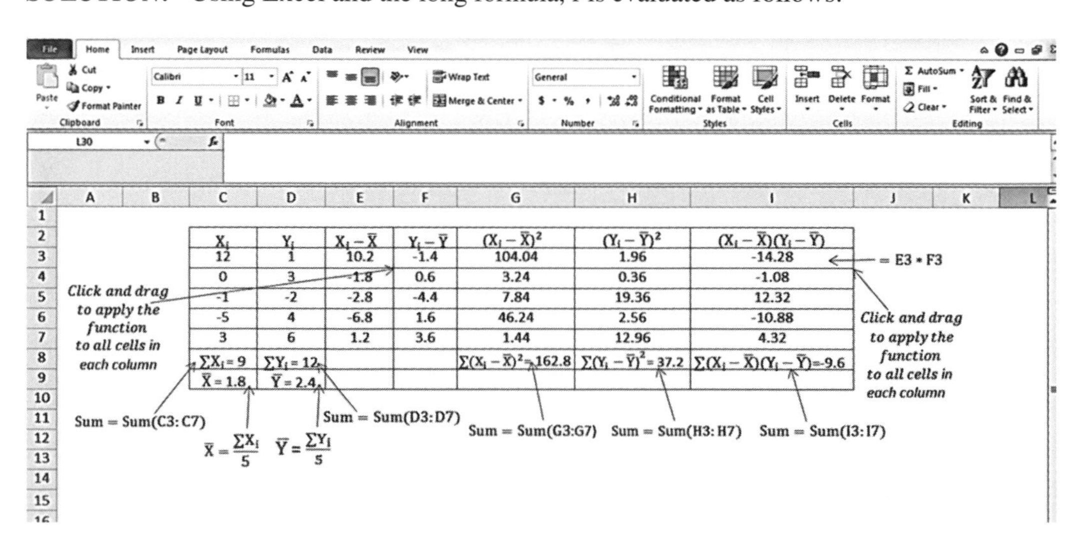

X_i	Y_i	$X_i - \bar{X}$	$Y_i - \bar{Y}$	$(X_i - \bar{X})^2$	$(Y_i - \bar{Y})^2$	$(X_i - \bar{X})(Y_i - \bar{Y})$
12	1	10.2	-1.4	104.04	1.96	-14.28
0	3	-1.8	0.6	3.24	0.36	-1.08
-1	-2	-2.8	-4.4	7.84	19.36	12.32
-5	4	-6.8	1.6	46.24	2.56	-10.88
3	6	1.2	3.6	1.44	12.96	4.32
$\Sigma X_i = 9$	$\Sigma Y_i = 12$			$\Sigma(X_i - \bar{X})^2 = 162.8$	$\Sigma(Y_i - \bar{Y})^2 = 37.2$	$\Sigma(X_i - \bar{X})(Y_i - \bar{Y}) = -9.6$
$\bar{X} = 1.8$	$\bar{Y} = 2.4$					

$$r = \sum_{i=1}^{5} \frac{(X_i - \overline{X})(Y_i - \overline{Y})}{\sqrt{\sum_{i=1}^{5} (X_i - \overline{X})^2 \sum_{i=1}^{5} (Y_i - \overline{Y})^2}}$$

$$= \frac{-9.6}{\sqrt{(162.8)(37.2)}} = -0.1104.$$

Using Excel and the short formula, r is evaluated as follows:

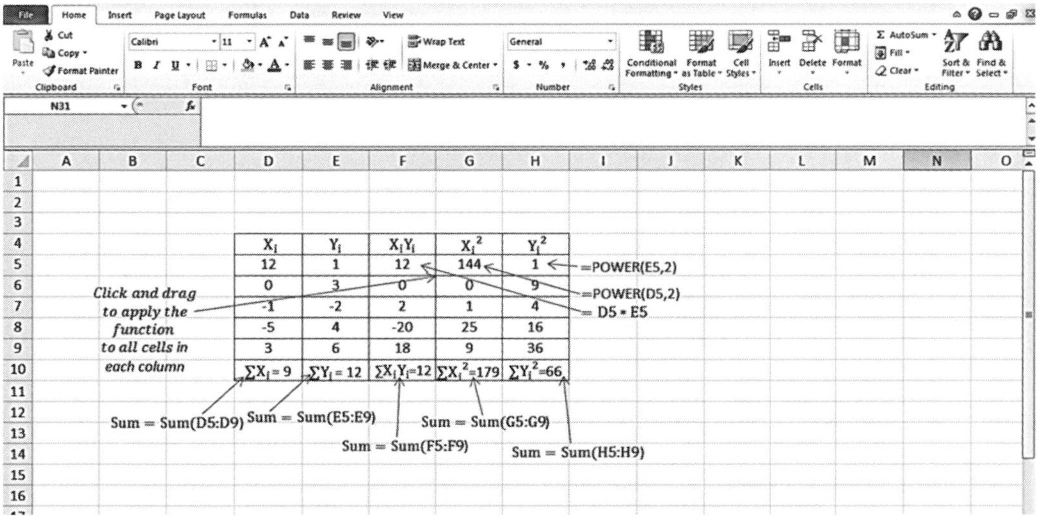

X_i	Y_i	X_iY_i	X_i^2	Y_i^2
12	1	12	144	1
0	3	0	0	9
-1	-2	2	1	4
-5	4	-20	25	16
3	6	18	9	36
$\sum X_i = 9$	$\sum Y_i = 12$	$\sum X_iY_i = 12$	$\sum X_i^2 = 179$	$\sum Y_i^2 = 66$

=POWER(E5,2)
=POWER(D5,2)
= D5 * E5

Click and drag to apply the function to all cells in each column

Sum = Sum(D5:D9) Sum = Sum(E5:E9) Sum = Sum(G5:G9)
Sum = Sum(F5:F9) Sum = Sum(H5:H9)

$$r = \frac{\sum X_iY_i - \frac{\sum X_i \sum Y_i}{n}}{\left[\left(\sum X_i^2 - \frac{(\sum X_i)^2}{n}\right)\left(\sum Y_i^2 - \frac{(\sum Y_i)^2}{n}\right)\right]^{1/2}}$$

$$= \frac{12 - \frac{(9)(12)}{5}}{\left[\left(179 - \frac{9^2}{5}\right)\left(66 - \frac{12^2}{5}\right)\right]^{1/2}}$$

$$= \frac{-9.6}{[(162.8)(37.2)]^{1/2}} = -0.1104.$$

■

EXAMPLE 1.31

Using the X and Y values given in Example 1.5, evaluate the slope coefficient b by the long formula with Excel. Also, predict the value of Y when X = 11, assuming that the linear relationship between X and Y is Y = a + bX.

SOLUTION: Using Excel, the slope coefficient b is evaluated by the long formula as follows:

X_i	Y_i	$X_i - \overline{X}$	$Y_i - \overline{Y}$	$(X_i - \overline{X})^2$	$(Y_i - \overline{Y})^2$	$(X_i - \overline{X})(Y_i - \overline{Y})$
12	1	10.2	-1.4	104.04	1.96	-14.28
0	3	-1.8	0.6	3.24	0.36	-1.08
-1	-2	-2.8	-4.4	7.84	19.36	12.32
-5	4	-6.8	1.6	46.24	2.56	-10.88
3	6	1.2	3.6	1.44	12.96	4.32
$\sum X_i = 9$	$\sum Y_i = 12$			$\sum (X_i - \overline{X})^2 = 162.8$	$\sum (Y_i - \overline{Y})^2 = 37.2$	$\sum (X_i - \overline{X})(Y_i - \overline{Y}) = -9.6$
$\overline{X} = 1.8$	$\overline{Y} = 2.4$					

Click and drag to apply the function to all cells in each column

= E4 * F4

Sum = Sum(C4:C8) Sum = Sum(D4:D8) Sum = Sum(G4:G8) Sum = Sum(H4:H8) Sum = Sum(I4:I8)

$$b = \frac{\sum_{i=1}^{5} (X_i - \overline{X})(Y_i - \overline{Y})}{\sum_{i=1}^{5} (X_i - \overline{X})^2}$$

$$= \frac{-9.6}{162.8} = -0.058968.$$

$$a = \overline{Y} - b\overline{X}$$

$$= 2.4 - (-0.058968)(1.8)^{\circ} = 2.5061424.$$

$$Y = a + bX$$

$$= 2.5061424 + (-0.058968)(11) = 1.8574944.$$

Using Excel, the slope coefficient b is evaluated by the short formula below:

X_i	Y_i	X_iY_i	X_i^2	Y_i^2
12	1	12	144	1
0	3	0	0	9
-1	-2	2	1	4
-5	4	-20	25	16
3	6	18	9	36
$\sum X_i = 9$	$\sum Y_i = 12$	$\sum X_iY_i = 12$	$\sum X_i^2 = 179$	$\sum Y_i^2 = 66$

Click and drag to apply the function to all cells in each column

=POWER(E5,2)
=POWER(D5,2)
= D5 * E5

Sum = Sum(D5:D9) Sum = Sum(E5:E9) Sum = Sum(F5:F9) Sum = Sum(G5:G9) Sum = Sum(H5:H9)

$$b = \frac{\sum X_i Y_i - \frac{\sum X_i \sum Y_i}{n}}{\left[\sum X_i^2 - \frac{\left(\sum X_i\right)^2}{n}\right]}$$

$$= \frac{12 - \frac{(9)(12)}{5}}{179 - \frac{9^2}{5}}$$

$$= \frac{-9.6}{162.8} = -0.058968.$$

■

EXAMPLE 1.32

Solve the following simultaneous linear equations using the Solver Add-in in Excel.

$$3P + Q = 26 \text{ and } P - Q = 2.$$

SOLUTION: INSTRUCTIONS:

A. Excel spread sheet setup of the problem:
Enter data in Excel spread sheet.

a. Show variables P in B1 and Q in C1. Trial value 1 in B2 and 1 in C2. Type "equation 1" in A3 and enter the coefficient 3 of the variable P in B3 and 1 of the variable Q in C3. In the cell D3, enter the Array formula "=Sumproduct (highlight the trial values 1 and 1, highlight 3 and 1) and hit Enter. You will see 4 in D3, which is the left-hand side value of the equation 1 with the given trial values. In the cell E3, type "equal" (not the symbol = since Excel expects a formula to follow) and enter 26 in cell F3, which is the right-hand side value of equation 1.

b. Repeat the above step for equation 2 in cell A4. This will result in 0 in D4, "equal" in E4 and 2 in F4.

c. Click any empty cell below the setup table as follows:

	A	B	C	D	E	F
1	Variables	P	Q			
2	Trial values	1	1			
3	Equation 1	3	1	4	equal	26
4	Equation 2	1	-1	0	equal	2

B. Using Solver Add-In*

Click on DATA in the menu bar and then Solver in the extreme right side of the ribbon at the top of your screen.

a. In the Solver parameter dialog box, at set objective, select Target Cell D3, where 4 is obtained by the Array formula.
b. At "to" select the button "value of" and enter 26 in white bar.
c. Under "By Changing Variable Cells," click on the white bar and highlight both the trial value cells at the same time.
d. Click inside the lower bigger rectangular box and click on "Add" on the right-hand side to get "Add Constraint" dialog box.
e. Under Cell Reference: Highlight Cell, D4 where "0" is displayed by the formula.
f. Change the middle symbol to "=".
g. Constraint: Highlight Cell where "2" is typed.
h. Click "OK" since you don't have a third constraint equation.
i. Uncheck "Make Unconstrained Variables Non-Negative" box and select solving method as "Simplex LP."
j. Click on "Solve"; you see a circle marked "Keep Solver Solution" in the Solver Results dialog box.
k. Click OK to see the solution as in the table below:

	A	B	C	D	E	F
1	Variables	P	Q			
2	Trial values	7	5			
3	Equation 1	3	1	26	equal	26
4	Equation 2	1	-1	2	equal	2

Note: The trial values of variables P and Q are changed by solver program to $P=7$ and $Q=5$, which satisfy both the equations. Therefore, the final solution of the two equations is $P=7$ and $Q=5$.

* If you do not find Solver when you click on "Data" in the menu bar in Microsoft Excel (Office 2007 and onward),

1. Click *the MICROSOFT ICON or FILE* (at the top left of your screen).

2. Click *Excel Options* (at the bottom of the menu).

3. Click *ADD-INS.*

4. Click *GO* (at the bottom of the window).

5. Check the box with "Solver Add-in."

6. Click on *OK* and see *"Solver"* at the top right of your screen. ■

EXAMPLE 1.33

Solve the following simultaneous linear equations using the Solver Add-in in Excel.

$$2X + 3Y = 18 \quad \text{and} \quad X + 3Y = 15.$$

SOLUTION: Let $2X+3Y=18$ be equation 1 and $X+3Y=15$ be equation 2.
INSTRUCTIONS:

A. Excel spread sheet setup of the problem:
Enter data in Excel spread sheet.

a. Show variables X in B1 and Y in C1. Trial value 1 in B2 and 1 in C2.
Type "equation 1" in A3 and enter the coefficient, 2, of the variable X in B3 and 3 of the variable Y in C3. In the cell D3, enter the Array formula "=Sumproduct(highlight the trial values 1 and 1, highlight 2 and 3)" and hit Enter. You will see 5 in D3, which is the left-hand side value of equation 1 with the given trial values. In the cell E3, type "equal" (not the symbol = since Excel expects a formula to follow) and enter 18 in cell F3, which is the right-hand side value of equation 1.
b. Repeat the above step for equation 2 in cell A4. This will result in 4 in D4, "equal" in E4 and 15 in F4.
c. Click any empty cell below the setup table as follows:

	A	B	C	D	E	F
1	Variables	X	Y			
2	Trial values	1	1			
3	Equation 1	2	3	5	equal	18
4	Equation 2	1	3	4	equal	15

B. Using Solver Add-In*
Click on DATA in the menu bar and then Solver in the extreme right side of the ribbon at the top of your screen.

a. In the Solver parameter dialog box, at "set objective," select Target Cell D3, where 5 is obtained by the Array formula.
b. At "to" select the button "value of" and enter 18 in white box.
c. Under "By Changing Variable Cells," click on the white box and highlight both the trial value cells at the same time.
d. Click inside the lower bigger rectangular box and click on "Add" on the right-hand side to get "Add Constraint" dialog box.
e. Under Cell Reference: Highlight Cell, D4 where "4" is displayed by the formula.
f. Change the middle symbol to "=".
g. Constraint: Highlight Cell, where "15" is typed.
h. Click "OK" since you do not have a third constraint equation.
i. Uncheck "Make Unconstrained Variables Non-Negative" box and select solving method as "Simplex LP."
j. Click on "Solve"; you see a circle marked "Keep Solver Solution" in the Solver Results dialog box.
k. Click OK to see the solution as in the table below:

	A	B	C	D	E	F
1	Variables	X	Y			
2	Trial values	3	4			
3	Equation 1	2	3	18	equal	18
4	Equation 2	1	3	15	equal	15
5						

Note: The trial values of variables X and Y are changed by solver program to X = 3 and Y = 4, which satisfy both the equations. Therefore, the final solution of the two equations is X = 3 and Y = 4.

* If you do not find "Solver" when you click on "Data" in the menu bar in Microsoft Excel (Office 2007 and onward),

1. Click *the MICROSOFT ICON or FILE* (at the top left of your screen).
2. Click *Excel Options* (at the bottom of the menu).
3. Click *ADD-INS.*
4. Click *GO* (at the bottom of the window).
5. Check the box with "Solver Add-in."
6. Click on *OK* and see *"Solver"* at the top right of your screen. ■

CHAPTER 1 REVIEW

You should be able to:

1. Distinguish between a linear and nonlinear function.
2. Distinguish between dependent and independent variables.
3. Distinguish between the slope and Y intercept of a linear function.
4. Solve two simultaneous linear equations in two unknowns using the method of elimination.
5. Distinguish between consistent and dependent linear equation systems.
6. Explain the notation $\sum_{i=1}^{n} X_1$.
7. Calculate a Pearson correlation coefficient r.
8. Distinguish between the empty set and the universal set.
9. Identify the complement of a set.
10. Distinguish between the set operations of union and intersection.
11. Distinguish between open and closed intervals.
12. Distinguish between the domain and range of a function.
13. Determine if a particular graph represents a function using the vertical line test.

14. Graph a linear function using the intercept method.
15. Give an intuitive definition of a continuous function.
16. Apply the power, product, and quotient rules of differentiation.
17. Characterize the concavity of a function in terms of the sign of its second derivative.
18. Distinguish between a definite and indefinite integral.
19. Explain why marginal values are slopes of functions.

Key Terms and Concepts:

basic points, 19
closed interval, 15
complement, 13
concavity, 27
consistency test, 5
continuous, 22
countably infinite, 13
definite integral, 28
dependent linear equations, 6
dependent variable, 2, 15, 18
derivative, 22-3
domain, 15
element, 12-13
element inclusion, 13
function, 15-16
fundamental theorem of integral calculus, 30
graph, 15-16
horizontal intercept, 19
image, 15
indefinite integral, 28, 30-1
independent variable, 1-2, 15, 18
intersection, 14-15
limit, 22-3
linear function, 2
marginal change, 23
method of elimination, 4
nth order derivative, 27
null set, 13
one-to-one, 15
open interval, 15
Pearson correlation coefficient, 10
range, 15
rules of differentiation, 23
rules of integration, 31
second derivative, 27
set, 12-13
set inclusion, 13
slope, 2
summation notation, 6-12
third derivative, 27
uncountably infinite, 13
union, 14-15
universal set, 13
vertical intercept, 19
vertical line test, 16-17

EXERCISES

Solve the following.

1. $\frac{1}{3}X + 7 = 51.$
2. $\frac{2}{3X} + 9 = 12.$
3. $2.3X - 8.9 = -4.3.$

Solve the following sets of simultaneous equations:

4. $$\begin{aligned} 7X + 3Y &= 14, \\ X - Y &= 12. \end{aligned}$$

5. $$\begin{aligned} 2X + 3Y &= 18, \\ X + 3Y &= 15. \end{aligned}$$

6. $$\begin{aligned} 5X - 7Y &= 28, \\ 2X + 5Y &= 19. \end{aligned}$$

7. $$\begin{aligned} 2X_1 - X_2 &= 3, \\ 7X_1 + 2X_2 &= 27. \end{aligned}$$

8. **a.** $$\begin{aligned} 3P + Q &= 26, \\ P - 5Q &= 2. \end{aligned}$$

b. $$\begin{aligned} -6X + 3Y &= 0, \\ 3X + Y &= 0. \end{aligned}$$

c. $$\begin{aligned} P + 2Q &= 10, \\ 4P + 8Q &= 20. \end{aligned}$$

Suppose that there are six observations for the variables X and Y as given below:

$$\begin{array}{llllll} X_1 = 2, & X_2 = 1, & X_3 = 3, & X_4 = -5, & X_5 = 1, & X_6 = -2 \\ Y_1 = 4, & Y_2 = 0, & Y_3 = -1, & Y_4 = 2, & Y_5 = 7, & Y_6 = -3 \end{array}$$

For Exercises 9–19, compute the following:

9. $\sum_{i=1}^{6} X_i$.

10. $\sum_{i=1}^{6} Y_i$.

11. $\sum_{i=1}^{6} X_i^2$.

12. $\sum_{i=1}^{6} Y_i^2$.

13. $\sum_{i=1}^{6} X_i Y_i$.

14. $\sum_{i=1}^{6} (X_i + Y_i)$.

15. $\sum_{i=1}^{6} (X_i - Y_i)$.

16. $\sum_{i=1}^{6} (X_i - 3Y_i + 2X_i^2)$.

17. $\sum_{i=1}^{6} cX_i$, where $c = -1$.

18. $\sum_{i=1}^{6} (X_i - 3Y_i + c)$, where $c = 3$.

19. $\sum_{i=1}^{6} (X_i - Y_i)^2$
$= \sum_{i=1}^{6} X_i^2 - \sum_{i=1}^{6} 2X_i Y_i + \sum_{i=1}^{6} Y_i^2$.

20. Write set notation for the set of all whole numbers less than 8.

21. Write set notation for the set of odd whole numbers greater than 0.

22. Write set notation for the solution set of equation $(x - 2)(x + 3) = 0$.

Determine whether true or false in Exercises 23–25.

23. $2 \in \{-3,2,4,5,7\}$.

24. $3 \in \{0,2,4,6,8,\ldots\}$.

25. $\{0,1,7\} \subset \{0,1,8,13\}$.

26. Find $\{0,3,5,7\} \cap \{0,1,3,11\}$.

27. Find $\{3,5,9\} \cup \{1,6,7\}$.

28. If U = {3 red marbles, 7 blue marbles, 8 white marbles} and A = {8 white marbles}, find $\overline{A}$. What is $\overline{\overline{A}}$ (the complement of $\overline{A}$)?

29. Using the information in Exercise 28, find $A \cup \overline{A}$, $A \cap \overline{A}$, $\overline{U}$, and $\overline{\phi}$.

30. Determine which of the following graphs are also functions:

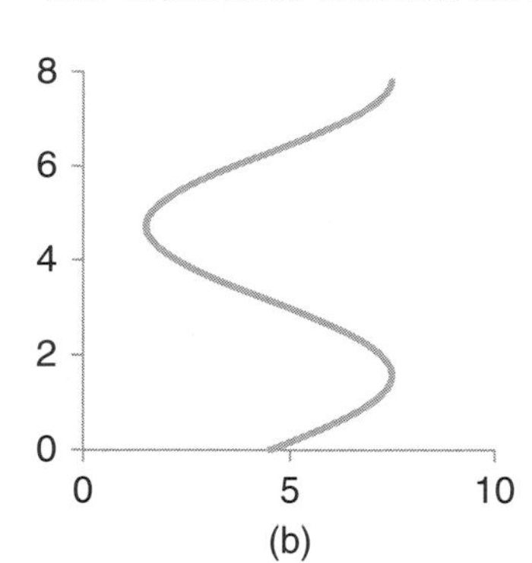

(b)

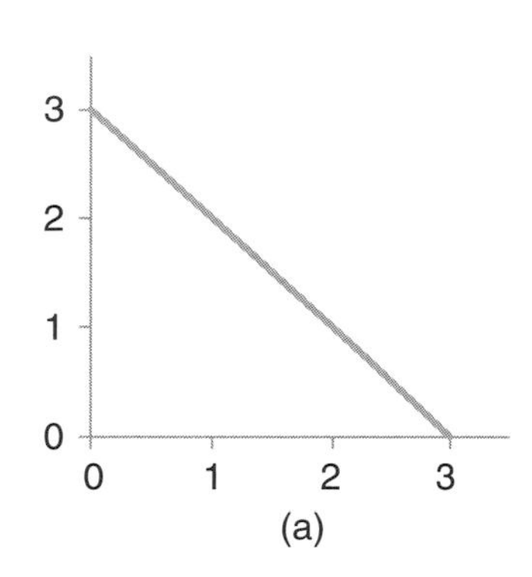

(a)

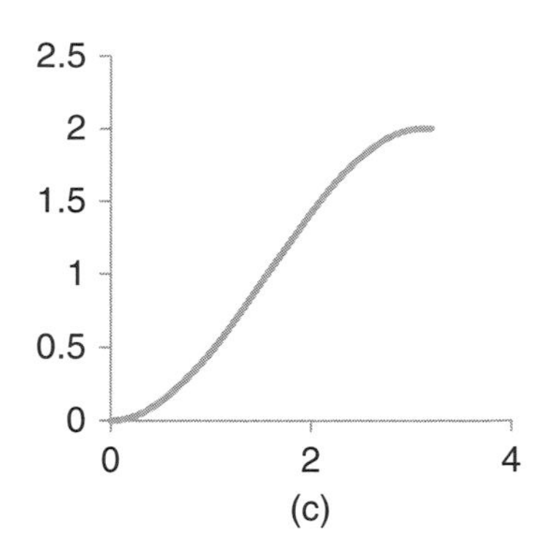

(c)

Graph the following functions:

31. $2x + y - 10 = 0.$

32. $2y = \frac{1}{x}, \quad x \neq 0.$

33. $y - 3\sqrt{x} = 5.$

Find dy/dx for the functions given in Exercises 34–43.

34. $y = x^{3/4}.$

35. $y = \frac{1}{x^3}, \quad x \neq 0.$ Evaluate $\frac{dy}{dx}$ at $x = 1$.

36. $y = 15.$

37. $y = \left(\frac{1}{2}\right).$

38. $y = 5x^3$. Evaluate $\frac{dy}{dx}$ at $x = 2$.

39. $y = \left(\frac{1}{5}\right)x^5.$

40. $y = 6x + 7.$

41. $y = -0.5x + 3.$

42. $y = 3x^2 + 4\left(\frac{1}{x^2}\right), x \neq 0.$

43. $y = 5x^3 + 4x^2 + 3x + 2.$

44. **a.** If $f(x) = 2x^2 + 1$, $g(x) = 7x^3$, and $y = f(x)g(x)$, find $\frac{dy}{dx}$.

b. If $f(x) = 5x^3$, $g(x) = 6 + 4x^5$, and $y = f(x)g(x)$, find $\frac{dy}{dx}$.

45. **a.** If $f(x) = 3x - 6$, $g(x) = 4x^2 + x$, and $y = f(x)/g(x)$, find $\frac{dy}{dx}$.

b. If $f(x) = 5x^3 - 10$, $g(x) = -x^2 + x$, and $y = \frac{f(x)}{g(x)}$, find $\frac{dy}{dx}$.

46. **a.** If $f(x) = 5x^7 + 2x^2$, find $f'''(x)$.

b. If $f(x) = 9x^2 - 11x^5 + 7x$, find $f'''(x)$.

47. $\int \frac{1}{x^2} dx.$

48. $\int x^3 dx.$

49. $\int \left(3x^2 - 4x + 8\right) dx.$

50. $\int xe^{2x}dx$.

(*Hint:* Use integration by parts.)

51. $\int_a^b \frac{1}{(b-a)} dx$.

52. $\int_0^t 100e^{-0.1x}dx$.

53. Given

$$X_1 = -2, \quad X_2 = -3, \quad X_3 = 0, \quad X_4 = 5, \quad X_5 = 9$$
$$Y_1 = 5, \quad Y_2 = 9, \quad Y_3 = 6, \quad Y_4 = 8, \quad Y_5 = 7$$

a. Evaluate the Pearson correlation coefficient

$$r = \sum_{i=1}^{5} \frac{(X_i - \overline{X})(Y_i - \overline{Y})}{\sqrt{\sum_{i=1}^{5} (X_i - \overline{X})^2 \sum_{i=1}^{5} (Y_i - \overline{Y})^2}}, \quad (1 \leq r \leq 1),$$

where $\overline{X} = \frac{\sum X_i}{5}$ and $\overline{Y} = \frac{\sum Y_i}{5}$. How is r interpreted?

b. Also, reevaluate r using the following "short formula":

$$r = \frac{\sum X_i Y_i - \frac{\sum X_i \sum Y_i}{n}}{\left(\left(\sum X_i^2 - \frac{(\sum X_i)^2}{n}\right)\left(\sum Y_i^2 - \frac{(\sum Y_i)^2}{n}\right)\right)^{\frac{1}{2}}}.$$

(*Hint:* To facilitate your calculations, construct a table.)

54. Using the data and the results in Exercise 53, compute $\frac{r \times \sqrt{3}}{\sqrt{1-r^2}}$.

55. Using the data and results in Exercise 53,

a. Compute the slope value, $b = \frac{\sum_{i=1}^{5} (X_i - \overline{X})(Y_i - \overline{Y})}{\sum_{i=1}^{5} (X_i - \overline{X})^2}$, and the intercept value, $a = \overline{Y} - b\overline{X}$, of the linear equation $Y = a + bX$. Use this estimated equation to predict the value of Y when $X = 11$.

b. Also, reevaluate the slope value b using the following "short formula":

$$b = \frac{\sum X_i Y_i - \frac{\sum X_i \sum Y_i}{n}}{\left[\sum X_i^2 - \frac{(\sum X_i)^2}{n}\right]}.$$

56. Repeat Exercise 53 with the following values of the variables:

$$X_1 = 7, \quad X_2 = 1, \quad X_3 = 4, \quad X_4 = 8$$
$$Y_1 = 5, \quad Y_2 = 0, \quad Y_3 = 10, \quad Y_4 = 1$$

57. Find the derivative of $y = f(x)g(x)$, where

a. $f(x) = 9x+7$, $g(x) = 2x^3 + x$.

b. $f(x) = 4x^2 - 3x$, $g(x) = x^7 - 7$.

58. If $f(x) = 3x^6 - 2x^2$, $g(x) = 4x^3 + x^2$, and $y = f(x)\,g(x)$, find $\frac{dy}{dx}$.

59. Find the derivative of $y = \frac{f(x)}{g(x)}$, where

a. $f(x) = x^2 - 9$, $g(x) = x + 3$.

b. $f(x) = \frac{1}{x} + x$, $x \neq 0$, $g(x) = 2x^3 + 3$.

60. a. If $f(x) = 6x^9 - 5x^{-2}$, find $f'(x)$, $f''(x)$, and $f'''(x)$.

b. If $f(x) = 8x^3 - 7x^2$, find $f'(x)$, $f''(x)$, and $f'''(x)$.

EXCEL APPLICATIONS

61. Answer Exercise 53, both parts (a) and (b), by formulating appropriate Excel spread sheets.

62. Also answer Exercise 55, both parts (a) and (b), by formulating appropriate Excel spread sheets.

63. Solve the simultaneous equations given in Exercises 4, 6, and 8 (a) earlier using the Solver Add-in in Excel.

APPENDIX 1.A: A REVIEW OF BASIC MATHEMATICS

1.A.1 Order of Arithmetic Operations

In simplifying arithmetic expressions, it is conventional to follow the order given below:

1. Evaluate items in parentheses (•) first, items in braces {•} second, and items in large brackets [•] next.
2. Exponents.
3. Multiplication or division, whichever comes first, left to right.
4. Addition or subtraction, whichever comes first, left to right.

EXAMPLE A.1

$$\text{Solve}: \left[\{(20-12)\div(7-5)\}+2\right]\times 7+\sqrt{36}+9^2+11-4\times\left(\frac{6}{\sqrt{144}}\right)$$

$$= \left[\{8\div 2\}+2\right]\times 7+\sqrt{36}+9^2+11-4\times\left(\frac{6}{\sqrt{144}}\right)\text{(parentheses first)}$$

$$= [4+2]\times 7+6+81+11-4\times\frac{6}{12}\text{(powers and/or square roots next)}$$

$$= 6\times 7+6+81+11-4\times\frac{1}{2}\text{(large brackets next)}$$

$$= 42+6+81+11-2\text{(multiplication)}$$

$$= 138\text{(addition and subtraction, left to right).}$$

Note : Define the positive square root (denoted $\sqrt{\cdot}$) of a positive number as positive.

For example, $\sqrt{25} = 5$.

■

EXAMPLE A.2

$$\text{Solve}: \left[(8^2) \times 4\right] \times 7 + 16$$
$$= [64 \times 4] \times 7 + 16$$
$$= 256 \times 7 + 16$$
$$= 1792 + 16$$
$$= 1808.$$

■

1.A.2 Exponents

An *exponent* or *power* is a symbol placed at the upper right of the *base* (a number or variable), which indicates the number of times the latter is to be multiplied by itself. For example, $3^4 = 3 \times 3 \times 3 \times 3 = 81$. Here, 3 is called the base and 4 is called the power or exponent. In general, $a^p = a \times a \times a \times a \cdots$ (p times). With base e, the exponential function $y = e^x$ has several applications. Some of these are shown later in Chapter 2. Here, e is an irrational number (it can be expressed as a nonrepeating and unending decimal) introduced by Leonhard Euler (1707–1783). It is expressed as

$$e = 1 + \left(\frac{1}{1}\right) + \left(\frac{1}{1 \times 2}\right) + \left(\frac{1}{1 \times 2 \times 3}\right) + \left(\frac{1}{1 \times 2 \times 3 \times 4}\right) + \cdots$$

and approximately equals to 2.7182818. Figure A.1 shows $y = e^x$.

Exponents are extremely useful in simplifying expressions in multiplication and division involving a combination of algebraic and numerical values. There are several applications of exponents in economics and business dealing with growth rates, especially continuous compounding. Exponents are useful in simplifying both linear and nonlinear functions and comparing their graphs. They are also used in the smoothing of functions in forecasting.

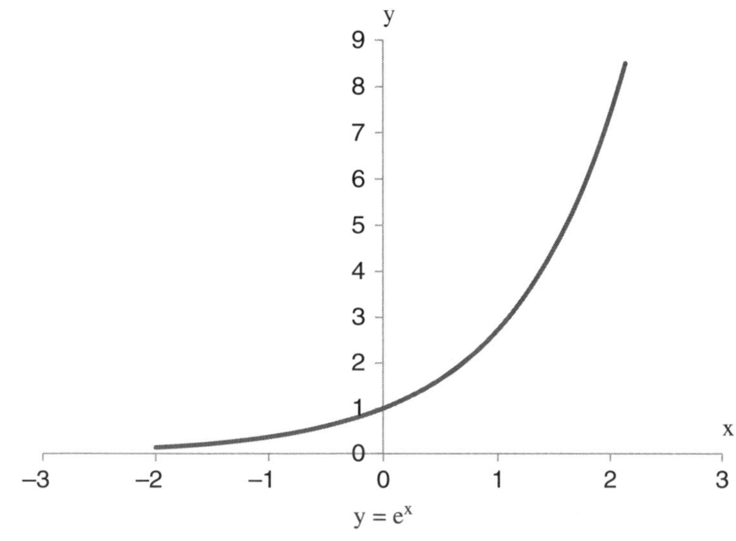

Figure A.1 Graph of the exponential function e^x.

Rules of Exponents

1. $a^p \times a^q = a^{(p+q)}$.
  Example: $3^3 \times 3^2 = 3^{3+2} = 3^5 = 3 \times 3 \times 3 \times 3 \times 3 = 243$.

2. $(a^p)^q = a^{p \times q}$.
  Example: $(3^3)^2 = 3^{(3\times2)} = 3^6 = 3 \times 3 \times 3 \times 3 \times 3 \times 3 = 729$.

3. $(a \times b)^p = a^p \times b^p$.
  Example: $(2 \times 4)^3 = 2^3 \times 4^3 = 8 \times 64 = 512$.

4. $\left(\frac{a}{b}\right)^p = \frac{a^p}{b^p}$.
  Example: $\left(\frac{2}{3}\right)^3 = \left(\frac{2^3}{3^3}\right) = \left(\frac{8}{27}\right) = 0.2963$.

5. $\frac{a^p}{a^q} = a^{p-q}$.
  Example: $\left(\frac{6^4}{6^2}\right) = 6^{4-2} = 6^2 = 36$.

6. $a^{1/p} = \sqrt[p]{a}$
  Example: $64^{1/6} = \sqrt[6]{64} = 2$.
  Note: If p is even, a has to be positive to get a real number. If a is negative, $\sqrt{a}$ is not a real number. If p is odd, then $\sqrt[p]{a}$ is defined for both positive and negative values of a. For example, $\sqrt[6]{-64}$ is not a real number, while $\sqrt[3]{-64} = -4$.

7. $a^{p/q} = \sqrt[q]{a^p}$.
  Example: $(64)^{5/6} = \sqrt[6]{64^5} = \left(\sqrt[6]{64}\right)^5 = 2^5 = 32$.

8. $a^{-p} = \frac{1}{a^p}$.
  Example: $3^{-3} = \left(\frac{1}{3^3}\right) = \left(\frac{1}{27}\right) = 0.0370$.

9. $a^1 = a$.
  Example: $6^1 = 6$.

10. $a^0 = 1 (a \neq 0)$.
  0° is not defined.
  Example: $10^\circ = 1$.

11. $1^p = 1$.
  Example: $1^4 = 1$.

EXAMPLE A.3

$$\text{Simplify}: \frac{(a^3)^2 \times (b^{(4/5)})^3}{a^4 \times b^4}$$

$$= \left(\frac{a^6 \times b^{(12/5)}}{a^4 \times b^4}\right)$$

$$= a^{6-4} \times b^{(12/5)-4}$$

$$= a^2 \times b^{-((8/5))}$$

$$= \frac{a^2}{\sqrt[5]{b^8}} \quad \text{or} \quad \frac{a^2}{b\sqrt[5]{b^3}}.$$

■

EXAMPLE A.4

$$\text{Simplify}: \frac{\left(x^3\right)^{1/9} \times y^{3/4}}{\left(\frac{x}{y}\right)^{3/4}}$$

$$= \frac{x^{3/9} \times y^{3/4}}{x^{3/4} \times y^{-(3/4)}}$$

$$= x^{(3/9)-(3/4)} \times y^{(3/4)+(3/4)}$$

$$= x^{-(15/36)} \times y^{6/4}$$

$$= \frac{y^{3/2}}{x^{5/12}}.$$

■

1.A.3 Logarithms

The *logarithm* of a positive number y is the power to which the base must be raised to equal the given number. In general, if $y = a^x$, thenx $= \log_a y$, where a is the base. For example, the logarithm of 8 to the base 2 is 3, because 3 is the power to which 2 must be raised to produce 8 ($2^3 = 8, \log_2 8 = 3$). If $y = 1$, $\log y = 0\left(1 = a^0 \text{ or } x = 0\right)$. Clearly, a logarithm is a fancy way to write an exponent.

The base a cannot be 1. If $a = 1$, the logarithm is still defined but is trivial. If we want to solve $x = \log_1 y$, we can convert the log to exponential form, $y = 1^x$. In this case, y must be 1 because any real power of 1 is 1, and x can be any real number. So, there is no certain answer when $a = 1$.

There is a one-to-one relationship between the logarithmic functions and exponential functions. Table A.1 shows some examples of this relationship.

Special Cases in Logarithms

Common logs:	If the base a is 10, the logarithms are called *common logs.*
Natural (Napierian) logs:	If the base a is the mathematical constant e, the logarithms are called *natural logs* and denoted "ln" for short. These logs are used in the natural and social sciences to represent growth or decay.

Table A.1 Equivalence of Exponential and Logarithmic Functions: Some Examples

Exponential	Logarithmic
$y = a^x$	$x = \log_a y$
$8 = 2^3$	$3 = \log_2 8$
$16 = \left(\frac{1}{2}\right)^{-4}$	$-4 = \log_{1/2} 16$
$\frac{1}{25} = 5^{-2}$	$-2 = \log_5 \left(\frac{1}{25}\right)$
$1000 = 10^3$	$3 = \log_{10} 1000$
$81 = 3^4$	$4 = \log_3 81$

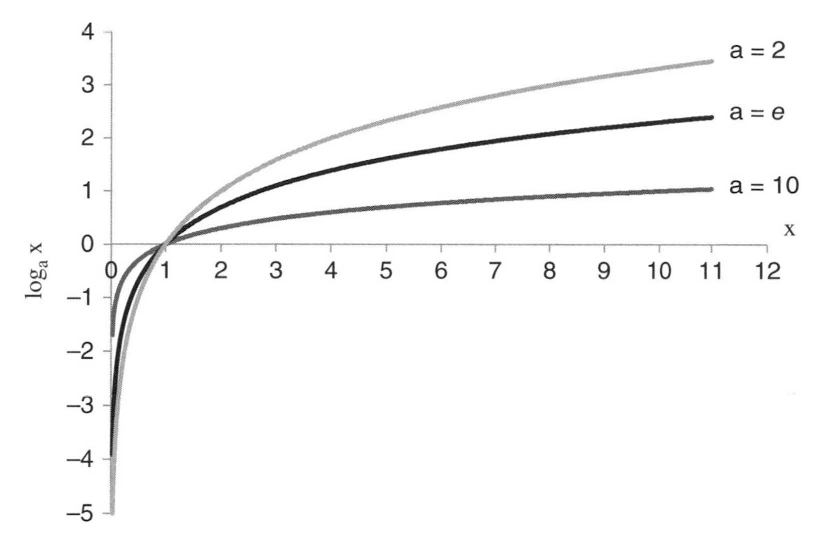

Figure A.2 Graphs of logarithmic functions with three different bases.

Binary (Base 2) logs: If the base a is 2, the logarithms are called *binary logs* and denoted "lb." Binary logs are commonly used in computer science and information theory.

Figure A.2 depicts common, natural, and binary logs.

Most calculators can be used to evaluate common logs as well as natural logs. For example, we can find the common log of 0.00127 using a calculator's "log function" as $\log_{10} 0.00127 = -2.896$. Scientific calculators typically have both "log" and "ln" buttons. For example, if we want to find the natural logs of 157.2 and 0.2759, we can use "ln function." So, $\log_e 157.2 = \ln 157.2 = 5.0575$. Similarly, $\ln 0.2759 = -1.2877$.

Tables provided in some textbooks can also be used to find the logs of numbers or numbers from logs (also called antilogs).

Rules of Logarithms

1. *Product Rule:* If x and y are positive numbers, then $\log_a xy = \log_a x + \log_a y$

The logarithm of a product of two numbers equals the sum of logarithms of the individual numbers. This is similar to the rule that the exponent of the product of two numbers to the same base equals the sum of the exponents of the same base. (See rule 1, Section 1.2.)

EXAMPLE A.5

a. $\log_{10}(3 \times 7) = \log_{10}3 + \log_{10}7$.

b. $\log_4 7 + \log_4 5 = \log_4(7 \times 5) = \log 35$.

c. $\log_7 7x = \log_7 7 + \log_7 x = 1 + \log_7 x$, since $\log_7 7 = 1$.

d. $\log_{10}(357 \times 269) = \log_{10}357 + \log_{10}269$.

■

2. *Quotient Rule:* If x and y are positive numbers, then $\log_a(x/y) = \log_a x - \log_a y$.

EXAMPLE A.6

a. $\log_4(7/3) = \log_4 7 - \log_4 3$.

b. $\log_3(y/8) = \log_3 y - \log_3 8$. ■

3. *Power Rule:* $\log_a x^k = k \log_a x$.

This rule follows from the product rule above. For example, if $y = x$ in the product rule, then

$$\log_a xy = \log_a x^2 = \log_a x + \log_a x = 2 \log_a x.$$

EXAMPLE A.7

a. $\log_3 x^7 = 7 \log_3 x$.

b. $2\log_a x^n = 2n\log_a x$. ■

4. *Rule of Addition or Subtraction:* $\log_a(x \pm y) \neq \log_a x \pm \log_a y$.

This rule follows the one-to-one relationship between logs and exponents: $a^{x+y} \neq a^x + a^y$.

EXAMPLE A.8

Find the common log of 100.

Any positive number can be represented as a power of 10. Thus, $100 = 10^2$. Therefore, $\log_{10} 100 = 2$. ■

1.A.4 Prime Factorization

Factoring to prime numbers is a process by which a number is written as a product of its prime factors. A *prime number* is a natural number that can only be divided by 1 and itself without a remainder. Prime factorization is a widely used process in dealing with fractions. The following are the smallest prime numbers less than 50:

$$2,3,5,7,11,13,17,19,23,29,31,37,41,43,47.$$

The number 2 is the only even prime number because all other nonzero even numbers can be divided by 2.

EXAMPLE A.9

Find the prime factors of 18, 44, 54, and 120.

$$18 = 2 \times 3^2$$
$$44 = 2^2 \times 11$$
$$54 = 2 \times 3^3$$
$$120 = 2^3 \times 3 \times 5$$

■

1.A.5 Factoring

Monomial Factoring

Monomial factoring is a process by which an expression such as (ab + ac) is written as a product of its factors a and (b + c). For example, to factorize the expression 8xy − 4x, first we need to find the greatest common factor of 8xy and 4x, which is 4x. Then rewrite the expression as (4x)(2y − 1). Similarly, the factors of the expression 5ax − 20bx are 5x and (a − 4b).

EXAMPLE A.10

Factorize the following expressions:

a. $3x^2 + 5xy$.

b. $6x^3y^2 - 24x^2y + 18x^4y^3$.

SOLUTIONS:

a. $3x^2 + 5xy = x(3x + 5y)$.

b. $6x^3y^2 - 24x^2y + 18x^4y^3 = 6x^2y(xy) - 6x^2y(4) + 6x^2y(3x^2y^2) = 6x^2y(xy - 4 + 3x^2y^2)$.

Here, $6x^2y$ is the greatest common factor. ■

Binomial Factoring

Binomial factoring is used when the expression is a quadratic. A *quadratic expression* is a second-degree polynomial equation that has the form $y = ax^2 + bx + c$. When a is 1, the expression has the form of $y = x^2 + bx + c$, and factoring is simpler. In this case, find two integers, K and L, such that KL = c and K + L = b. In case a is not equal to 1, find two integers K and L such that KL = ac and K + L = b.

For example, if we want to factorize $y = x^2 - 5x + 6$, we need find two integers, K and L, such that KL = 6, and K + L = −5. In this case, factors are (x + K) and (x + L), such that $(x + K)(x + L) = x^2 + (K + L)x + KL$. By trial and error, we find K = −2 and L = −3. Thus, $x^2 - 5x + 6 = (x - 2)(x - 3)$.

EXAMPLE A.11

a. Factorize the expression $y = 2x^2 + 6x - 8$.

The expression can be written as $y = 2(x^2 + 3x - 4)$. As $x^2 + 3x - 4$ has the form of $y = x^2 + bx + c$, we need to find two integers, K and L, such that KL = −4 and K + L = 3. In this case, K and L are −1 and 4, respectively. So we can write the expression as $y = 2(x - 1)(x + 4)$. Note that $y = 2x^2 + 6x - 8$ has three factors.

b. Find the factors of $2x^2 + x - 6$.

We need to find two integers, K and L, such that KL = 2(−6) = −12, the product of the coefficient of x^2, a, and the constant c, and K + L = 1, the coefficient of x, b. By trial and error, we see that K = 4 and L = −3. Then, we rewrite the given expression as $2x^2 + 4x - 3x - 6$ and factor by grouping $2x(x + 2) - 3(x + 2)$ or $(x + 2)(2x - 3)$. ■

EXAMPLE A.12

Find the factors of $y = 3x^2 - 11x - 4$.

First, we need to find two integers K and L, such that $KL = 3 \times -4 = -12$ and $K + L = -11$. In this case, K and L are −12 and +1 ($(-12) \times 1 = -12$ and $(-12) + 1 = -11$). So, we can rewrite the expression as

$$\begin{aligned} y &= 3x^2 + x - 12x - 4, \\ y &= x(3x + 1) - 4(3x + 1), \\ y &= (3x + 1)(x - 4). \end{aligned}$$

■

1.A.6 Fractions: Ratios

A *fraction* is a ratio of the form x/y, where y is different from zero and x and y are both whole numbers. x is known as the *numerator* and y the *denominator*. Ratios $\frac{13}{4}$, $\frac{4}{9}$, and $\frac{-6}{7}$ are examples of such fractions. If the numerator of a fraction is less than the denominator, the fraction is called a *proper fraction*. When the numerator is greater than or equal to the denominator, the fraction is called an *improper fraction*. Thus, $\frac{4}{9}$ and $\frac{-6}{7}$ are proper fractions and $\frac{13}{4}$ is an improper fraction. Note that the ratio becomes a percent when the denominator becomes 100. For example, 67% means a ratio of $\frac{67}{100}$.

1.A.6.1 Greatest Common Factor The *greatest common factor* (GCF), also known as the greatest common denominator or greatest common divisor, is the largest number that is a common divisor of two or more non-zero numbers with zero remainder. GCF is used to reduce fractions to their lowest terms, which are called *irreducible fractions*. When the fractions are irreducible, the greatest common factor is 1.

To find the greatest common factor, prime factors of each number should be listed. Then the numbers that are common should be multiplied to find the GCF.

EXAMPLE A.13

Reduce the fraction 36/48.

To reduce, we need to find the greatest common factor of 36 and 48.

$$\begin{aligned} &\text{Prime factors of 36:} \quad 2 \times 2 \times 3 \times 3. \\ &\text{Prime factors of 48:} \quad 2 \times 2 \times 2 \times 2 \times 3. \end{aligned}$$

Here, $2 \times 2 \times 3$ is common in both numbers. So, $2 \times 2 \times 3 = 12$ is the greatest common factor.

$$\text{GCF}(36, 48) = 12.$$

So, the fraction can be written as

$$\frac{36}{48} = \frac{(3 \times 12)}{(4 \times 12)} = \frac{3}{4}.$$

■

1.A.6.2 Least Common Multiple The *least common multiple* (LCM) or least common denominator is the smallest positive number that is the common multiple of two or more nonzero numbers. LCM of two or more numbers can be divided by these numbers with no remainder. It is required to find the lowest common denominator when adding or subtracting fractions.

There are several ways to find the least common multiple. One way is to list the multiples of the numbers and then find the smallest common one. The other way is to prime factorize the numbers, and then multiply all the primes at their highest powers.

EXAMPLE A.14

Find the least common multiple of 6 and 8.

Method 1
Multiples of 6: 0,6,12,18,24,30,36, . . .
Multiples of 8: 0,8,16,24,32,40,48, . . .

Here, 24 is the lowest nonzero multiple for both numbers; 24 can be divided by both 6 and 8. So,

$$\mathrm{LCM}(6,8) = 24.$$

Method 2
Prime factors of 6: 2×3
Prime factors of 8: $2 \times 2 \times 2 = 2^3$.

Here, there are two primes, 2 and 3. To find the LCM, we should take the primes at their highest powers and multiply them. So, we have 2^3 and 3^1. If we multiply these, we will find

$$\mathrm{LCM}(6,8) = 2^3 \times 3 = 8 \times 3 = 24.$$

■

1.A.6.3 Addition and Subtraction of Fractions If the denominator is common to the fractions, which are called *like fractions*, add or subtract the numerators and divide by the common denominator to obtain the result.

EXAMPLE A.15

a. $\frac{15}{19} + \frac{6}{19} = \frac{15+6}{19} = \frac{21}{19}.$

b. $\frac{15}{19} - \frac{6}{19} = \frac{15-6}{19} = \frac{9}{19}.$

However, if the denominator is not common to the fractions (we have *unlike fractions*), then the following procedure is used to obtain the least common denominator (LCD). First, note that multiplying the numerator and denominator by the same nonzero number does not change the value of the fraction. Second, the usual procedure for obtaining the common denominator is to multiply the numerator and the denominator of each fraction by the ratio of the LCM and the denominator of each fraction. ■

EXAMPLE A.16a

Evaluate $\frac{7}{6} + \frac{11}{4}$.

We first find LCD.

Step 1: Factor each denominator as a product of primes.

Step 2: Figure out the greatest number of times that each prime factor occurs in any single factorization.

Step 3: LCM is the product of the factors found in step 2.

Thus, $6 = 3 \times 2$ and $4 = 2 \times 2$.

$$\text{LCM} = 3 \times 2 \times 2 = 12.$$

Thus, 12 is the LCM. In relation to fractions, note that LCD is the same as LCM.

With $12 \div 6 = 2$, the fraction $\frac{7}{6} = \frac{7 \times 2}{6 \times 2} = \frac{14}{12}$.

And with $12 \div 4 = 3$, the fraction $\frac{11}{4} = \frac{11 \times 3}{4 \times 3} = \frac{33}{12}$.

$$\frac{7}{6} + \frac{11}{4} = \frac{14}{12} + \frac{33}{12} = \frac{47}{12}.$$

■

EXAMPLE A.16b

Find $\frac{8}{6} + \frac{21}{18} - \frac{11}{12}$.

To solve this expression, we need to find the least common multiple of 6, 18, and 12.

$$\begin{aligned} 6 &= 3 \times 2 \\ 18 &= 3 \times 3 \times 2 \\ 12 &= 3 \times 2 \times 2 \end{aligned}$$

In this case, $\text{LCD} = \text{LCM} = 36 = 3 \times 3 \times 2 \times 2$, so that

$$\begin{aligned} \frac{8}{6} &= \frac{8 \times 6}{6 \times 6} = \frac{48}{36}, \\ \frac{21}{18} &= \frac{21 \times 2}{18 \times 2} = \frac{42}{36}, \\ \frac{11}{12} &= \frac{11 \times 3}{12 \times 3} = \frac{33}{36}. \end{aligned}$$

$$\text{Thus, } \frac{8}{6} + \frac{21}{18} - \frac{11}{12} = \frac{48}{36} + \frac{42}{36} - \frac{33}{36} = \frac{48 + 42 - 33}{36} = \frac{57}{36}.$$

■

1.A.6.4 Multiplication and Division of Fractions The product of any two fractions is obtained by the ratio of the product of the two numerators and the product of the two denominators.

When dividing a fraction by another fraction, take the reciprocal of the second action and multiply it by the first fraction, following the previous procedure.

EXAMPLE A.17

a. $\frac{8}{5} \times \frac{10}{3} = \frac{8 \times 10}{5 \times 3} = \frac{80}{15} = \frac{16}{3}$.

b. $\frac{-9}{12} \times \frac{2}{3} = \frac{(-9) \times 2}{12 \times 3} = \frac{-18}{36} = \frac{-1}{2}$.

■

EXAMPLE A.18

a. $\frac{2}{7} \div \frac{9}{5} = \frac{2}{7} \times \frac{5}{9} = \frac{2 \times 5}{7 \times 9} = \frac{10}{63}$.

b. $\left(\frac{4}{7}\right) \div \left(\frac{3}{(-5)}\right) = \left(\frac{4}{7}\right) \times \left(\frac{(-5)}{3}\right) = \left(\frac{4 \times (-5)}{7 \times 3}\right) = \left(\frac{-20}{21}\right)$.

c. $10 \div \left(\frac{3}{2}\right) = \left(\frac{10}{1}\right) \times \left(\frac{2}{3}\right) = \left(\frac{10 \times 2}{1 \times 3}\right) = \left(\frac{20}{3}\right)$. ■

1.A.6.5 Mixed Fractions A *mixed fraction* is the sum of a whole number and a fraction less than 1.

EXAMPLE A.19

Express in terms of thirds.

$$4\frac{2}{3} = 4 + \frac{2}{3} = \frac{4}{1} + \frac{2}{3} = \frac{4 \times 3}{1 \times 3} + \frac{2 \times 1}{3 \times 1} = \frac{12}{3} + \frac{2}{3} = \frac{12 + 2}{3} = \frac{14}{3}.$$ ■

EXAMPLE A.20

Express $2\frac{3}{7}$ as ratio.

An easy way to do this is to multiply 2 by 7 and add 3 to get the numerator of the ratio and the denominator remains the same. Thus,

$$2\frac{3}{7} = \frac{(2 \times 7) + 3}{7} = \frac{17}{7}.$$

When adding or subtracting mixed fractions, whole parts and fractional parts can be added or subtracted separately, as shown in the following example. ■

EXAMPLE A.21

Simplify, $$7\frac{1}{3} + 2\frac{3}{4} + \frac{1}{2} - 1\frac{7}{12}.$$

The expression can be written as $\left(7 + \frac{1}{3}\right) + \left(2 + \frac{3}{4}\right) + \left(\frac{1}{2}\right) - \left(1 + \frac{7}{12}\right)$.

If we move the whole numbers, $(7 + 2 - 1) + \left(\frac{1}{3} + \frac{3}{4} + \frac{1}{2} - \frac{7}{12}\right)$,

LCD for 3,4,2, and 12 is 12. So, $8 + \left(\frac{4}{12} + \frac{9}{12} + \frac{6}{12} - \frac{7}{12}\right)$

$$= 8 + \frac{12}{12} = 8 + 1 = 9.$$ ■

1.A.7 Decimals

A *decimal* is a number that involves the following three parts:

- **i.** Whole number
- **ii.** Decimal point
- **iii.** Numbers whose denominators are powers of 10

EXAMPLE A.22

Show the position of each digit in 15.231.

Here, 15 is the whole number and "." is the decimal point. The numbers after decimal are, respectively, 2 in "tenths", $\left(\frac{1}{10}\right)$, 3 in "hundredths", $\left(\frac{1}{100}\right)$, and 1 in "thousandths," $\left(\frac{1}{1000}\right)$. Thus,

$$\begin{aligned}15.231 &= 15 + \left(\frac{2}{10}\right) + \left(\frac{3}{100}\right) + \left(\frac{1}{1000}\right)\\ &= (1 \times 10) + (5 \times 1) + 2 \times \left(\frac{1}{10}\right) + 3 \times \left(\frac{1}{100}\right) + 1 \times \left(\frac{1}{1000}\right)\\ &= 10 + 5 + 0.2 + 0.03 + 0.001.\end{aligned}$$

■

1.A.7.1 Operations with Decimals

Addition or Subtraction These operations are carried out after aligning the numbers in their appropriate positions.

EXAMPLE A.23

- **a.** $175.31 + 21.5692 = 175.31 + 21.5692 = 196.8792.$
- **b.** $80.05 - 108.5381 = -108.5381 + 80.05 = -28.4881.$

■

Multiplication

To multiply two numbers with decimals, first obtain the product of the two numbers ignoring the decimal points. Next, count the "total" number of positions after the decimal points in both the numbers and place the decimal point in the product number after the "total" number of positions counting from right to left.

EXAMPLE A.24

- **a.** Simplify 3.13×0.002.

Step 1: $313 \times 2 = 626$.
Step 2: Total number of positions after the decimal in both the numbers, $2 + 3 = 5$.
Step 3: 5 positions, 0.00626.

b. Simplify 7.47×5.0239.

Step 1: $747 \times 50239 = 37528533$.
Step 2: Total number of positions after the decimal in both the numbers, $2 + 4 = 6$.
Step 3: 6 positions, 37.528533. ■

Division To divide a decimal by another, first multiply both the numerator and the denominator by a power of 10 until they become whole integers. The power of 10 is decided based on the maximum number of digits after the decimal. The following example demonstrates this particular procedure.

EXAMPLE A.25

Simplify $125,488 \div 2.48$.

To find $\frac{125,488}{2.48}$, solve $\frac{12,548,800}{248} = 50,600$. Then,

$125,488 \div 2.48 = 50,600$. ■

Rounding Off Decimals Sometimes, when there are a large number of decimal places in a number and an exact answer is not needed, you can round off decimals to a specified place, depending upon the desired accuracy needed.

Rule: If the digit to the right of the digit to be rounded off is 5 or more, then you round off the digit by 1 and drop the digits to the right of the rounded off digit. Thus, 3.142857 rounded off to the third decimal place is 3.143, since the fourth number after the decimal, 8, is greater than 5, and the other digits beyond 8 are dropped. Similarly, if the digit to the right of the digit to be rounded off is less than 5, do not change the digit to be rounded off and drop all the digits to the right of the digit to be rounded off. Thus, 3.142857 rounded off to two decimal places is 3.14, since the decimal number after the second decimal, 2, is less than 5, and all digits to the right of second decimal place are dropped.

EXAMPLE A.26

Round off the following to four decimal places:

a. 1.4142136, Ans : 1.4142.

b. 2.3178623, Ans : 2.3179. ■

1.A.8 Percent

Percent is a ratio of two numbers, where the denominator is 100. A percent is shown with the symbol "%", which means $1/100$. For example, 20% means $20/100 = 1/5$. So 1 out of 5 units is 20%. Similarly, -1.5% is $-1.5/100 = -15/1000 = -3/200$. This concept is useful in the comparison of several situations involving data. It is to be noted, however, that the denominator 100 as a base is arbitrarily used for convenience, and the actual percentage figures are based on the actual data values.

EXAMPLE A.27

In a class of 45, there are 27 male and 18 female students. What are the percentages of the female and male students in the class?

The percentage of female students is

$$\frac{18}{45} = 0.4 = \frac{40}{100} = 40\%.$$

Similarly, the percentage of male students is

$$\frac{27}{45} = 0.6 = \frac{60}{100} = 60\%.$$

Note that sum of the percentages of male and female students is $40\% + 60\% = 100\%$. ■

1.A.8.1 Finding Percent (%) of a Number To determine percent of a number, change the percent to a fraction or decimal (whichever is easier for you) and multiply this fraction (or the decimal) with the number. Remember, the word "of" means multiply.

Note: Remember that the symbol % means $\frac{1}{100}$.

You may want to adapt the following rules to find answers. Turn the question word-for-word into an equation. For "what" substitute the letter x; for "is", substitute an equal sign; and for "of", substitute a multiplication sign. Change percent to decimals or fractions, whichever you find easier. Then solve the equation.

EXAMPLE A.28

a. What is 20% of 80?

$$\frac{20}{100} \times 80 = \frac{1600}{100} \text{ or } 0.20 \times 80 = 16.00 = 16.$$

b. What is 12% of 50?

$$\frac{12}{100} \times 50 = \frac{600}{100} = 6 \quad \text{or} \quad 0.12 \times 50 = 6.00 = 6.$$

c. What is $\frac{1}{2}$% of 18?

$$\frac{1/2}{100} \times 18 = \frac{1}{200} \times 18 = \frac{18}{200} = \frac{9}{100} \quad \text{or} \quad 0.005 \times 18 = 0.09.$$

d. What is 0.4% of 80?

$$\frac{0.4}{100} \times 80 = 0.004 \times 80 = 0.32.$$

e. 18 is what percent of 90?

Suppose x is the percent of 90. Then

$$18 = \frac{x}{100}(90), \quad x = \left(\frac{18}{90}\right) \times 100 = \frac{100}{5} = 20.$$

f. 10 is 0.5% of what number?

$$10 = \frac{0.50}{100}(x), \quad \left(\frac{1000}{0.50}\right) = x, \quad 2000 = x.$$

g. What is 15% of 60?

$$x = \left(\frac{15}{100}\right) \times 60 = \left(\frac{90}{10}\right) = 9 \quad \text{or} \quad 0.15(60) = 9.$$

■

1.A.8.2 Finding Percentage Increase (+) or Percentage Decrease (−) To find the *percentage change* (increase or decrease), use the following formula:

$$\frac{\text{Actual change}}{\text{Starting value}} \times 100\% = \text{percentage change} = \left(\frac{\Delta X}{X_1}\right) \times 1,$$

where $\Delta X = X_2 - X_1$ denotes the actual change in X, X_1 = starting value, and X_2 = end value.

EXAMPLE A.29

f. What is the percentage change of a $500 item that is sold for $400?

Here $X_1 = \$500$,

$X_2 = \$400$,

$\Delta X = \$400 - \$500 = -\$100$.

$$\text{Percentage change} = \left(\frac{-100}{500}\right) \times 100\% = -\left(\frac{1}{5}\right) \times 100\% = -20\%.$$

Therefore, X exhibits a decrease of 20%.

g. What is the percentage change of Jon's salary if it went from $150 a week to $200 a week?

Here $X_1 = \$150$,

$X_2 = \$200$,

$\Delta X = \$200 - \$150 = \$50$.

$$\text{Percentage change} = \left(\frac{50}{150}\right) \times 100\% = \left(\frac{1}{3}\right) \times 100\% = 33.33\%.$$

■

1.A.9 Conversions between Decimals (D), Fractions (F), and Percents (P)

1.A.9.1 Changing Decimals to Fractions (D → F) To change a decimal with two decimal places to a fraction:

Step 1: Move the decimal point two places to the right.

Step 2: Put that number over 100 as the numerator.

Step 3: Reduce if necessary.

EXAMPLE A.30

a. $0.65 = \left(\frac{65}{100}\right) = \left(\frac{13}{20}\right).$

b. $0.05 = \left(\frac{5}{100}\right) = \left(\frac{1}{20}\right).$

c. $0.75 = \left(\frac{75}{100}\right) = \left(\frac{3}{4}\right).$

This rule can be generalized to a number with any number of decimal places. For example, with three decimals, 0.075 becomes $0.075 \times \left(10^3/10^3\right) = (75/1000) = (3/40)$. The procedure can be summarized as shown in the following example:

Read it:	0.8
Write it:	8/10
Reduce it:	4/5

■

1.A.9.2 Changing Fraction to Decimals: (F → D) Any fraction can be converted to an equivalent decimal. To change a fraction to a decimal, simply do what the operation says. Since the fraction (a/b) means a ÷ b, we can divide the numerator of a fraction by its denominator to convert the fraction to a decimal. For example, to convert 3/8 to a decimal, divide 3 by 8 by long division as follows: $3 \div 8 = 0.375$.

EXAMPLE A.31

a. Thefraction $\frac{13}{20}$ means 13 dividedby 20(insertdecimalpointsandzerosaccordingly).

$$\text{So,} \quad 13.00 \div 20 = 0.65.$$

b. $5.000 \div 8 = 0.625.$

■

1.A.9.3 Changing Decimals to Percents (D → P) To change a decimal to a percent:

Step 1: Move the decimal point two places to the right.

Step 2: Insert a percent sign.

EXAMPLE A.32

a. $0.75 = 75\%$.

b. $0.05 = 5\%$. ■

1.A.9.4 Changing Percents to Decimals (P → D)

To change percent to decimals:

Step 1: Divide by 100.

Step 2: Move the decimals to the left by two places.

EXAMPLE A.33

a. $64\% = \frac{64}{100} = 0.64$.

b. $2.5\% = \frac{2.5}{100} = 0.025$.

c. $0.1\% = \frac{0.1}{100} = 0.001$(one tenth of 1%). ■

1.A.9.5 Changing Fractions to Percents (F → P)

To change a fraction to a percent:

Step 1: Multiply by 100.

Step 2: Insert a percent sign.

EXAMPLE A.34

Change the following fractions to percents:

a. $\frac{1}{2}$

b. $\frac{2}{5}$

c. $\frac{3}{4}$

d. $\frac{7}{5}$

SOLUTIONS:

a. $\frac{1}{2} \times 100 = \frac{100}{2} = 50\%$.

b. $\frac{2}{5} \times 100 = \frac{200}{5} = 40\%$.

c. $\frac{3}{4} \times 100 = \frac{300}{4} = 75\%$.

d. $\frac{7}{5} \times 100 = \frac{700}{5} = 140\%$. ∎

1.A.9.6 Changing Percents to Fractions (P → F) To change percents to fractions:

Step 1: Divide the percent by 100.

Step 2: Eliminate the percent sign.

Step 3: Reduce if necessary.

EXAMPLE A.35

a. $60\% = \left(\frac{60}{100}\right) = \left(\frac{3}{5}\right)$.

b. $13\% = \left(\frac{13}{100}\right)$.

c. One tenth of 1% : $0.1\% = \left(\frac{0.1}{100}\right) = 0.001$.

d. Two tenths of 1% : $0.2\% = \left(\frac{0.2}{100}\right) = 0.002$. ∎

1.A.10 Ratio and Proportion

A *proportion* is, by definition, the equality of two ratios. For example, $\frac{3}{5} = \frac{6}{10}$.

In general, $\frac{x}{y} = \frac{z}{w}$, where y and w are each different from zero.

Multiplying both sides by the common denominator, yw, we obtain

$$yw \cdot \left(\frac{x}{y}\right) = yw \cdot \left(\frac{z}{w}\right), \quad \text{which in turn gives } wx = yz.$$

This result can also be obtained by multiplying diagonally, known as *cross-multiplication*. The terms wx and yz are known as *cross-products*. Using any three values in the proportion, we can always find the fourth value.

For example, if $\frac{3}{4} = \frac{x}{12}$, then $4x = 3 \times 12$ and $x = \frac{3 \times 12}{4} = \frac{36}{4} = 9$.

EXAMPLE A.36

A day trip to a Stock Exchange for a class of 25 students costs $1250. How much does it cost for the entire class of 150 freshmen to visit the Stock Exchange?

SOLUTION: Let us assume it costs $x, then using the concept of proportion, we obtain

$$\frac{25}{1250} = \frac{150}{x}.$$

Cross-multiplication produces, $25x = 150 \times 1250$.

Dividing both sides by 25,

$$x = \frac{150 \times 1250}{25} = 7500 \text{ dollars.}$$ ■

EXAMPLE A.37

In a business school, the student to faculty ratio is found to be 1100 to 77. If the university on the whole maintains a student body of 2900, find the actual number of faculty needed in order to keep the same ratio as in the business school.

SOLUTION: Let x stand for the number of faculty needed at the university level to maintain the same student faculty ratio as in the business school. Then

$77/1100 = x/2900$. Cross-multiplying, we obtain $1100x = 2900 \times 77$

Dividing both sides by 1100, $x = 2900 \times 77/1100 = 203$. ■

Exercises

Simplify the following:

A.1 $(18 \div 12) \times 8^2 - 88$.

A.2 $20 - 6 \times 12 + 20^2 + (12 + 2) \times 6$.

A.3 $[(4)^2 \times 4] \times 16 + 10$.

A.4 $\frac{18}{12} \times 3^2 - 10$.

A.5 $X^2YXY^3X^2Y^2$.

A.6 $\frac{X^2Y^3}{\sqrt{X}}$.

A.7 $\left(\frac{X^3}{-64Y^6}\right)^{1/3}$.

A.8 $\left[4\sqrt{\frac{625}{81}}\right]3$.

A.9 $\left[3\sqrt{\frac{1}{27}}\right]^2$.

A.10 $40 - 1.96\frac{20}{\sqrt{16}}$.

A.11 $0.30 + 1.645\sqrt{\frac{0.3(1-0.3)}{100}}$.

A.12 $(0.1)^3$.

A.13 $\{(0.3)xyz\}^0$.

A.14 $\left(\frac{0.03}{0.004}\right)^4$.

A.15 $\frac{(0.3)^4}{(0.2)^3} \times 10^3$.

A.16 $0.027^{2/3}$.

A.17 $\left(\frac{1.24(0.25)}{0.62}\right)^{1/2}$.

A.18 $x \cdot x \cdot x^4$.

A.19 $x \cdot x \cdot y^2$.

A.20 $\frac{x^2 \cdot x^4}{y^4 \cdot y^2}$.

A.21 $(x^4)^4$.

A.22 $\left((x^3)^4\right)^5$.

A.23 $(4 - 2^2)^0$.

A.24 $\frac{x^4}{x^3}$.

A.25 $\left(\frac{x^3}{y^4}\right)^5$.

A.26 $\left(\frac{x^{-3}}{y^{-4}}\right)^{-6}$.

A.27 $\left(\frac{x^4}{x^3}\right)^5$.

A.28 $\left(\frac{15}{3q}\right)^4$.

A.29 $\frac{12^3}{12^0}$.

A.30 $\frac{x^{-5}}{x^3}$.

A.31 $\left(\frac{9}{16}\right)^{1/2}$.

A.32 $\frac{1}{3}y^3(2x^{-3})^4$.

A.33 $27^{4/3}$.

A.34 $\frac{5}{3}\left(2\frac{5}{3}\right)^{-1}$.

A.35 $\frac{x^8/y^4}{8/y^2}$.

A.36 $\frac{(625)^{1/4}}{25^2}$.

A.37 $\left(\frac{1}{3}\right)^3$.

A.38 $\frac{-(2.5)(1.2)-(3.4)(0.2)}{30}$.

A.39 Evaluate $y = \left(\frac{1}{x}\right)^{-(4/6)}$, at $x = 27$.

A.40 Find the prime factors of 36, 48, 112, and 192.

Factor the following:

A.41 $x^2 - 7x + 12$.

A.42 $x^2 + 5x - 24$.

A.43 $x^2 + x - 12$.

A.44 $3x^2 - 4x + 1$.

A.45 $5x^2 + 7x + 2$.

A.46 $2x^2 + x - 6$.

A.47 $x^2 + 5x + 6$.

Evaluate the following expressions:

A.48 $\frac{1}{3}+\frac{1}{5}+\frac{1}{2}$.

A.49 $\frac{7}{8}-\frac{3}{24}$.

A.50 $-\frac{5}{8}+\frac{1}{6}-\frac{3}{2}$.

A.51 $\frac{1}{2}-\frac{1}{4}$.

Solve the following:

A.52 $\left(\frac{7}{3}-\frac{2}{3}+\frac{1}{5}\right)/20$.

A.53 $\frac{(1/3)-(1/5)}{(2/3)-(1/6)}$.

A.54 $\frac{2}{7}\times\frac{14}{4}$.

A.55 $\frac{3}{5}\times\frac{-2}{7}$.

A.56 $\frac{1}{4}\div\frac{2}{3}$.

A.57 $\frac{-2}{3}\div\frac{3}{5}$.

A.58 $0\div\frac{1}{3}$.

A.59 $\frac{0}{5}\div\frac{5}{6}$.

Simplify the following:

A.60 $3\frac{5}{8}+2\frac{1}{4}-7\frac{2}{3}$.

A.61 $2\frac{3}{5}+\frac{1}{3}-\frac{2}{7}$.

A62 $\left(1\frac{1}{4}\right)\left(\frac{3}{4}\right)$.

A.63 $\left(2\frac{3}{4}\right)\left(4\frac{1}{2}\right)$.

A.64 $\frac{1}{3}\div\frac{2}{3}$.

A.65 $2\frac{3}{4}\div4\frac{1}{3}$.

Show the position of each digit in the following:

A.66 4.76.

A.67 0.873.

A.68 125.007.

Find:

A.69 $1.273 + 3.2$.

A.70 $0.25 + 0.5$.

A.71 $0.34 - 0.07 + 1.7$.

Evaluate:

A.72 0.02×0.7.

A.73 $0.1 \times 0.01 \times 0.005$.

A.74 $0.09 \div 0.1$.

A.75 $\frac{(1.5)(-0.6)-(2.5)(1.7)}{20}$.

A.76 $\frac{-(2.5)(1.2)-(3.4)(0.2)}{30}$.

A.77 $\frac{0.4}{0.04}-\frac{6(3.6)}{0.072}$.

Round off the resulting decimal to the given place:

A.78 $\frac{1}{3}$, to the nearest tenth.

A.79 $\frac{5}{7}$, to the nearest hundredths

A.80 $\frac{2}{3}$, to the nearest thousandth

A.81 $\frac{22}{7}$, to the nearest ten thousandth.

A.82 An investor bought 125 shares of a company stock at a price of $\$79\frac{5}{16}$ a share. What is the cost of the stock? If this represents (3/5) of the total amount of investment, find the total amount of investment.

A.83 In a business school, the ratio of female to male undergraduate students is 1 to 4. After 200 female students are admitted, the ratio of female to male students became $\frac{2}{3}$. Find the total number of students after admitting 200 female students in the school.

A.84 A U.K. store sells an item of jewelry for $1725.00, which includes all taxes at the rate of 17.5%. For export purposes, to earn foreign exchange, if the government exempts taxes on all exports to other countries, what price should be paid for the jewelry piece by a foreign visitor if taxes are excluded?

A.85 Mr. John Smith receives $1400.00 as salary every week after taxes. If he is in 30% tax bracket, what is his salary per week before taxes?

A.86 Movie tickets are sold for $7.50 per adult and $6.50 per child. A party of 10 paid a total of $71.00. How many adults and children are there in the party?

A.87 Fill in the blanks:

Row	Fraction	Decimal	%, Percentage
1	...	0.02	...
2	...	0.125	...
3	...	3.750	...
4	$\frac{1}{16}$	...	...
5	$\frac{1}{8}$	...	...
6	$\frac{3}{2}$	...	...
7	...	...	30
8	...	...	50
9	...	...	10
10	...	...	200
11	$\left(\frac{1}{4}\right)^2$	...	...
12	$\left(\frac{1}{10}\right)^5$	...	...

A.88 The *Wall Street Journal* reported on Friday, May 25, 2012 a high of 12,539.59 for the Dow Jones Industrial Average Index and a low of 12,419.63. Find the percentage increase from low to high.

A.89 The *Wall Street Journal*, Friday, May 25, 2012, reported that the 52 week crude oil high and low prices per barrel were, respectively, $109.77 and $75.67. Find the percentage decrease from high price to low price.

Chapter 2

Applications of Linear and Nonlinear Functions

2.1 INTRODUCTION

The concept of a linear function was introduced in Chapter 1. This chapter deals with the application of linear functions in business and economics. There are many instances where a linear function is used for decision making. For example, in determining the market equilibrium with demand and supply functions, in determining the break-even level of output with revenue and cost functions, and in identifying the intervals of dissaving and saving using aggregate consumption and income relationships.

2.2 LINEAR DEMAND AND SUPPLY FUNCTIONS

Here we specify the quantity demanded or the quantity supplied of a product (or service) as a linear function of the price of the product (or service). While there are other determinants of quantity demanded or supplied, we shall assume that they are being held constant. We demonstrate, by means of specific examples, how the concept of slope can be used in a meaningful way to depict the marginal effect of changes in price on quantity. The concept of marginality plays a dominant role in business, as in hiring and pricing decisions, and as a public policy guide. The hiring of factors of production as well as the determination of wages and prices (rents) are based on marginal productivities. This process is generated from the theory that in a competitive mixed economy, factors of production are paid according to their marginal products. Also, the concept of marginality plays a fundamental role in the formulation of fiscal, as well as monetary, policy in the economy. For example, often tax cuts are based on the marginal rate of consumption of an "average consumer." Likewise, the Federal Reserve Bank alters the Federal Funds Rate to control inflation and achieve a level of desired economic growth based on the elasticity effects on the economy. We illustrate these points by specific examples as shown below.

Introduction to Quantitative Methods in Business: With Applications Using Microsoft® Office Excel®, First Edition. Bharat Kolluri, Michael J. Panik, and Rao Singamsetti.

Companion website: www.wiley.com/go/Kolluri/QuantitativeMethods

EXAMPLE 2.1

$Q = 40 - 2P + 0.5Y$ is called a *demand function*, where Q is the quantity demanded of a product, P is the price of the product in dollars, and Y is the family income in thousands of dollars.

a. Draw the demand curve (a graphical relationship between Q and P for fixed Y) for $Y = 50$ at price levels of $P = 0, 5, 10, 15,$ and 20.

b. Repeat the above for $Y = 100$, for the same price levels on the same graph.

c. Comment on the demand curve in part (b) in relation to the demand curve obtained in part (a).

Note: In economics, generally the independent variable P is shown on the vertical axis and the dependent variable Q appears on the horizontal axis.

SOLUTIONS:

a. If $Y = 50$, then $Q = 40 - 2P + 0.5(50) = 40 - 2P + 25 = 65 - 2P$ (Table 2.1a).

Table 2.1a Demand Schedule at Income Level $50,000

P	Q
0	65
5	55
10	45
15	35
20	25

b. If $Y = 100$, then $Q = 40 - 2P + 0.5(100) = 40 - 2P + 50 = 90 - 2P$ (Table 2.1b).

Table 2.1b Demand Schedule at Income Level $100,000

P	Q
0	90
5	80
10	70
15	60
20	50

Figure 2.1 illustrates the graphs for parts (a) and (b). The lower line depicts the demand curve for part (a). The top line shows the demand curve for part (b).

c. It is evident from the graph that the demand curve in part (b) is to right of the demand curve in part (a). This is due to an increase in income. This is known as a "shift in the demand curve," meaning that a larger quantity is demanded at the same price due to an increase in income.

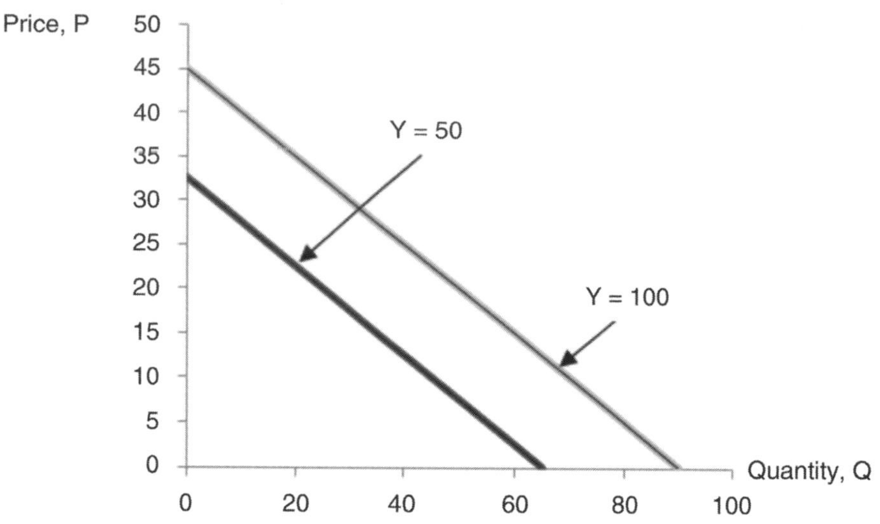

Figure 2.1 Demand curves at different income levels.

■

EXAMPLE 2.2

Let us express the (aggregate) *consumption function* as $C = 10 + (3/4)Y$, where C stands for consumption expenditure in thousands of dollars, Y stands for disposable income in thousands of dollars, the slope $\frac{3}{4}$ is the *marginal propensity to consume* (MPC) (meaning the increase in consumption expenditure due to a 1 unit increase in income or $\frac{\Delta C}{\Delta Y} = \frac{3}{4}$), and the intercept (10,000) is known as the "subsistence level" of consumption expenditure when income $Y = 0$.

a. Verify that $\text{MPC} = \frac{3}{4}$.

b. Find consumption expenditure at $Y = 20$ and at $Y = 60$.

c. Find the "break-even level" of income at which income and consumption expenditure are equal.

SOLUTION: The graph of the consumption function $C = 10 + \frac{3}{4}Y$ and the graph of $C = Y$ (the 45° line) appear in Figure 2.2. We measure disposable income in thousands of dollars on the Y-axis and consumption expenditure in thousands of dollars on the C-axis.

From $C = 10 + \frac{3}{4}Y$, we can readily obtain

$$C + \Delta C = 10 + \frac{3}{4}(Y + \Delta Y) = 10 + \frac{3}{4}Y + \frac{3}{4}\Delta Y = C + \frac{3}{4}\Delta Y.$$

Subtracting C from both sides of this equation yields $\Delta C = \frac{3}{4}\Delta Y$ or $\frac{\Delta C}{\Delta Y} = 3/4$, the MPC. (Note that from $C = 10 + \frac{3}{4}Y$ we could just simply calculate $dC/dY = 3/4$.)

At $Y = 20$, $C = 10 + \frac{3}{4}(20) = 25$ (a dissaving of \$5000 occurs since consumption exceeds income or $C - Y = 25 - 20 = 5 > 0$.)

At $Y = 60$, $C = 10 + \frac{3}{4}(60) = 55$ (now a positive saving of \$5000 occurs since consumption is lower than income or $C - Y = 55 - 60 = -5 < 0$.)

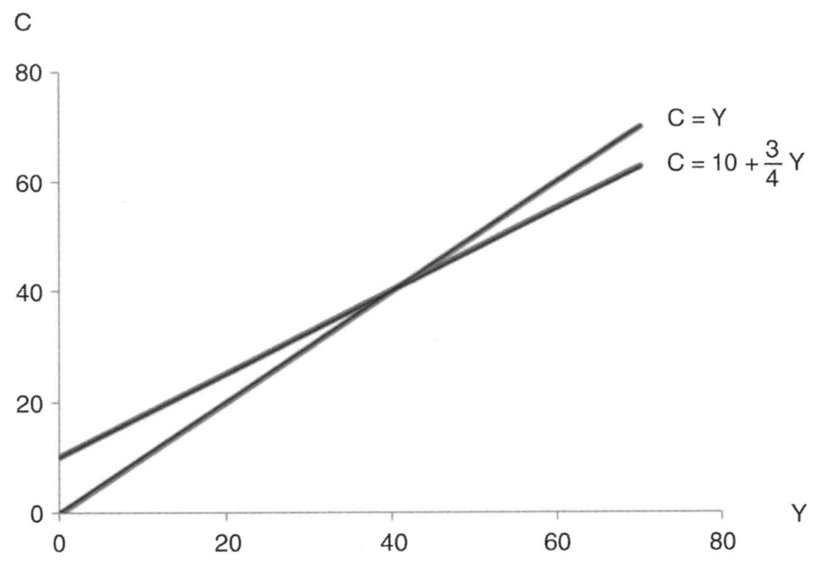

Figure 2.2 Consumption relationship and equality of consumption and income.

To calculate break-even income, set C = Y, that is,

$$10 + \frac{3}{4}Y = Y$$

or $10 = \frac{1}{4}Y$ so that Y = 40. ■

2.3 LINEAR TOTAL COST AND TOTAL REVENUE FUNCTIONS

The *total cost* function is TC = FC + VC × Q, where FC is *fixed cost*, VC stands for *unit variable cost*, and Q is quantity in units.

The *total revenue* function TR is linear and of the form TR = P × Q for a given price P.

(What sort of market does the firm operate in if P is fixed?)

EXAMPLE 2.3

Suppose that the fixed cost of production for a commodity is \$10 and the price is \$4 per unit. The variable cost is \$2 per unit.

a. Find the break-even quantity Q_e, where TR = TC.

b. Find the loss or gain for the company at production levels of Q = 3 and Q = 7 units of output.

SOLUTION: TC = 10 + 2Q and TR = 4Q, where TC stands for total cost, TR stands for total revenue, and Q stands for output in units. A plot of the revenue and cost schedules (Table 2.2) is shown in Figure 2.3.

Table 2.2 Revenue and Cost Schedules

Quantity (Q)	Total Revenue (TR = 4Q)	Total Cost (TC = 10 + 2Q)
2	8	14
4	16	18
6	24	22
8	32	26

a. To find the break-even output level, equate TR and TC. To this end:

$$\text{TR} = \text{TC}$$

$$4\text{Q} = 10 + 2\text{Q}$$

$$2\text{Q} = 10 \text{ and thus}$$

$$\text{Q} = \text{Q}_e = 5 \text{ units.}$$

b. At Q = 3, TR = 12, and TC = 16,

$$\text{Profit} = \text{ total revenue} - \text{ total cost} = 12 - 16 = -4.$$

Loss = $4 (total cost exceeds total revenue).

c. At Q = 7, TR = 28, and TC = 24,

$$\text{Profit} = 28 - 24 = 4.$$

Gain = $4 (total cost falls short of total revenue).

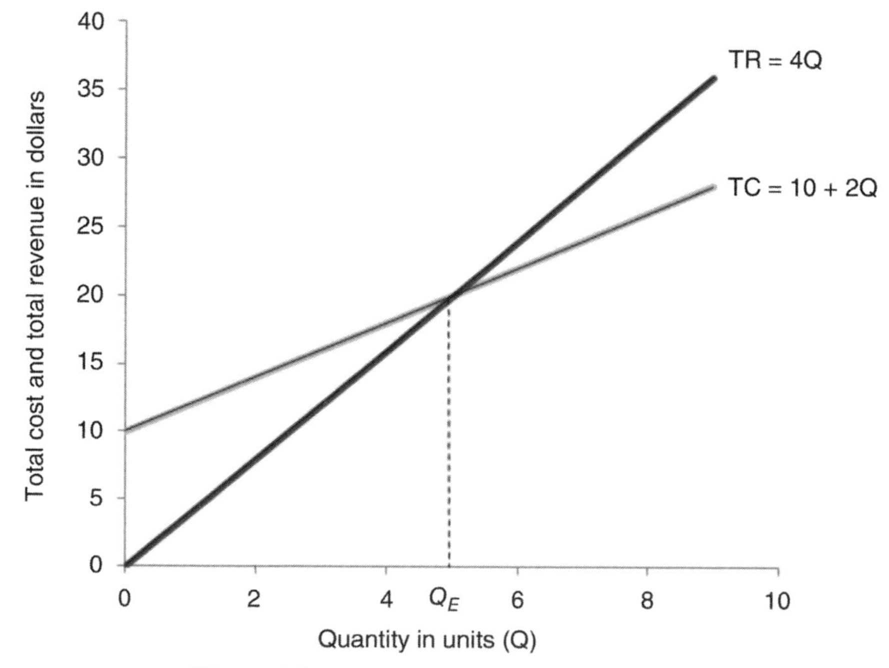

Figure 2.3 Total revenue and total cost functions.

■

EXAMPLE 2.4

A firm plans to sell a standard family-size box of cereal for \$2.40. Production estimates have shown that the variable cost of producing one box is \$2.16. Fixed cost of production is \$3600. What is the break-even volume of sales?

SOLUTION: The total revenue function TR is linear and of the form $TR = P \times Q = 2.40Q$

The total cost function TC is $TC = FC + VC \times Q = 3600 + 2.16Q$, where FC is fixed cost and VC stands for unit variable cost.

Break-even quantity Q_e is established by equating the total revenue and total cost functions and solving for $Q = Q_e$. Thus, at Q_e,

$$TR = TC,$$

$$2.40Q = 3600 + 2.16Q,$$

$$0.24Q = 3600,$$

so

$Q = Q_e = 15,000$ units.

All the intermediate steps are shown to demonstrate the relation between fixed cost, product price, and variable cost per unit of output. Revenue and cost corresponding to this quantity can be found by substituting Q_e into either the total revenue or total cost function since the dollar figures are equal. This is computed below using the total revenue function:

$$TC = TR = \$2.40 \times Q_e = \$2.40 \times 15000.$$

$$\text{Thus, } TC = TR = \$36,000 \text{ at } Q_e.$$

This analysis has determined that the firm will earn a positive profit after the sale of the first 15,000 boxes. Total revenue and total cost at 15,000 boxes will be each equal to \$36,000 and losses will be incurred at quantities below 15,000 boxes. These results apply unless changes occur in the price and cost data. ■

2.4 MARKET EQUILIBRIUM

How should we describe the workings of a particular market? Our approach will be to use a *model*: a system of structural equations that mirror the behavior of the market. (*Structural equations* describe the basic structure of the market.) For instance, consider the following demand and supply functions;

$$Q_D = f(P_D) \quad \text{and} \quad Q_S = g(P_S).$$

Here, the subscript "D" stands for the demand for the product by consumers and subscript "S" stands for the supply of the product offered by producers. Q is the quantity and P is the unit price.

These functions are the structural equations of the model. Moreover, they are "behavioral" in nature since they generally reflect the behavior of individuals, households, and firms; they describe how these entities react to market conditions, for example, to a change in market price or even in expectations.

To depict market equilibrium, we need an *equilibrium condition*: a condition that brings quantity demanded into alignment with quantity supplied; a condition that eliminates excess market demand or excess market supply. In this regard, a market equilibrium is a point where the quantity demanded and the quantity supplied are equal. Obviously, at this equilibrium, the demand price and the supply price are also equal. Hence, our market model is made up of the two structural equations, and the equilibrium (market-clearing) condition or

$$\begin{aligned} Q_D &= f(P_D), \\ Q_S &= g(P_S), \\ Q_D &= Q_S. \end{aligned}$$

Let us denote the equilibrium point by (Q_e, P_e), where $Q_D = Q_S = Q_e$ and $P_D = P_S = P_e$. Then, by solving the demand and supply functions simultaneously, we obtain the market equilibrium: a position from which, once attained, there will be no "net" tendency to depart.

2.5 GRAPHICAL PRESENTATION OF EQUILIBRIUM

Generally, supply has a positive slope and demand has a negative slope (Figure 2.4).

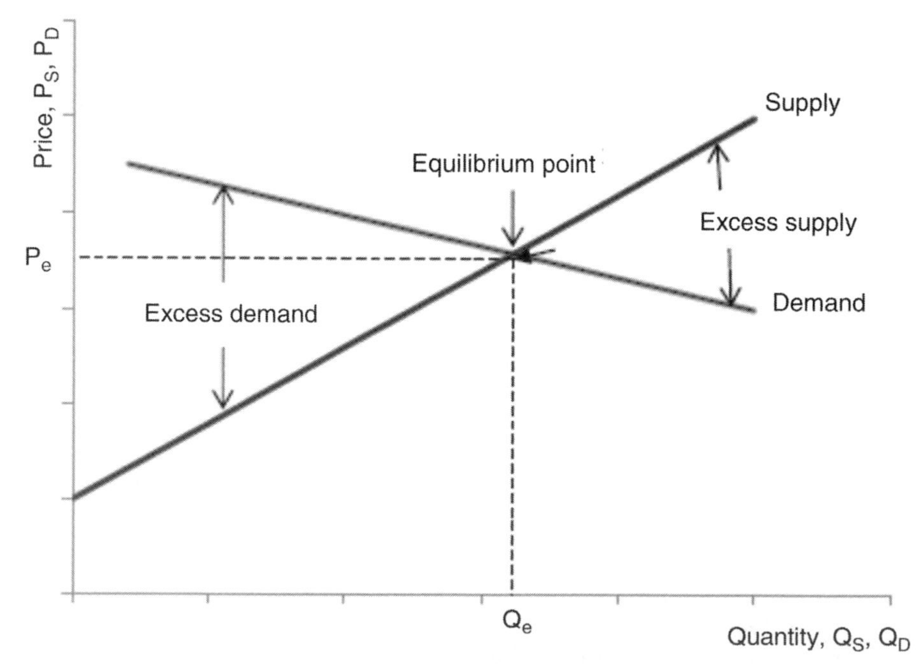

Figure 2.4 Supply and demand relationships.

EXAMPLE 2.5

Given the following demand and supply equations for automobiles, find the equilibrium price and quantity:

$$P_D = 16000 - 20Q_D,$$

$$P_S = 6000 + 30Q_S.$$

Note: At equilibrium, $P_D = P_S = P_e$ and $Q_D = Q_S = Q_e$.

SOLUTION: From the demand and supply equations, we have, respectively,

$$P_e = 16000 - 20Q_e$$

and

$$P_e = 6000 + 30Q_e.$$

Subtracting the second equation from the first,

$$0 = 10,000 - 50Q_e,$$

$$50Q_e = 10,000,$$

and thus

$$Q_e = 200.$$

Substituting $Q_e = 200$ into the demand equation, we get the equilibrium price:

$$P_e = 16,000 - 20(200) = 16,000 - 4000 = 12,000.$$

Thus, the equilibrium price and quantity are, respectively, \$12,000 and 200 units. ■

2.6 APPLICATIONS OF NONLINEAR FUNCTIONS

A *nonlinear function*, by definition, is an algebraic expression involving both variables and parameters (constants), in which at least one variable and/or a parameter appears with an exponent other than 1. For example, $y = A_0 + A_1x_1 + A_2x_2^2 + A_3x_3^3 + A_4^2x_4 - A_5x_1x_2$ is a nonlinear function. Additionally, the exponential functions and logarithmic functions introduced in Chapter 1 are also nonlinear functions. This section deals with some applications of nonlinear functions in business.

One of the most common applications of exponential functions involves the growth of a variable. For example, if the variable P accumulates over time t on a continuous basis, the appropriate formula is $P_t = P_0e^{Rt}$. This can be justified as follows. If the initial amount P_0 grows at the rate of R% per annum for t periods, the accumulated value is $P_t = P_0(1 + R/n)^{nt}$, where n is the number of times the increment is added per period (frequency of compounding within each period). As n approaches infinity, this function assumes the form $P_t = P_0e^{Rt}$. This formula is derived from the calculations below. Specifically,

$$P_1 = P_0\left(1+\frac{R}{n}\right)^n,$$

$$P_2 = P_0\left(1+\frac{R}{n}\right)^{2n},$$

$$\vdots$$

$$P_t = P_0\left(1+\frac{R}{n}\right)^{tn}.$$

Let $m = n/R$ so that

$$\left(1+\frac{R}{n}\right)^{tn} = \left(1+\frac{R}{n}\right)^{(n/R)Rt} = \left(1+\frac{1}{m}\right)^{mRt} = \left(\left(1+\frac{1}{m}\right)^m\right)^{Rt}.$$

As n goes to infinity (continuous compounding), m also goes to infinity since $m = n/R$. Hence,

$$\lim_{m\to\infty}\left(1+\frac{1}{m}\right)^m = e.$$

Similarly,

$$\lim_{n\to\infty}\left(1+\frac{R}{n}\right)^{tn} = e^{Rt}.$$

So,

$$P_t = P_0e^{Rt}.$$

EXAMPLE 2.6

Suppose you deposit \$1000.00 in a certificate of deposit (CD) in a bank. Find how much it appreciates (future value) at a 10% interest rate per annum after 2 years if interest is added (compounded)

a. once a year,

b. semiannually,

c. quarterly,

d. monthly, and

e. continuously.

Comment on the value of your deposit after 2 years as the frequency of compounding per year increases.

SOLUTION: As explained above, the future value after t periods, P_t, is given by $P_t = P_0(1 + (R/n))^{nt}$, where P_0 is the initial value, R is the rate of interest per annum, and n is the number of times the interest is added per year. (*Hint:* Use the y^x key on your calculator.)

a. Here $P_0 = \$1000.00$, $n = 1$, $R = 10\%$ or 0.10, and $t = 2$. So,

$$P_2 = 1000\left(1 + \frac{0.10}{1}\right)^{(1\times 2)} = 1000(1.1)^2 = \$1210.$$

b. Here n takes on the value 2 and it is the only change from (a) above. So,

$$P_2 = 1000\left(1 + \frac{0.10}{2}\right)^{(2\times 2)} = 1000(1 + 0.05)^4 = 1000(1.05)^4 = 1000(1.21551) = \$1215.51.$$

c. Here n assumes the value 4 and it is the only change from (a) above. So,

$$P_2 = 1000\left(1 + \frac{0.10}{4}\right)^{(4\times 2)} = 1000(1 + 0.025)^8 = 1000(1.21840) = \$1218.40.$$

d. Now n takes on the value 12 and it is the only change from (a) above. So,

$$P_2 = 1000\left(1 + \frac{0.10}{12}\right)^{(12\times 2)} = 1000(1 + 0.00833)^{24} = 1000(1.22029) = \$1220.29.$$

Alternatively, taking natural logs on both sides yields

$$\begin{aligned}\ln P_2 &= \ln 1000 + 24\ln(1.00833) = 6.9077553 + 24(0.0082955)\\ &= 6.9077553 + 0.199092 = 7.1068473\end{aligned}$$

and taking antilogs using the INV button and LN button on your calculator yields the number for P_2 as 1220.29.

It can also be evaluated by taking logs to the base 10. Try it.

e. As noted in Section 2.2, in the case of continuous compounding, we use the formula

$$P_t = P_0 e^{Rt} = 1000e^{0.1\times 2} = 1000e^{0.2} = 1000(2.71828)^{0.2} = 1000(1.22140) = \$1221.40.$$

Here, e is approximated as 2.71828. Alternatively, as in part (d) above, P_2 can also be found by taking natural logs. *Comment:* We can see from the above calculations that the future value is higher as the frequency of compounding increases due to the fact that the process generates interest on interest more frequently.

■

EXAMPLE 2.7

Graph the following *exponential growth function* $P_t = P_0 e^{Rt}$, where R is the exponential growth rate, t is the time period, P_t is the value of the function at time t, P_0 is the value of P at $t = 0$, and e is a mathematical constant.

SOLUTION: You are given

$$P_0 = 8,$$

$$e \cong 2.71828,$$

$$R = 0.02.$$

Thus, the growth function has the form $P = 8e^{0.02t}$, with the growth values given in Table 2.3.

Table 2.3 Time and Growth Values

t	P
−40	3.5946
−20	5.3626
0	8.0000
20	11.9346
40	17.8043
60	26.5609
80	39.6243
100	59.1125
120	88.1854

■

A graphical description of the tabular values is shown in Figure 2.5.

EXAMPLE 2.8

In 1985, the population of the United States was 234 million and the exponential growth rate was 0.8% per year. What will be the U.S. population in the year 2020?

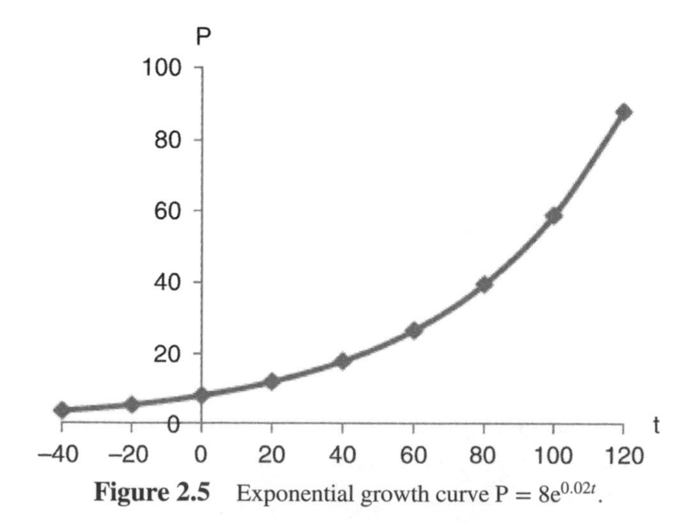

Figure 2.5 Exponential growth curve $P = 8e^{0.02t}$.

SOLUTION: At $t = 0$ (1985), the population was $P_0 = 234$ million. Since $R = 0.008$, the exponential growth equation is $234e^{0.008t}$ and, thus, for the year 2020, we have $t = 35$, $P_{2020} = 234e^{0.008(35)} = 234(1.3231) = 309.6$ million. ■

EXAMPLE 2.9

Suppose the marginal propensity to consume of a household is 0.7. Derive the associated consumption function.

SOLUTION: Let C(Y) represent the consumption function of the household, where Y represents the income of the household. Then,

$$C(Y) = \int MPCdY = \int 0.7dY = 0.7Y + c,$$

where c is the constant of integration that gives the subsistence level of consumption when the income level is zero. ■

EXAMPLE 2.10

Assume that the MPC of a typical household is 3/4 or 75% and the subsistence level of consumption is 16,000. With these facts, find the consumption function of the household.

SOLUTION: By integrating the constant MPC with respect to income (Y), we obtain $C(Y) = \int(3/4)dY = (3/4)Y + c$, where c is the constant of integration.

The subsistence level of consumption occurs at $Y = 0$, and this is given as 16,000. Hence, $C(0) = (3/4) \times 0 + c = 0 + 16{,}000$, and thus $c = 16{,}000$. Now the consumption function of the household is $C(Y) = (3/4)Y + 16{,}000$. ■

EXAMPLE 2.11

Suppose the marginal revenue function is $MR = 3Q^{-0.7}$. Find the total revenue when $Q = 5$.

SOLUTION: $TR(Q) = \int 3Q^{-0.7}\,dQ = (3Q^{-0.7+1})/(-0.7 + 1) + c = (3Q^{0.3})/(3/10) + c = 10Q^{0.3} + c$, where c is a constant. To find c, we set $Q = 0$. Then, $TR(0) = 10(0^{0.3}) + c = 0$ or $c = 0$. Hence, $TR(5) = 10 \times 5^{0.3} + 0 = 10 \times 1.621 = 16.21$. ■

EXAMPLE 2.12

Suppose the marginal cost of producing Q units of output is given by $MC = 10 + 2Q^{-1/2}$. Find the total cost function if the fixed cost is \$1000.

SOLUTION: By integrating the above MC function, we get $C(Q) = \int(10 + 2Q^{-1/2})dQ = 10Q + 4Q^{1/2} + c$. Then, fixed cost $= C(0) = 10 \times 0 + 4 \times 0^{1/2} + c = 1000$.

Therefore, $c = 1000$ and the total cost function is $C(Q) = 10Q + 4Q^{1/2} + 1000$. ■

EXAMPLE 2.13

It is known that, by definition, total investment is the rate of change in capital stock. Thus, $I(t) = dk(t)/dt$, where $I(t)$ is the investment (a flow concept) in millions of dollars and $K(t)$ stands for capital stock (a stock concept) at any time t. To derive the capital stock at any time, we have to integrate the investment function $I(t)$ with respect to t. Thus, $K(t) = \int I(t)dt$. More specifically, in a fixed interval (t_1, t_2), we get total capital accumulation during this period as $K(t_2) - K(t_1)$. Find the capital stock accumulation of a company with $I(t) = 10t^{3/4}$ during the period: (a) $t_1 = 0$ and $t_2 = 1$; and (b) $t_1 = 5$ and $t_2 = 10$.

SOLUTION: Integrating the given investment function yields

$$K(t) = \int 10t^{3/4}dt = 10t^{(3/4)+1}/(3/4+1) + c = 10t^{7/4}/(7/4) + c = (40/7)\left(t^{7/4}\right) + c.$$

(a) Capital accumulation during the period from $t_1 = 0$ to $t_2 = 1$ is $K(1) - K(0) = (40/7) \times 1^{7/4} + c) - \left((40/7) \times 0^{7/4} + c)\right) = 40/7$ millions of dollars.

(b) Similarly, capital accumulation during the period $t_1 = 5$ to $t_2 = 10$ is $K(10) - K(5) = (40/7)\ 10^{7/4} + c) - ((40/7)\ 5^{7/4} + c)) = (40/7)\ 10^{7/4} - (40/7)\ 5^{7/4}) = (40/7)\ (10^{7/4} - 5^{7/4}) = (40/7)\ (56.23413 - 16.71851) = (40/7)\ (39.51562) = 225.80354$ millions of dollars. ■

2.7 PRESENT VALUE OF AN INCOME STREAM

It can be shown that the *present value of an income stream*, $f(t)$, obtained continuously over time t is given by $P = \int f(t)e^{-rt}dt$ from $t = 0$ to $t = X$, where X stands for the number of years the income stream lasts and r is the interest rate (in percent per year) compounded continuously over time.

EXAMPLE 2.14

Find the present value of an employee's salary and bonus in a financial company that pays a continuous income stream at the rate of \$300,000 per year, assuming that the employment contract lasts for 5 years at the current rate of interest of 3% per year.

SOLUTION: The present value of the income stream of the employee is

$$P = \int 300000e^{-0.03t}\,dt \text{ from } t = 0 \text{ to } t = 5. \text{ Then,}$$

$$P = 300,000\left[\frac{e^{-0.03t}}{-0.03}\right]_0^5 = \left(\frac{300,000}{-0.03}\right)\left[e^{-0.03\times 5} - e^{-0.03\times 0}\right] = -10,000,000[0.861 - 1]$$
$$= -10,000.000[-0.139] = \$1,390,000.$$ ■

EXAMPLE 2.15

Find the present value of a small business company generating a perpetual income stream of \$300 million per year, assuming a 3% continuous rate of interest per year.

SOLUTION: The present value of the income stream of the small business company is

$$P = \int 300e^{-0.03t}dt \text{ from } t = 0 \text{ to } t = \infty. \quad \text{Then,}$$

$$P = 300\left[\frac{e^{-0.03t}}{-0.03}\right]_0^{\infty} = \left(\frac{300}{-0.03}\right)\left[e^{-0.03x\infty} - e^{-0.03x0}\right] = -10,000[0-1]$$

$= 10000$ in millions of dollars (same as 10 billion \$). ■

2.8 AVERAGE VALUES

In general, if $y = f(x)$, then the *average of y* is defined as $\frac{y}{x} = \frac{f(x)}{x}$. Specifically, *average cost* is defined as

$$\text{Average cost (AC)} = \frac{\text{total cost}}{Q},$$

where Q denotes quantity, and *average revenue* is defined as

$$\text{Average revenue (AR)} = \frac{\text{total revenue}}{Q} = \frac{\text{price} \times Q}{Q} = \text{price} = P.$$

EXAMPLE 2.16

Find the average values for functions 1, 2, and 3 given below and evaluate each at $Q = 100$. Comment on (2) and (3).

1. $C(Q) = 200 + 3Q$, where $C(Q)$ is the total cost function and Q is the level of production.

2. $R(Q) = 5Q$, where $R(Q)$ is total revenue when Q units are transacted at a constant price of $P = 5$.

3. $R(Q) = P(Q)Q$, where $P(Q) = 300 - 2Q$ is a demand (inverse) function showing the dependence of price on quantity demanded Q.

SOLUTION:

1. $C(Q) = 200 + 3Q$, where $C(Q)$ is the total cost and Q is the level of production. Average cost (AC) $= \frac{\text{total cost}}{Q}$, so that

$AC = \frac{200 + 3Q}{Q}$. At $Q = 100$,

$$AC = \frac{200 + 3(100)}{100} = \frac{500}{100} = 5.$$

2. $R(Q) = 5Q$, where $R(Q)$ is the total revenue when Q units are transacted at a constant price of $P = 5$.

$$\text{Average revenue (AR)} = \frac{\text{total revenue}}{Q} = \frac{\text{price} \times Q}{Q} = \frac{5Q}{Q} = 5 = \text{price}.$$

In this case, average revenue is equal to price and is independent of the value of Q.

3. $R(Q) = P(Q)Q$, where $P(Q) = 300 - 2Q$ is the demand (inverse) function showing the dependence of price on quantity demanded Q.

$$\text{Average revenue (AR)} = \frac{\text{total revenue}}{Q} = \frac{R(Q)}{Q} = \frac{(300-2Q)Q}{Q} = 300 - 2Q = P(Q).$$

At $Q = 100$, $AR = 300 - 2(100) = 100$.

Comment on 2 and 3:

Average revenue is always the same as price, no matter whether the price is fixed or related to the quantity Q. ■

2.9 MARGINAL VALUES

In general, if $y = f(x)$, then the marginal value of y with respect to x is defined as $\frac{dy}{dx} = \frac{d}{dx}f(x) = f'(x)$. This represents the *slope* of the function f(x) at a given x value. It represents the *rate of change* in f(x) with respect to x. For example, *marginal cost* is defined as

$$\text{Marginal cost (MC)} = \frac{d}{dQ}C(Q), \text{ where } C(Q) = \text{total cost}.$$

Marginal cost is the incremental cost of producing one more unit of output. Similarly, *marginal revenue* is the additional revenue obtained by selling one more unit of output. Marginal revenue is defined as

$$\text{Marginal revenue (MR)} = \frac{d}{dQ}R(Q), \text{ where } R(Q) = \text{total revenue}.$$

EXAMPLE 2.17

Find the marginal values for each of the following functions a, b, and c, at $Q = 100$:

a. $C(Q) = 200 + 3Q$.

b. $R(Q) = 5Q$.

c. $R(Q) = (300 - 2Q)Q$.

SOLUTION:

a. $C(Q) = 200 + 3Q$.

$$\text{Marginal cost (MC)} = \frac{d}{dQ}C(Q),$$

$$MC = \frac{d}{dQ}(200 + 3Q) = 0 + 3 = 3.$$

At $Q = 100$, $MC = 3$. (In fact MC is 3 at any value of Q.)

Note that since total cost = total variable cost (TVC) plus total fixed cost (TFC) or $C(Q) = TVC(Q) + TFC$, it follows that

$$MC = \frac{dC(Q)}{dQ} = \frac{dTVC(Q)}{dQ}, \text{ that is,}$$

marginal cost is also the derivative of total variable cost with respect to output Q.

b. $R(Q) = 5Q$.

$$\text{Marginal revenue (MR)} = \frac{d}{dQ}R(Q) = \frac{d}{dQ}(5Q) = 5.$$

At $Q = 100$, $MR = 5$. (Again MR is 5 at any value of Q.)
Note: If price is constant, AR and MR both are equal to price.

c. $R(Q) = (300 - 2Q)Q$

$$\text{Marginal revenue (MR)} = \frac{d}{dQ}R(Q),$$

$$MR = \frac{d}{dQ}(300 - 2Q)Q = \frac{d}{dQ}\left(300Q - 2Q^2\right) = 300 - 4Q.$$

At $Q = 100$, $MR = 300 - 4(100) = -100$.

Note: In general, if $y = f(x) = a + bx$ is a linear function of x, then $\frac{dy}{dx}$ is b, the slope of the linear equation.

Comparing the demand (inverse) function $P = 300 - 2Q$ with the MR function, $MR = 300 - 4Q$, we see that the MR function is twice as steep as the P function, even though each has the same vertical intercept (300). This observation breaks down if the P function is nonlinear in Q. ■

2.10 ELASTICITY

In general, the *point elasticity* of $y = f(x)$ at a given x is defined as the ratio of percentage change in y to percentage change in x. If E denotes elasticity, then

$$E = \frac{dy/y}{dx/x} = \frac{d \log y}{d \log x} = \frac{x}{y} \cdot \frac{dy}{dx}$$

It is to be noted that point elasticity is independent of the units of measurement of the variables.

EXAMPLE 2.18

Find price elasticities of demand given the demand function $Q = 400 - 4P$ at the prices levels ($) 90, 70, 50, 30, and 10. Explain your results. (Since $dQ/dP < 0$, the price elasticity will be calculated as $E = -\frac{dQ}{dP}\frac{P}{Q}$.)

SOLUTION: Details of the requisite computations are given in Table 2.4 using the definition of point elasticity. First, we use the demand equation $Q = 400 - 4P$ to produce the demand schedule and note that $\frac{dQ}{dP} = -4$. Then the price elasticities of demand are derived at various price levels, as shown in the last column of Table 2.4.

Table 2.4 Computation of Price Elasticities of Demand

P in $	Q in units	Price Elasticity $= E = -\frac{dQ}{dP}\frac{P}{Q} = 4\frac{P}{Q}$
90	40	9.00
70	120	2.33
50	200	1.00
30	280	0.43
10	360	0.11

Incidentally, the reason why these elasticity values are called "point elasticities" is that they hold at specific points on a demand curve. ■

POINTS TO BE NOTED

- The sign of the slope coefficient dQ/dP indicates the type (negative or positive) of relationship between the dependent variable Q and the independent variable P.
- Demand elasticities can also be computed with respect to a variety of independent variables, such as income, interest rate, and prices of other goods.
- If the numerical value of elasticity of demand is greater than unity, then demand is said to be *elastic*. This means that a 1% increase (decrease) in price will result in a greater than 1% decrease (increase) in quantity demanded. Correspondingly, total revenue will decrease (increase).
- If the numerical value of point elasticity of demand is equal to unity, then demand is said to be of *unitary elasticity*. This means that a 1% increase (decrease) in price will result in exactly a 1% decrease (increase) in quantity demanded. In this case, total revenue will remain the same.
- And if the numerical value of point elasticity of demand is less than unity, then demand is said to be *inelastic*. This means that a 1% increase (decrease) in price will result in less than a 1% decrease (increase) in quantity demanded. Correspondingly, total revenue will increase (decrease).
- The point elasticity of demand coefficient is not generally of the same value at each point on a demand curve (see Table 2.4). If the demand function is linear, then point elasticity can be readily determined graphically (Figure 2.6). In this regard, point elasticity can be determined geometrically as AB/OA.
- A demand curve that has the same elasticity value at each price–quantity combination has the form $Q = aP^b$, $b < 0$. Here, $\log Q = \log a + b \log P$ so that

$$E_p = \text{price elasticity} = -\frac{d \log Q}{d \log P} = -b > 0,$$

the *constant price elasticity* value at each (Q, P).

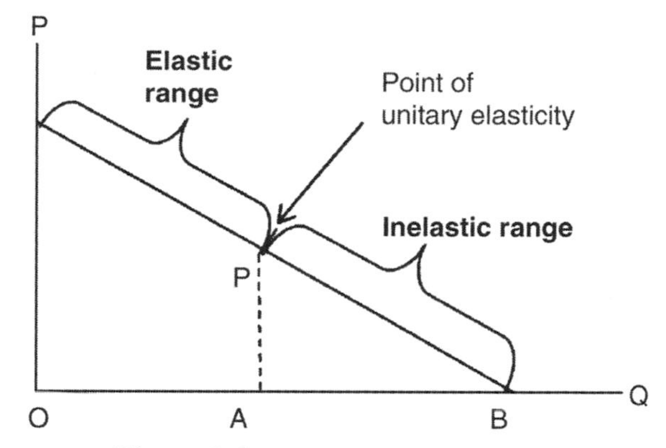

Figure 2.6 Price elasticity of demand.

EXAMPLE 2.19

a. Find the elasticity of demand at P = 30 for the demand function 2P + Q = 100.

b. Find how a price increase from \$30 to \$32 will affect total revenue, and by how much?

c. Find how revenue will be affected by a decrease in price from \$30 to \$28, and by how much? Comment on your results.

SOLUTION:

a. Rewriting the demand function as Q = 100 – 2P, $\frac{dQ}{dP} = -2$. Then

$$E = -\frac{P}{Q}\cdot\frac{dQ}{dP} = \frac{-30}{[100-2(30)]}\cdot[-2] = \frac{30}{40}\cdot(2) = \frac{60}{40} = \frac{3}{2} = 1.5 > 1.$$

Thus demand is elastic.

b. At P = \$30,

$$Q = 100 - 2P = 100 - 2(30) = 40 \text{ units. Then,}$$

$$TR = P \times Q = 30 \times 40 = \$1200.$$

At P = \$32,

$$Q = 100 - 2P = 100 - 2(32) = 36 \text{ units. Thus,}$$

$$TR = P \times Q = 32 \times 36 = \$1152.$$

Therefore, revenue declined by \$1200 – \$1152 = \$48.

c. At P = \$28,

$$Q = 100 - 2P = 100 - 2(28) = 44 \text{ units so that}$$

$$TR = P \times Q = 28 \times 44 = \$1232$$

Thus, total revenue increased by \$1232 – \$1200 = \$32.

Comment: Since demand is elastic, a price increase resulted in a decline in revenue; a price decrease resulted in an increase in revenue. ■

2.11 SOME ADDITIONAL BUSINESS APPLICATIONS

In this section, we present some applications of integrals.

EXAMPLE 2.20

Find the total cost function in a production process, where the marginal cost of production is \$3 per unit. Assume that the fixed cost of production is \$1000.

SOLUTION: Let us assume that the total cost of units produced is denoted by C(Q), where Q stands for the number of units. By integrating marginal cost, we get function $C(Q) = \int 3dQ = 3Q + c$, where c is the constant of integration. The value of c is obtained as $C(0) = 3 \times 0 + c = 1000$, which is the given fixed cost value. ■

EXAMPLE 2.21

Find the total revenue function when marginal revenue is $MR = 2000 - 4Q$, where Q stands for quantity demanded. Also derive the demand function.

SOLUTION: By integrating marginal revenue $MR = 2000 - 4Q$, we get the total revenue function, $R(Q) = \int (2000 - 4Q)\ dQ = 2000Q - 4Q^2/2 + c$. The constant of integration c is obtained as follows. Note first that when demand Q is zero, revenue is zero, that is, $R(0) = 2000 \times 0 - 4 \times 0^2/2 + c$ so that $c = 0$. Then, $R(Q) = 2000Q - 2Q^2$.

Since $R(Q) = \text{price} \times \text{quantity} = P \times Q$, where P stands for price and Q stands for quantity demanded, we see that quantity demanded is $Q = R(Q)/Q = (2000Q - 2Q^2)/Q = 2000 - 2Q$. ■

2.12 EXCEL APPLICATIONS

EXAMPLE 2.22

Assume you opened and deposited \$1000.00 into a savings account that pays 4% per annum.

a. If the bank compounds interest annually, how much will you have in your account at the end of 3 years (assuming no deposits or withdrawals are made for 3 years)?

b. Find the balance if the bank compounds interest quarterly under the same conditions.

c. Find the balance if the bank compounds interest *continuously* under the same conditions.

d. Comment on what will happen to the balance at the end of 3 years as the frequency of compounding per year increases.

Instructions for Using Appropriate Excel Formulas:

a. = principal*(1 + i)^t

b. = principal*$(1 + \frac{i}{4})$^(t*4)

c. = principal*EXP(i*t)

d. Comment

Note: i = annual percentage rate of interest expressed in decimals = 0.04,

t = time in years = 3

Principal = 1000

Open Excel spreadsheet to obtain the solution for a). Type = 1000*(1 + 0.04) ^3 in cell B3 and hit enter. This gives the solution 1124.864 as shown below. Similarly, for (b) type =1000*(1 + .04/4) ^ (3*4) in cell B4 and hit enter to find the solution 1126.82503 as shown below.

Finally, to answer (c) type =1000*EXP (0.04*3) in cell B5 and hit enter to find the solution 1127.496852 as shown below. (d) *Comment:* The balance at the end of 3 years increases as the frequency of compounding per year increases.

1	A	B
2	Solution	
3	(i)	1124.864
4	(ii)	1126.82503
5	(iii)	1127.496852

■

EXAMPLE 2.23

Exponential Growth

Graph the following exponential growth function $P_t = P_0e^{Rt}$, where R is the exponential growth rate, t is the time period in years, P_t is the value of the function at time t, P_0 is the value of P at t = 0, and e is a mathematical constant.

Using the information $P_0 = 8$, e = 2.71828, and R = 0.02, generate an excel table for values of P_t, where time (t) ranges from −40 to 120 at 20 unit intervals. Plot a graph in Excel for those values of P.

SOLUTION: Open Excel spreadsheet and type t in A1 and P in B1. Type −40,−20, . . . , 120 in A2, A3, . . . , A10 cells. In B2, type =8*exp(0.02*highlight A2 cell) and hit Enter. Then you see 3.5946 (approximately) in B2 cell. Go back to cell B2 and point the cursor to bottom right corner of cell B2 (fill handle). When the cursor becomes black, cross, click and drag down to copy the formula from B3 to B10 cells. The result is as shown below:

	A	B
1	t	P
2	−40	3.5946
3	−20	5.3626
4	0	8.0000
5	20	11.9346
6	40	17.8043
7	60	26.5609
8	80	39.6243
9	100	59.1124
10	120	88.1854

To plot a graph of the values of P, highlight values of "t" and "P" in the above table, go to "Insert" on menu bar of the Excel spreadsheet, and select "scatter" on the ribbon and choose curve option and click to get a chart as shown below:

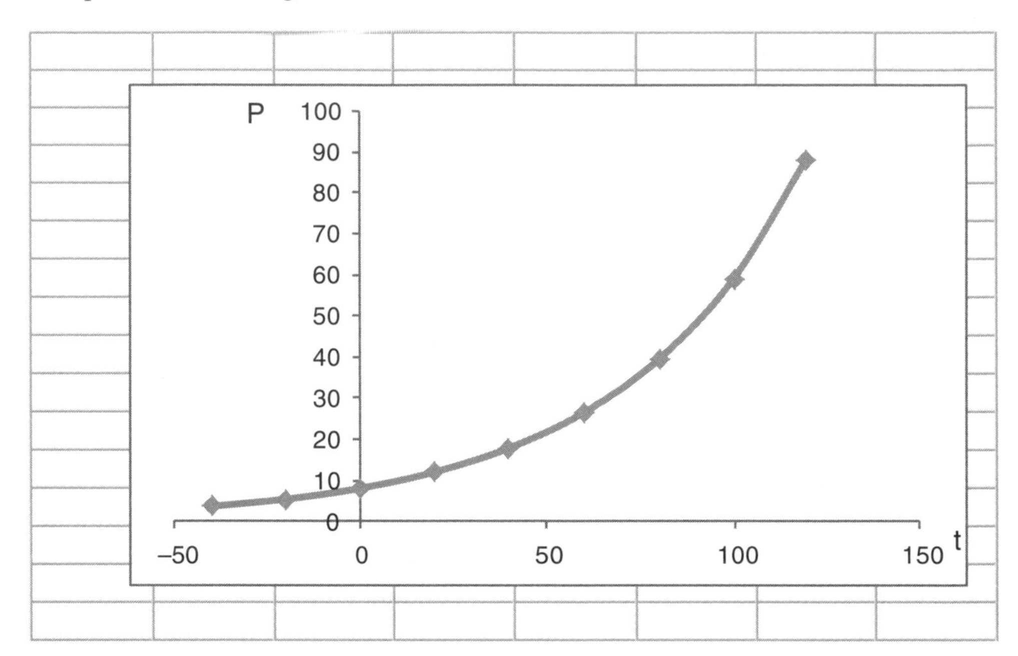

■

CHAPTER 2 REVIEW

You should be able to:

1. Distinguish between linear demand and supply functions.
2. Discuss the (two) key components of a total cost function.
3. Discuss the role of price in determining the shape of a total revenue function.
4. Determine the break-even quantity given the total cost and total revenue functions.
5. Discuss the attainment of market equilibrium.
6. Distinguish between excess demand and excess supply.
7. Comment on the role of the frequency of compounding in calculating the future value of an initial investment.
8. Distinguish between average cost and marginal cost, and between average revenue and marginal revenue.
9. Discuss the concept of price elasticity of demand.
10. Distinguish between elastic demand, inelastic demand, and unitary elasticity of demand.

Key Terms and Concepts:

EXERCISES

1. Determine the demand curve for the price–quantity data in the following table for Q = f(P). Also show it graphically.

P in $	0	10	20	30	40
Q in 1000's	8	6	4	2	0

2. $Q = -5 + 10P - 2R$ is an example of a supply function, where Q is the quantity supplied of a product, P is the price in $ of the product, and R is the interest rate in percent terms.
   a. Draw the supply curve (a graphical relationship between Q and P for a fixed R) for $R = 5$, at price levels of $P = 2, 4, 6,$ and 8.
   b. Repeat the above for $R = 7$ and for the same price levels in the same graph.
   c. Comment on the supply curve in part (b) in relation to the supply curve in part (a).
3. For the consumption function $C = 20 + \frac{2}{3}Y$:
   a. Find the break-even income level.
   b. Find the consumption expenditure at income levels of 40 and 80. (In this exercise, assume C and Y are measured in thousands of dollars.)
   c. Show the answers to parts (a) and (b) graphically and identify the ranges of income corresponding to dissaving and saving.
   d. Plot the *savings function* $S = Y - C$. What is the *marginal propensity to save* (MPS)? Does MPC + MPS = 1? At $Y = 20$, does $Y = C + S$?
4. Suppose the fixed cost of production for a commodity is $45,000. The variable cost is 60% of the selling price of $15.00 per unit. Find the break-even level of output.
5. **a.** Find the break-even output level with the following information.

Selling price = $5.00

Unit variable cost = $2.00

Fixed cost = $3000

b. Suppose a particular firm is producing and selling fig products in Toronto. The selling price of a 10 oz jar of fig jam is $9. Fixed cost of producing the fig jam is $540 and unit variable cost of a 10 oz jar of fig jam is $6. Find the break-even quantity.

6. A furniture company produces folding wooden tables. The selling price of a table is $80. Fixed production cost is $2000 and the unit variable cost is $60. The manager wants to know how many tables the company should produce and sell to break even. Find the loss or gain if the company plans to produce $Q = 50$ tables or $Q = 150$ tables.

7. In Example 2.5, find the quantity demanded and the quantity supplied if the price is artificially fixed at $8000. (Note that at this price of $8000, which is lower than the equilibrium price, the quantity demanded far exceeds the quantity supplied. Thus, there will be product shortage.)

8. Suppose the demand curve for a particular brand of shoes is linear, and is of the form $P_D = 84 - 0.14Q_D$; and the linear supply function has the form $P_S = 18 + 0.02Q_S$.

a. Find the equilibrium price and quantity for the firm's shoes. *Note:* P_D and P_S are measured in dollars and Q_D and Q_S indicate the number of pairs of shoes.

b. If the demand function shifts to $P_D = 96 - 0.14\,Q_D$, find the new equilibrium price and quantity. Describe the change in demand by comparing the demand functions in parts (a) and (b).

9. The following are the supply and demand functions for a portable music player. Draw the supply and demand graphs for the music player and find the equilibrium price and quantity.

$$Q_D = 560 - 10\,P_D$$

$$Q_S = 80 + 5\,P_S$$

10. The U.S. Post Office first introduced the "Forever Stamp" in 1998 with a value of 33 cents without printing the face value on it. Now, in 2014, the price of the "Forever Stamp" is 49 cents. Assuming that the value increases at an annual growth rate of 2.5%, find the value of the "Forever Stamp" in 2018.

11. Find the future value of an investment of $1000.00 in 10 years if the interest rate of 10% per year is compounded continuously.

12. Suppose the grandparents in a family with three grandchildren decided to provide financial support at a rate of $10,000.00 per each grandchild per year for a period of 20 years, starting in the year 2015. Find the present value of this income stream at the market interest rate of 4% per year.

13. A company offers an employee a pension plan providing payment of $60,000 per year beginning at the age of 65 along with a one-time cash payment of $100,000. Find the present value of the total amount, assuming an average interest rate of 5% per year.

14. Find the present value of an income stream of $10,000 forever at a constant interest rate of 10% per year.

15. Find the savings function for an individual whose marginal savings rate is a constant at 10% of disposable income Y (income minus taxes). Assume that the amount of savings is zero when the disposable income for the person is zero but the individual needs $15,000 as minimum consumption for survival.

16. Suppose the investment $I(t)$ of a technology firm grows exponentially by $I(t) = 1,000,000e^{0.4t}$, where t stands for time and 0.4 stands for a 40% growth rate in investment per year. Find the capital accumulation in 5 years from now.

17. Suppose the marginal revenue function of an auto manufacturing company is given by MR = 100,000-2 Q, where Q is the number of autos produced. Derive the total revenue function as well as the demand function for the autos.

18. Suppose the marginal cost of production in a firm producing cereal boxes is $2.15 per box. Find the total cost function, assuming that the fixed cost of production is $3600.

19. Find the average values for functions a, b, and c given below and evaluate each at Q = 100.

a. $C(Q) = 540 + 6Q$, where C(Q) is the total cost function and Q is the level of production.

b. $R(Q) = 9\,Q$, where R(Q) is the total revenue when Q units are transacted at a constant price of P = $9.

c. $R(Q) = P(Q)Q$, where $P(Q) = 40 - 0.1Q$, is a demand (inverse) function showing the dependence of price on quantity demanded Q.

20. Find the marginal values for functions a, b, and c given below and evaluate each at Q = 100.

a. $C(Q) = 540 + 6Q$, where C(Q) is the total cost function and Q is the level of production.

b. $R(Q) = 9Q$, where R(Q) is the total revenue when Q units are transacted at a constant price of P = $9.

c. $R(Q) = P(Q)Q$, where $P(Q) = 40 - 0.1Q$, is a demand (inverse) function showing the dependence of price on quantity demanded Q.

21. Find the price elasticity of demand for the following demand function at price levels of $2, $3.33, and $4, and interpret price elasticity at each price level. $Q = 20 - 3P$.

22. Find the price elasticity of supply for the supply function $Q = -30 + 5P$ at a price level of $12. (*Hint:* Calculate $\frac{dQ}{dP}\frac{P}{Q} > 0$.)

23. Find the price elasticity of the supply function $Q = 11 + 7P$ at $P = 10$.

24. Given the total revenue and total cost graphs presented below, comment on points A and B and regions S_1 and S_2. What do you call the slopes of the total revenue and total cost lines in economics or business terms?

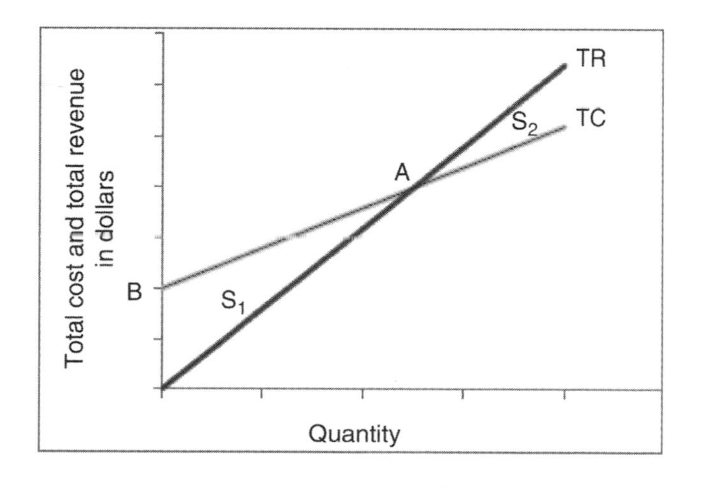

Excel Applications

25. This exercise deals with the case where price is not a constant but related to quantity demanded. The demand (price) function is $P = 50 - 2.5Q$, where Q is the quantity demanded. The revenue function is $TR = Q \times P = 50Q - 2,5Q^2$ and the total cost function is $TC = 25 + 25Q$. Using Excel, develop a table showing columns for Q, P, TR, TC, and Profit (TR-TC). Graph the total cost and total revenue functions at the values of Q: 0, 1, 2, 3, 4, 5, 6, 7, 8, and 9. Indicate approximately on the graph the break-even output levels where total cost equals total revenue. Distinguish between these two output levels.

26. Assume you opened and deposited \$100,000.00 into a savings account that pays 2% per annum.

Use Excel to find solutions to the following questions.

a. If the bank compounds interest annually, how much you will have in your account at the end of 3 years (assuming no deposits or withdrawals are made for 3 years).

b. Find the balance if the bank compounds interest quarterly under the same conditions.

c. Find the balance if the bank compounds interest *continuously* under the same conditions.

d. Comment on what will happen to the balance at the end of 3 years as the frequency of compounding per year increases.

27. Using Excel, draw the curve of the following function: $Y = \frac{1}{\sqrt{2\pi}} \exp\left(-\frac{1}{2}x^2\right)$, where $\pi = \frac{22}{7}$. Consider the following values of x: −3,−2,−1, 0,1,2,3 for preparing a table and drawing the graph.

Chapter 3

Optimization

3.1 INTRODUCTION

Optimization is the process of obtaining the most desirable outcome in any given decision situation. It is especially important in business to maximize profit or, equivalently, minimize cost under a variety of conditions. It is obvious that either at the firm level or for industry as a whole, management has to take into consideration both predictable and nonpredictable events by being sensitive to the legal, social, and global issues that confront them in order to achieve the *most desirable outcome*. More specifically, in this chapter we focus on procedures to arrive at maximum profit or, alternatively, to minimize total cost with or without constraints. First, we start with models of profit or revenue maximization in production. Second, we deal with cost minimization issues, especially in the case of inventory maintenance in business. Finally, this chapter deals with constrained optimization, maximization, and minimization, using linear programming models.

3.2 UNCONSTRAINED OPTIMIZATION

3.2.1 Models of Profit and Revenue Maximization

Many business decisions are based on marginal analysis. For example, should an additional machine be purchased to maximize profit, or an additional advertising dollar be spent in order to maximize sales? Decisions such as these can be made using marginal analysis. In order to understand the role of marginal analysis in optimization, we revisit the concepts of marginal cost (MC) and marginal revenue (MR) introduced in Chapter 2. We follow the profit-maximizing rule that states that a firm's profit is maximized at an output level where MR = MC. This is a necessary condition because, as long as MR > MC, the firm continues to make increasing profits by producing more output; and when MR < MC, the firm experiences declining profits. Therefore, the firm's profits are at a maximum when MR = MC. In a similar vein, revenue (sales) is maximized when MR = 0. It is equally important to

Introduction to Quantitative Methods in Business: With Applications Using Microsoft® Office Excel®, First Edition. Bharat Kolluri, Michael J. Panik, and Rao Singamsetti.

Companion website: www.wiley.com/go/Kolluri/QuantitativeMethods

maximize sales (revenue) since (a) salesmen are interested in maximizing their own individual performance by maximizing sales; and (b) firms often try to maximize their market shares independent of cost considerations.

EXAMPLE 3.1

Management has at its disposal the following information:

Demand (price) function: $P = 50 - 2.50\,Q$,
Revenue function: $R(Q) = Q \cdot P = Q(50 - 2.50\,Q) = 50\,Q - 2.50\,Q^2$,
Total cost function: $C(Q) = 25 + 25\,Q$.

(Note that both price and costs are measured in dollars.)
Using the above functions, determine the following:

a. Profit maximizing output
b. Profit-maximizing price
c. Maximum profit value
d. Revenue-maximizing output.

Note that each of the preceding functions has been written in terms of quantity or output level Q. This is because Q is chosen to be the firm's decision or control variable, that is, to maximize profit, the firm must determine the level of Q that does so.

Obtain the solutions to the above questions by both the trial and error method and the calculus approach. ■

3.2.2 Solution by Trial and Error (Approximate) Method

This method seeks to find the quantity that maximizes profit by tracking profit at different levels of quantity. Thus, referring to the completed Table 3.1, obtained by using Excel, we find that profit is at a maximum at an equilibrium output level of $Q_e = 5$; and revenue is maximized at $Q = 10$. Note that at the maximum profit level, MR = MC; and at the maximum revenue point, MR = 0. As explained in Section 2.1, marginal revenue can be obtained as the ratio of the change in revenue to the change in quantity. Similarly, marginal cost is obtained as the ratio of the change in total cost to the change in quantity.

Note that production continues to increase as long as $MR - MC > 0$. Maximum total profit will be 37.5, when $MR - MC = 0$. (Note that it is just a coincidence that price and total profit both are equal to 37.5.) Profits decline after the $Q = 5$ level of output. For example, at $Q = 6$, we have $MR - MC < 0$. As output increases from 4 to 6, MR – MC changes in sign from positive to negative, and in between assumes a zero value, which ensures (guarantees) maximum total profit. In this case, total profit is maximized at $Q_e = 5$. Also note that the optimum value of $Q = 5$ (where MR = MC) is easily obtained by a trial and error process with Excel. Starting from, say, an output level below $Q = 5$, we can monitor MR – MC through increasing values of Q until the value of MR becomes identical to the value of MC or $MR - MC = 0$.

Table 3.1 Computation of Profit, MR, and MC Using Excel

=50–2.5*A2	=50*A2–2.5*A2^2	=25+25*A2	=C2–D2	=(C3–C2)/(A3–A2)	=(D3–D2)/(A3–A2)

B2 =50-2.5*A2

	A	B	C	D	E	F	G	H	I
1	Q	P	TR	TC	Total Profit	MR	MC	Tracking	Criterion
2	0	50	0	25	-25	NA	NA	NA	NA
3	1	47.5	47.5	50	-2.5	47.5	25	MR>MC	MR-MC>0
4	2	45	90	75	15	42.5	25	MR>MC	MR-MC>0
5	3	42.5	127.5	100	27.5	37.5	25	MR>MC	MR-MC>0
6	4	40	160	125	35	32.5	25	MR>MC	MR-MC>0
7	5	37.5	187.5	150	37.5	27.5	25	MR>MC	MR-MC>0
8	**5.000001**	37.5	187.5	150	37.5	24.99999752	25	**MR=MC**	**MR-MC=0**
9	6	35	210	175	35	22.4999975	25	MR<MC	MR-MC<0
10	7	32.5	227.5	200	27.5	17.5	25	MR<MC	MR-MC<0
11	8	30	240	225	15	12.5	25	MR<MC	MR-MC<0
12	9	27.5	247.5	250	-2.5	7.5	25	MR<MC	MR-MC<0
13	10	25	250	275	-25	2.5	25	MR<MC	MR-MC<0
14	**10.00001**	24.99998	250	275.0003	-25.00025	-2.49997E-05	25	**MR<MC**	**MR-MC<0**
15	11	22.5	247.5	300	-52.5	-2.500025	25	MR<MC	MR-MC<0
16	12	20	240	325	-85	-7.5	25	MR<MC	MR-MC<0

The profit-maximizing and revenue-maximizing output levels can also be observed from the TR and TC, and from the MR and MC graphs (Figure 3.1).

Sometimes businesses aim at revenue (sales) maximization without cost consideration for short periods of time in order to promote their product or service. From Table 3.1 you can see that total revenue is maximized when the quantity sold is Q = 10, and the maximum total revenue value is 250. Thus, profit maximization and revenue maximization will not lead to the same output decision.

3.2.3 Solution Using the Calculus Approach

Using the basic rules of differentiation of functions described in Chapter 1, one can also obtain the optimal level(s) of output. But first we need some specialized tools.

Suppose we decide to maximize or minimize a continuously differentiable function $y = f(x)$ over an open domain. Figure 3.2 sets the stage for this task. In panel (a), we have indicated that f attains a maximum at $x = x_0$. To the left of x_0, $f' > 0$; and to the right of x_0, $f' < 0$. Since f' is continuous at x_0, f' must equal zero there given that the sign of f' changes from positive to negative as x increases from left to right. Hence, we require that $f'(x_0) = 0$. (The line tangent to f at x_0 is horizontal.)

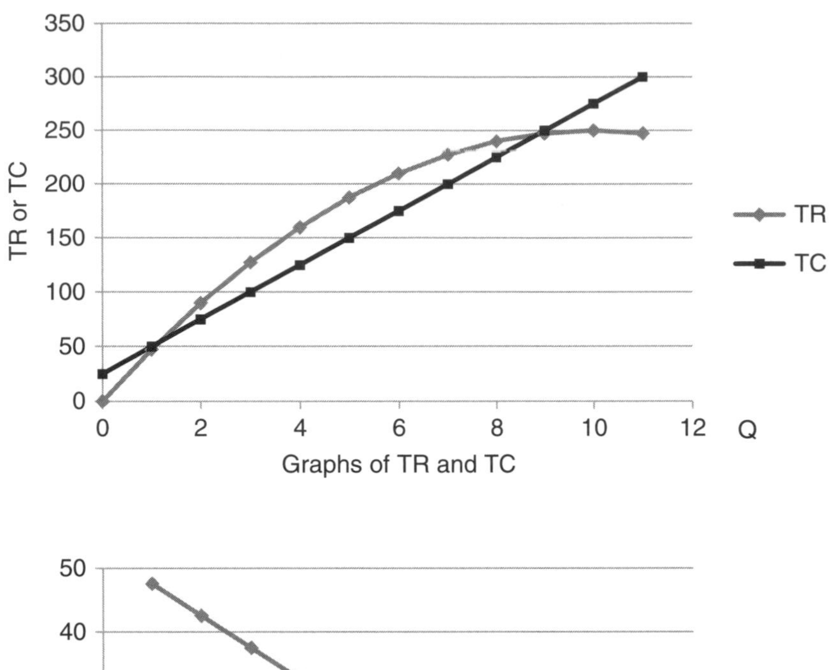

Figure 3.1 TR, TC and MR, MC graphs.

Looking to panel (b), we see that f has a minimum at $x = x_0$. To the left of x_0, $f' < 0$; and to the right of x_0, $f' > 0$. Again, under the continuity of f′ at x_0, the sign of f′ changes from negative to positive as x increases from left to right so that $f'(x_0) = 0$. (Here too the line tangent to f at x_0 is horizontal.) So whether f has a

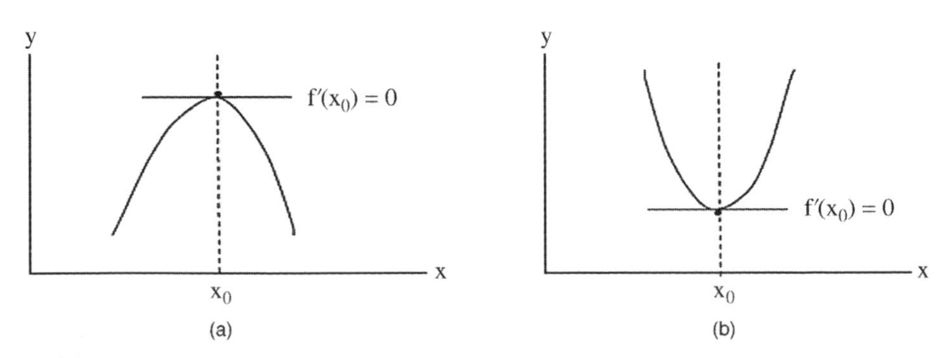

Figure 3.2 Finding the maximum or minimum of f. (a) $f'(x_0) = 0$, $f''(x_0) < 0$. (b) $f'(x_0) = 0$, $f''(x_0) > 0$.

maximum or a minimum at x_0, it must be the case that $f'(x_0) = 0$. The requirement $f'(x_0) = 0$ is called a *first-order condition* for a maximum or a minimum at x_0.

Suppose we find a value of x (call it x_0) where $f' = 0$. How do we distinguish a maximum from a minimum at this x value? The answer relies on the behavior of f'' at x_0. That is, if $f''(x_0) < 0$ (f is concave downward at x_0), then f has a maximum at x_0; if $f''(x_0) > 0$ (f is concave upward at x_0), then f has a minimum at x_0. To summarize:

1. If $f'(x_0) = 0$ and $f''(x_0) < 0$, then f has a maximum at x_0.
2. If $f'(x_0) = 0$ and $f''(x_0) > 0$, then f has a minimum at x_0.

The requirement that $f''(x_0) \neq 0$ is called a *second-order condition* for a maximum or a minimum at x_0.

Let us see how these criteria are applied. Does $y = f(x) = 2x^2 - 6x + 5$ possess an *extreme value*, that is, a maximum or a minimum? Our general procedure for answering questions such as this is as follows:

Step 1: Set $f' = 0$ and solve for the value of x that makes it vanish.

Step 2: Substitute the resulting x value into f'' so as to determine the sign of f''. To this end, we have

$$f' = 4x - 6 = 0 \quad \text{or} \quad x = 3/2.$$

Since $f'' = 4 > 0$, we conclude that f has a minimum at $x = 3/2$. (Note that since $f'' = 4$ is constant, f'' is positive for "any" x and thus it is positive for $x = 3/2$.)

What about the function $y = f(x) = 4x + x^{-1}$? (Any restriction on x?) Does it attain a maximum or a minimum? From

$$f' = 4 - x^{-2} = 0,$$

we get $x = \pm 1/2$. Hence, we have two values of x to test in f''. First, from $f'' = 2x^{-3}$, we see that $f''(-1/2) = 2/(-1/8) = -16 < 0$. Hence, f has a maximum at $x = -1/2$. Second, $f''(1/2) = 2/(1/8) = 16 > 0$. Thus, f has a minimum at $x = 1/2$.

As explained in Section 3.1, it is evident that total profit is maximized if MR = MC (this is also known as a *first-order condition for profit maximization*). First, we find MR and MC by differentiating the revenue and cost functions, respectively. Second, we solve for the output level by equating both. To this end,

$$\text{Total revenue } R = PQ = (50 - 2.5\,Q)Q = 50\,Q - 2.5\,Q^2,$$

$$MR = \frac{dR}{dQ} = R'(Q) = 50 - 2(2.5)Q = 50 - 5\,Q, \text{ and}$$

$$MC = \frac{dC}{dQ} = C'(Q) = 25.$$

By equating MR and MC, we obtain $50 - 5\,Q = 25$, so the answer to part (a) is $Q = 5$. Thus, we get the same solution as with the trial and error or approximation method. Then the corresponding price (the answer to part (b)) is $P = 50 - 2.5(5) = 50 - 12.5 = 37.5$, and maximum profit (the answer to part (c)) is total revenue − total cost = PQ − C(Q) = 37.5 × 5 − 25 + 255 = 37.5. The answer to part (d) is obtained by setting MR = 0. Hence, 50 − 5Q = 0 or 50 = 5Q and thus Q = 10. (Do the second-order conditions for a maximum hold at Q = 5 and Q = 10? We shall see how to check these requirements shortly.)

EXAMPLE 3.2

We again start with

Demand(price)function:	$P(Q) = 32 - Q$,
Revenue function:	$R(Q) = Q \cdot P = Q(32 - Q) = 32Q - Q^2$,
Total cost function:	$C(Q) = 200 + 2Q$.

(Note that both price and costs are measured in dollars.)
Using the above functions, determine the following:

a. Profit-maximizing output
b. Profit-maximizing price
c. Maximum total profit level
d. Revenue-maximizing output

Obtain the solutions to the above questions by both the trial and error method and by the calculus approach. ■

To accomplish the above tasks, we first use the trial and error (approximate) method and then the calculus approach.

3.2.4 Solution by Trial and Error (Approximate) Method

This technique seeks to find the quantity that maximizes profit by tracking profit at different levels of output. Thus, referring to the completed Table 3.2, obtained by using Excel, we find that profit is maximized at an output level of $Q_e = 15$ and revenue is at a maximum at Q = 16. Note that at maximum profit, MR = MC; and at maximum revenue, MR = 0. As before, marginal revenue is obtained as a ratio of the change in total revenue to the change in quantity. Similarly, marginal cost is obtained as the ratio of the change in total cost to the change in quantity.

From Table 3.2, note that production continues to increase as long as MR − MC > 0. Maximum total profit equals 25, and this occurs at $Q_e = 15$ when MR − MC = 0. Profits decline after that level of output since MR − MC < 0. As output expands, the change in the sign of MR − MC from positive to negative, and in between assuming a zero value, ensures (guarantees) maximum total profit. In this case, total profit is maximized at $Q_e = 15$. Thus, the answer to part (a) is that the profit-maximizing output level Q_e is 15 (approximately). For part (b), the profit-maximizing price is P = 17. For part (c), the maximum total profit level TP = 25. Looking to part (d) the revenue-maximizing output is Q = 16. Also note that the optimum output value of $Q_e = 15$ (where MR = MC) is easily obtained by the trial and error

Table 3.2 Computations of Profit, MR, MC Using Excel

B	C	D	E	F	G
=32–A2	=32*A2–A2^2	=200+2*A2	C2–D2	=(C3–C2)/(A3–A2)	=(D3–D2)/(A3–A2)

	A	B	C	D	E	F	G	H	I	J
1	Q	P	TR	TC	TP	MR	MC	Tracking	Criterion	
2	0	32	0	200	-200	NA	NA			
3	1	31	31	202	-171	31	2	MR>MC	MR-MC>0	
4	2	30	60	204	-144	29	2	MR>MC	MR-MC>0	
5	3	29	87	206	-119	27	2	MR>MC	MR-MC>0	
6	4	28	112	208	-96	25	2	MR>MC	MR-MC>0	
7	5	27	135	210	-75	23	2	MR>MC	MR-MC>0	
8	6	26	156	212	-56	21	2	MR>MC	MR-MC>0	
9	7	25	175	214	-39	19	2	MR>MC	MR-MC>0	
10	8	24	192	216	-24	17	2	MR>MC	MR-MC>0	
11	9	23	207	218	-11	15	2	MR>MC	MR-MC>0	
12	10	22	220	220	0	13	2	MR>MC	MR-MC>0	
13	11	21	231	222	9	11	2	MR>MC	MR-MC>	
14	12	20	240	224	16	9	2	MR>MC	MR-MC>0	
15	13	19	247	226	21	7	2	MR>MC	MR-MC>0	
16	14	18	252	228	24	5	2	MR>MC	MR-MC>0	
17	15	17	255	230	25	3	2	MR>MC	MR-MC>0	
18	15.001	17	255	230	25	2	2	MR=MC	MR-MC=0	
19	16	16	256	232	24	1	2	MR<MC	MR-MC<0	
20	16.001	15.999	256	232.002	23.998	0	2	MR<MC	MR-MC<0	
21	17	15	255	234	21	-1	2	MR<MC	MR-MC<0	
22	18	14	252	236	16	-3	2	MR<MC	MR-MC<0	
23	19	13	247	238	9	-5	2	MR<MC	MR-MC<0	
24	20	12	240	240	0	-7	2	MR<MC	MR-MC<0	
25	21	11	231	242	-11	-9	2	MR<MC	MR-MC<0	
26	22	10	220	244	-24	-11	2	MR<MC	MR-MC<0	
27	23	9	207	246	-39	-13	2	MR<MC	MR-MC<0	
28	24	8	192	248	-56	-15	2	MR<MC	MR-MC<0	
29	25	7	175	250	-75	-17	2	MR<MC	MR-MC<0	
30										

process with Excel by starting from an output level below Q = 15 and tracking MR – MC until MR – MC = 0.

The profit- and revenue-maximizing output levels can also be observed from the TR and TC, and from the MR and MC graphs in Figure 3.3.

3.2.5 Solution Using the Calculus Approach

Again using the basic rules of differentiation of functions described in Chapter 1, one can also obtain the optimal levels of output. But this time, instead of using the optimality criterion MR = MC, we shall maximize profit (π) directly. That is, from the *profit function*,

$$\pi(Q) = TR - TC = P(Q)Q - C(Q) = (32 - Q)Q - (200 + 2Q)$$
$$= 30Q - Q^2 - 2000,$$

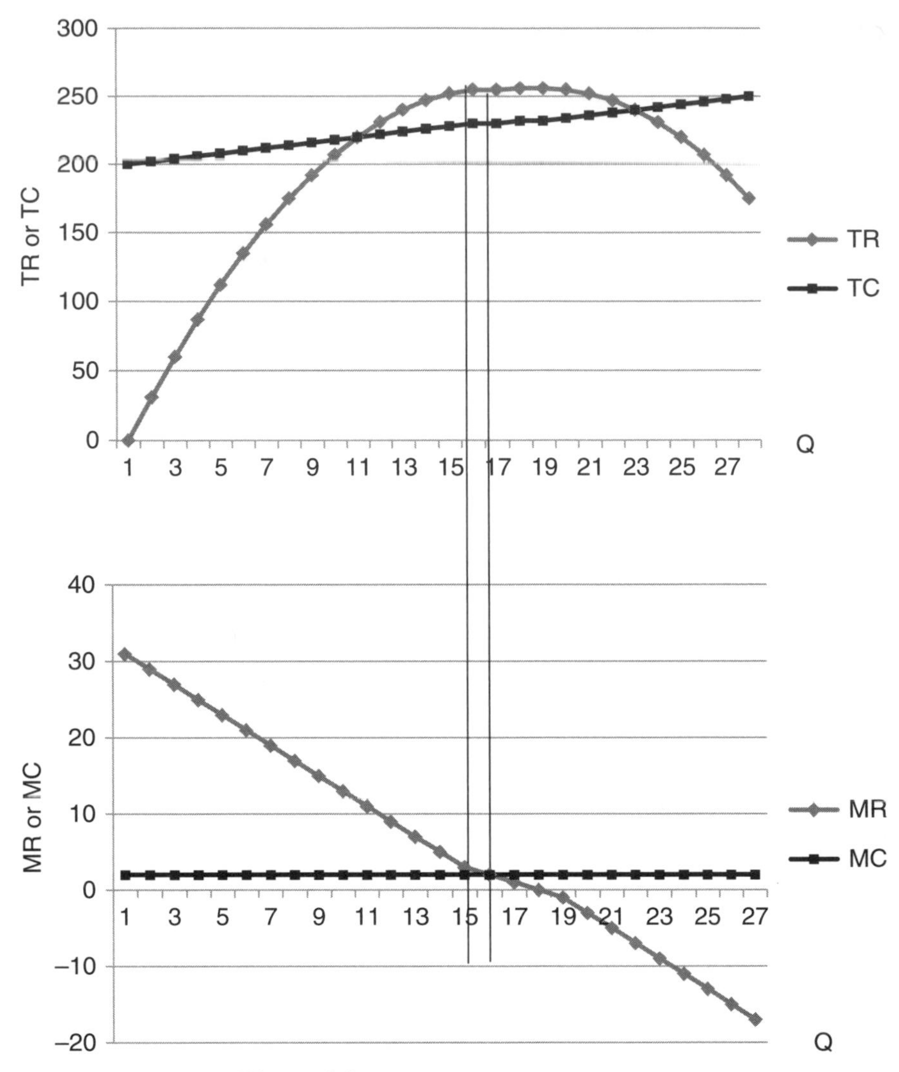

Figure 3.3 TR, TC and MR, MC graphs.

let us set

$$\frac{d\pi}{dQ} = 30 - 2Q = 0 \quad \text{or} \quad Q = 15.$$

And since $d^2\pi/dQ^2 = -2 < 0$, we see that $Q = Q_e = 15$ is the profit-maximizing level of output. (Note that Q_e is independent of total fixed cost.) This answers part (a).

At $Q_e = 15$, the corresponding price is $P = 32 - Q_e = 17$ (the part (b) solution).

To respond to part (c), we need to only evaluate $\pi(Q)$ at $Q_e = 15$ or maximum profit is $\pi(Q_e) = P(Q_e)Q_e - C(Q_e) = 17 \times 15 - [200 + 2(15)] = 25$.

Finally, the answer to part (d) is obtained by maximizing the total revenue function $TR = P(Q)Q = 32Q - Q^2$ directly. To this end, set

$$\frac{dTR}{dQ} = 32 - 2Q = 0 \quad \text{or} \quad Q = 16.$$

With $d^2TR/dQ^2 = -2 < 0$, we see that $Q = 16$ is the revenue-maximizing output level.

Note that working with the profit function directly implicitly incorporates the first-order condition $MR = MC$ since, from $\pi(Q) = TR - TC$, we obtain

$$\frac{d\pi}{dQ} = \frac{dTR}{dQ} - \frac{dTC}{dQ} = 0 \quad \text{or} \quad MR - MC = 0 \quad \text{or} \quad MR = MC \text{ as expected.}$$

3.3 MODELS OF COST MINIMIZATION: INVENTORY COST FUNCTIONS AND EOQ

It is often the case that when a firm receives orders for a certain amount of output or service, the dollar amount receivable is fixed by contract. In this instance, the firm is motivated to minimize the cost of producing the amount of output or service, thereby maximizing profit (the difference between the fixed dollar amount receivable and the cost of production).

Another example of cost minimization is when business organizations maintain inventories (stocks) of finished or unfinished goods and raw materials for future use. The reason why organizations maintain inventory is that it is not possible to predict exactly the uncertain demand, production times, and sales levels that ultimately materialize. Maintaining inventory is convenient for businesses to do, but it involves substantial costs. There are two important costs associated with maintaining inventory. One is holding cost and the other is ordering cost.

Holding cost refers to the costs associated with maintaining an inventory investment, such as insurance, taxes, and the cost of capital locked up in the inventory. This cost is usually expressed as a percentage of the inventory investment or, equivalently, as a percent of unit cost. It is also termed *carrying cost*. *Ordering cost* is a fixed cost associated with preparing and placing an order. It is independent of the size of the order. A tool that enables us to determine a firm's optimal inventory level is the *economic order quantity* (EOQ) *model*.

To derive the EOQ model, we first start with the total annual cost of maintaining inventory. This total annual cost is composed of two parts. These are annual holding cost (also called carrying cost) and the annual ordering cost. Let us use the following notation to develop the model. Assume that annual demand (D) is constant and there are no shortages. In addition, the rate of demand in each period T (which could be a week, month, or a quarter) is constant and the same for each period over the year. Let Q be the order quantity to be determined per period, with the cost per unit (C) or price of the product assumed constant for the entire year. Let C_h represent annual holding cost per unit, often expressed as a certain percentage of unit cost C or $C_h = K \times C$, where K is a given percentage. Also, let C_o stand for the fixed ordering cost per order, irrespective of the size of the order. Finally, let N represent the number of orders per year or $N = D/Q$. To summarize:

D: annual demand

T: inventory period (*cycle time*)

Q: order quantity

C: unit cost of the product

C_h: unit holding cost per annum ($C_h = KC$, K constant)

N: orders per year

C_o: fixed cost per order

Armed with the preceding nomenclature, the *annual ordering cost* (AOC) is given by

$$\mathrm{AOC(Q) = (D/Q)C_o}. \tag{3.1}$$

As Figure 3.4 illustrates, the *average inventory* held at any given point in time is $\frac{Q}{2}$ when demand is constant. For example, at the beginning of any period, the inventory level is Q units and at the end of the period, it is 0 units. Thus, the average is $(Q + 0)/2 = Q/2$. Therefore, the *annual holding cost* (AHC) is given by

$$\mathrm{AHC(Q) = (Q/2)C_h} \tag{3.2}$$

The *total annual inventory maintenance cost* (Y(Q)) is obtained by adding both of these cost elements described by (3.1) and (3.2). Thus,

$$\mathrm{Y(Q) = AHC(Q) + AOC(Q)} = \frac{Q}{2}C_h + \frac{D}{Q}C_o. \tag{3.3}$$

Since we are interested in determining the optimal order quantity (the quantity that minimizes the inventory maintenance cost), the annual purchase cost, which is fixed under the assumption that price is constant, is not added to the total cost of maintaining the inventory.

Our objective is to determine the order quantity Q that minimizes total inventory cost Y(Q). The resulting Q is called the EOQ or *economic order quantity*. This can be accomplished by trial and error and by the calculus methods, as illustrated by the following example.

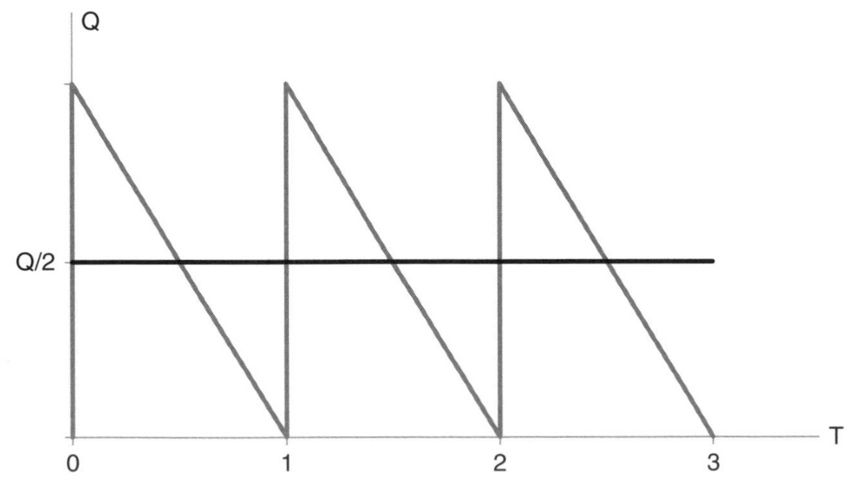

Figure 3.4 Use of inventory at a constant rate of demand.

EXAMPLE 3.3

A firm estimates the annual demand for one of its products to be 10,000 units. Each unit costs the firm $25.00. The inventory holding cost per unit is 4% of the unit cost. The fixed cost of placing an order is $50.00. The firm operates 360 days per year and it takes 5 days from the time of placing an order to the delivery (*lead time* L). Find the following:

a. EOQ

b. Average inventory level

c. The number of orders per year

d. Cycle time in days

e. Reorder point (R)

f. Annual holding cost

g. Annual ordering cost

h. Annual total cost without purchase cost ■

3.3.1 Solution by Trial and Error Method

Note: From Table 3.3, it is obvious that at $Q^* = 1000$, total cost is at a minimum, and at this level of Q, the holding and ordering costs both are equal to 500. Therefore, this is the desired level of order quantity or the optimal EOQ. These observations are reinforced by examining the plots of annual holding cost, annual ordering cost, and annual total cost (Figure 3.5).

It is evident from an examination of Table 3.3 and Figure 3.5 that, at this desired level of order quantity (1000), total cost is at a minimum and the annual holding cost (500) is equal to annual ordering cost (500). Since at the said minimum we must have $AHC(Q^*) = AOC(Q^*)$, let us use this requirement to algebraically solve for Q^*. To this end, we set

$$\frac{Q^*}{2} C_h = \frac{D}{Q^*} C_o$$

Table 3.3 Worksheet Showing Holding and Ordering Costs at Different Order Quantities

Trial Order Quantity, Q	Annual Holding Cost $= \frac{Q}{2} C_h = \frac{Q}{2} \times 1$	Annual Holding Cost $= \frac{D}{Q} C_o = \frac{10{,}000}{Q} \times 50$	Annual Total Cost
400	$200	$1250	$1450
600	300	833.33	1133.33
800	400	625	1025
1000 (=Q*)	**500**	**500**	**1000**
1200	600	416.67	1016.67
1400	700	357.14	1057.14
1600	800	312.5	1112.5

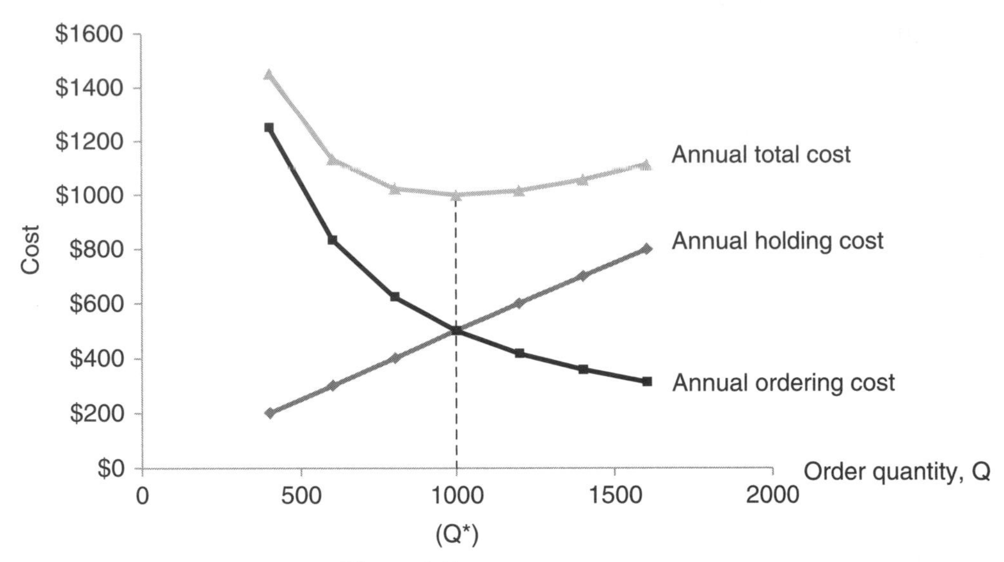

Figure 3.5 Annual cost curves.

or

$$(Q^*)^2 C_h = 2DC_o.$$

Hence,

$$(Q^*)^2 = \frac{2DC_o}{C_h}$$

and thus

$$Q^* = +\sqrt{\frac{2DC_o}{C_h}}, \quad \text{(optimal EOQ)}.$$

This is the optimal EOQ.

Using the EOQ formula, we can obtain the following solution values as requested at the outset in Example 3.3.

a.

$$Q^* = \sqrt{\frac{2(10,000)50}{1}} = 1000.$$

Knowing Q^*, we can readily find the following particulars of inventory management:

b. Average inventory level

$$\frac{Q^*}{2} = \frac{1000}{2} = 500.$$

c. Number of orders per year

$$N = \frac{D}{Q^*} = \frac{10,000}{1000} = 10.$$

d. Cycle time in days

$$\frac{360}{N} = \frac{360}{10} = 36 \text{ days}.$$

e. *Reorder point*: The inventory position at which a new order should be placed is R = L × daily demand, where L is the lead time. Thus,

$$R = 5 \times \frac{10,000}{360} = 138.9.$$

f. Annual holding cost from Table 3.3 when $Q^* = 1000$ is \$500.

g. Annual ordering cost from Table 3.3 when $Q^* = 1000$ is \$500.

h. Annual total cost = annual holding cost + annual ordering cost = \$500 + \$500 = \$1000.

3.3.2 Solution Using the Calculus Approach

At minimum cost, the first derivative of Y(Q) with respect to Q (marginal cost) must be set to zero. That is,

$$Y'(Q) = \frac{1}{2}C_h - \frac{D}{Q^2}C_o = 0$$

or

$$\frac{1}{2}C_h = \frac{D}{Q^2}C_o.$$

Note: This first-order condition automatically implies that the annual holding and ordering costs are equal at the optimal level of Q. Solving the preceding expression for Q yields

$$Q = \pm\sqrt{\frac{2DC_o}{C_h}}.$$

(We assume that Q in this relationship is positive since we rule out the possibility of back orders.) Furthermore, the second-order condition for a minimum of total cost at the above Q level requires that $Y''(Q)$ must be positive or

$$Y''(Q) = \frac{d}{dQ}\left[\frac{1}{2}C_h - \frac{D}{Q^2}C_o\right] = 0 + \frac{2DC_o}{Q^3} > 0$$

for any positive Q. We termed this particular level of Q, the EOQ, and denoted it as

$$Q^* = \sqrt{\frac{2DC_o}{C_h}} \quad \text{(Optimal EOQ)}.$$

Continuing as above:

a.
$$Q^* = \sqrt{\frac{2(10,000)50}{0.04(25)}} = \sqrt{\frac{1,000,000}{1}} = 1000.$$

Having found Q*, we can calculate the following:

b. Average inventory level

$$\frac{Q^*}{2} = \frac{1000}{2} = 500.$$

c. Number of orders per year

$$N = \frac{D}{Q^*} = \frac{10,000}{1000} = 10.$$

d. Cycle time in days

$$\frac{360}{N} = \frac{360}{10} = 36.$$

e. Reorder point (R), the inventory position at which a new order should be placed. By inventory position we mean the inventory on hand plus the inventory on order. Thus,

$$R = \text{lead time} \times \text{daily demand} = 5 \times \frac{10,000}{360} = 138.9.$$

f. Annual holding cost

$$\frac{Q^*}{2} \times C_h = \frac{1000}{2} \times 1 = 500.$$

g. Annual ordering cost

$$\frac{D}{Q^*} \times C_o = \frac{10,000}{1000} \times 50 = 500.$$

h. Annual total cost = annual holding cost + annual ordering cost = 500 + 500 = 1000.

EXAMPLE 3.4

A local auto repair shop predicted the annual demand for mufflers to be approximately 50,000. The fixed ordering cost is assumed to be $100 per order. The cost of each muffler is $50 and the holding cost is 2.5% of the muffler cost. Assume that there are 250 working days in the year and 4 days of lead time. Find the following:

a. EOQ

b. Average inventory level

c. Number of orders per year

d. Cycle time in days

e. Reorder point

f. Annual holding cost

g. Annual ordering cost

h. Annual total cost without the purchase cost

SOLUTION: We may consolidate the above information as follows:

$$D = 50,000,$$
$$C = \$50.00,$$
$$C_h = 0.025 \times \$50.00 = \$1.25,$$
$$C_o = \$100.00.$$

Then,

a. $\text{EOQ} = Q^* = \sqrt{\frac{2DC_o}{C_h}} = \sqrt{\frac{2 \times 50,000 \times 100}{1.25}} = 2828.43.$

b. Average inventory level $= \frac{Q^*}{2} = \frac{2828.43}{2} = 1414.22.$

c. Number of orders per year $= N = \frac{D}{Q^*} = \frac{50,000}{2828.43} = 17.68.$

d. Cycle time in days $= \frac{250}{N} = \frac{250}{17.68} = 14.14.$

e. Reorder point = lead time × daily demand $= 4 \times \frac{50,000}{250} = 800.$

f. Annual holding cost $= \frac{Q^*}{2} \times C_h = \frac{2828.43}{2} \times 1.25 = 1767.78.$

g. Annual ordering cost $= \frac{D}{Q^*} \times C_o = \frac{50,000}{2828.43} \times 100 = 1767.77.$

h. Annual total cost = annual holding cost + annual ordering cost = 1767.78 + 1767.77 = 3535.55. ■

3.4 CONSTRAINED OPTIMIZATION: LINEAR PROGRAMMING

In the process of business decision making, optimization generally involves constraints or side relations to which the variables must conform. For example, firms maximize profit subject to, among others, resource constraints; consumers maximize their utility subject to their budget constraints. Similarly, firms maximize return on a portfolio of investments subject to cash and other asset constraints.

Optimization also deals with the case of minimization subject to constraints. Households minimize total expenditure on food and other items subject to a budget (or other) constraint. Likewise, a manager would like to minimize the total cost of transporting goods from distribution centers to retail outlets subject to time, cash availability, and demand constraints at those outlets.

A commonly used constrained optimization model deals with maximizing or minimizing a linear objective function subject to linear inequality/equality constraints involving nonnegative variables. This is known as a *linear programming model*. Although there are many methods for solving this type of problem, we consider only two graphical methods involving two decision variables.

Let us examine the basic structure of a linear programming model. The following example problem will help us illustrate its salient features.

3.4.1 Linear Programming: Maximization

3.4.1.1 Solution by Graphical Method: First Approach

EXAMPLE 3.5

An investor would like to maximize total return with a portfolio of two mutual funds, A and B. Based on historical experience, it is estimated that the returns per share on these mutual funds are \$1.5 on fund A and \$1.00 on fund B. The price per share of fund A is \$1.00 and that of fund B is also \$1.00. The associated risk indices on these funds are given to be 2 on fund A and 1 on fund B. The investor is willing to invest \$5000.00 at a total risk index value of at most 7. Find the portfolio that maximizes total return subject to the given constraints.

SOLUTION: Given this information, our first step is to express the problem as a mathematical model. Specifically, let X_1 represent the number of shares of fund A (in thousands) and let X_2 depict the number of shares of fund B (also in thousands). Our objective is to maximize total return, in thousands of dollars, and this is given by $Z = 1.5X_1 + X_2$ (*the objective function*) subject to (abbreviated henceforth as "s.t.") the *constraint system*:

$$X_1 + X_2 \leq 5 \quad \text{(budget constraint in thousands of dollars),}$$
$$2X_1 + X_2 \leq 7 \quad \text{(risk constraint),}$$
$$X_1, X_2 \geq 0 \quad \text{(nonnegativity conditions).}$$

In general, the constraint system is made up of *structural constraints* (the budget and risk constraints) and *nonnegativity conditions*. Additionally, the values 5 and 7 are the *constraint capacities*.

Let us define the concept of a *solution set* as the set of points satisfying a linear inequality (or equality). For instance, the solution set for the budget constraint is shown in Figure 3.6a. Note that when strict equality holds, we just get the boundary line $X_1 + X_2 = 5$ (see the intercept method in Section 1.7.2 regarding the graphing of linear equations). And when we admit inequality ($X_1 + X_2 < 5$), we fall below the boundary line and end up within the shaded region proper. Hence, the solution set for the budget constraint involves all points on the boundary

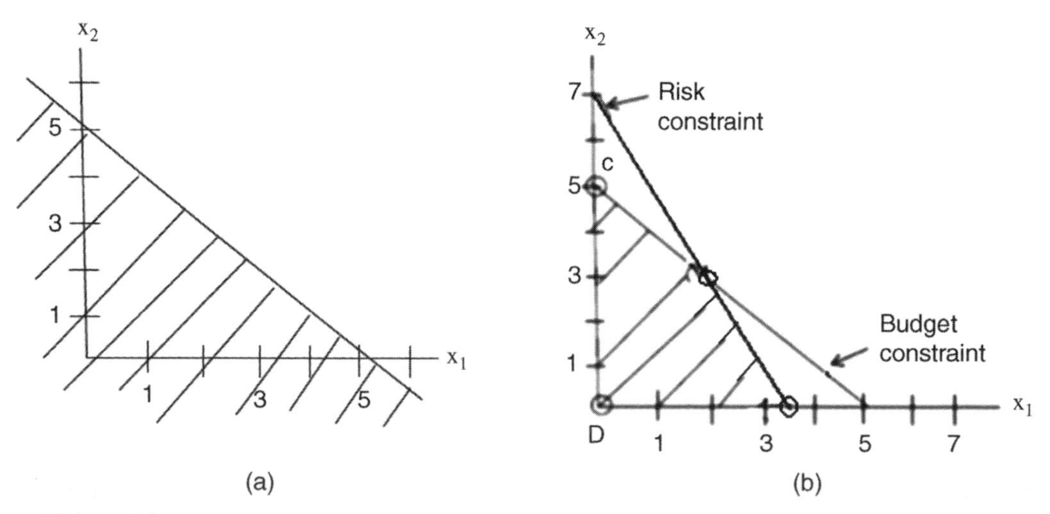

Figure 3.6 Solution sets. (a) Solution set for $X_1 + X_2 \leq 5$ (shaded). (b) Feasible region (shaded).

line and beneath it. (The reader should draw the solution sets for the risk constraint and for the nonnegativity conditions.) If all the constraints are satisfied simultaneously, then we obtain what is called the *feasible region* or region of admissible solutions to the linear programming problem. How is the feasible region formed? It is simply the intersection of the solution sets of all the constraints (Figure 3.6b). What are the characteristics of this feasible region? We may describe it as a "strictly bounded, closed, convex set with a finite number of extreme points." (Do not let this description "throw you for a loop"; it is not as daunting as it appears to be.) ■

Let us look at its particulars:

Strictly bounded: Its diameter is finite.

Closed: It contains all of its boundary points.

Convex: If you connect any two of its points by a straight line segment, that line segment lies entirely within the set.

Extreme (corner) point: A point that does not lie between any two other points of the feasible region (e.g., it occurs at the intersection of two (or more) of the constraint boundary lines).

Why are these concepts important? Simply because they (i) help specify the conditions under which the linear programming problem actually has an optimal solution; and (ii) tell us where to look for that solution. To this end, we state two important theorems:

Existence theorem: If the feasible region is a strictly bounded, closed, nonempty, and convex set, then the linear program always has an optimal solution.

Extreme point theorem: The objective function Z assumes its optimum at an extreme point of the feasible region (Figure 3.7a). If it assumes its optimum at more than one extreme point, then it takes on that same value all along an edge of the feasible region between those extreme points (Figure 3.7b).

As panel (b) of Figure 3.7 reveals, there are two optimal extreme point solutions (at A and B with Z(A) = Z(B)) and an infinity of optimal nonextreme point solutions between A and B (e.g., at C, Z(C) = Z(A) = Z(B)).

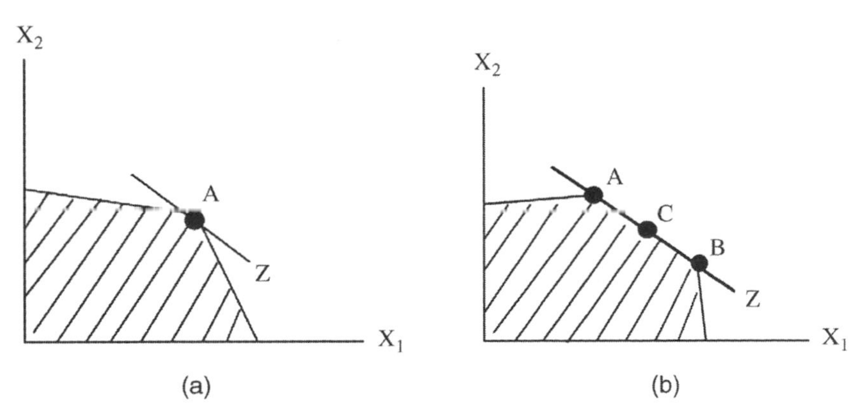

Figure 3.7 Illustration of the Extreme Point theorem. (a) Unique optimal solution (Z(A) = max Z). (b) Multiple optimal solutions (Z(A) = Z(B) + Z(C) = max Z).

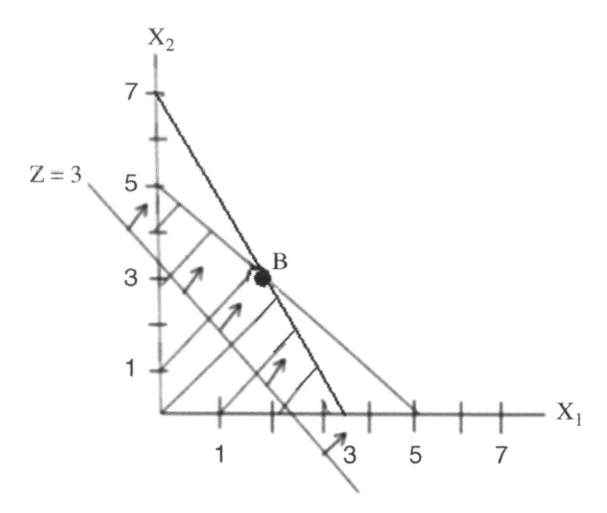

Figure 3.8 Optimizing Z.

To solve the investors problem, let us again consider the feasible region (Figure 3.6b) but now with the objective function $Z=1.5X_1+X_2$ introduced into our analysis. Suppose we choose an arbitrary (ballpark) value for Z, say, $Z=3$ and insert the objective function equation $3=1.5X_1+X_2$ into Figure 3.6b (call it Figure 3.8). Does $Z=3$ maximize total return? Obviously not, sincc a value of Z larger than 3 can be obtained by sliding $Z=1.5X_1+X_2$ parallel to itself up over the feasible region until tangency to the feasible region occurs at a unique extreme point or along an entire edge of the feasible region. Clearly, a maximal solution is obtained when $Z=1.5X_1+X_2$ is tangent to the feasible region at extreme point B. What are the coordinates of B? To answer this question, we need to only note that both the budget and risk constraints are *binding* at extreme point B in that they hold as strict equalities there. Hence, we need to solve only these binding constraints simultaneously in order to obtain the coordinates of this extreme point, that is, solving the linear system:

$$X_1 + X2 = 5,$$
$$2X_1 + X_2 = 7,$$

simultaneously yields $X_1=2$ and $X_3=3$. Hence, the maximum total return is

$$Z(B) = 1.5(2) + 1(3) = 6.$$

It was mentioned earlier that at extreme point B, both structural constraints are binding or hold as strict equalities. Clearly,

$$2 + 3 = 5,$$
$$2(2) + 3 = 7,$$

as expected. In this regard, there is no slack or excess capacity in either of these structural constraints. However, consider the feasible point with coordinates (1,3).

At this point the structural constraints yield

$$\left.\begin{array}{l} 1+3<5, \\ 2(1)+3<7, \end{array}\right\} \quad \text{or} \quad \left\{\begin{array}{l} 4<5, \\ 5<7. \end{array}\right.$$

Clearly, there is one unit of slack or unused capacity in the budget constraint. Denote it as $X_3=5-4=1$; and there are two units of slack or excess capacity in the risk constraint. Denote it as $X_4=7-5=2$. Here, X_3 and X_4 are termed *slack variables*: When positive, they account for the excess of the right-hand side over the left-hand side when dealing with "≤" constraints. We require that $X_3 \geq 0$, $X_4 \geq 0$. We can generally write the slack variables in terms of the "≤" constraints as

$$\left.\begin{array}{l} X_1+X_2+X_3=5, \\ 2X_1+X_2+X_4=7, \end{array}\right\} \quad \text{or} \quad \left\{\begin{array}{l} X_3=5-X_1-X_2, \\ X_4=7-2X_1-X_2. \end{array}\right.$$

We may summarize the solution to this linear programming problem as $X_1=2$ or 2000 shares of fund A, $X_2=3$ or 3000 shares of fund B, $X_3=0$ (all \$5000 invested), $X_4=0$ (risk tolerance limit is reached), and maximum return $Z=6$ or \$6000.

3.4.1.2 Solution by Graphical Method: Second Approach

We present here a second approach for finding the optimal solution (by an alternative graphical method) since it might be easier for the student to visualize the various steps involved. In both cases the graphical method breaks down in instances dealing with three or more decision variables. This is because the process deals with three or more dimensions, thus making it difficult to visualize the feasible region and objective function. These cases are dealt with by an algebraic approach such as the "simplex method." We use the simplex method of solution in the "Solver" program in Excel, as demonstrated later in the chapter.

We now turn to the second approach for finding the optimal solution for the same problem presented in Example 3.5.

After defining the decision variables and formulating the mathematical model, we describe below the alternative graphical method as a three-step procedure:

Step 1: Show the constraint equations graphically and identify the feasible region.

Step 2: Find the corner points (extreme points) of the feasible region.

Step 3: Find the value of the objective function at each one of the corner points and identify the optimal solution(s) as those corresponding to the maximum/minimum value of the objective function.

Based upon the nature of step 3, this second approach is often called the *complete enumeration of extreme points method*.

Example 3.5 problem has the form

$$\text{Maximize } Z = 1.5X_1 + X_2 \text{ s.t.}$$
$$X_1 + X_2 \leq 5 \quad \text{(budget constraint)}$$
$$2X_1 + X_2 \leq 7 \quad \text{(risk constraint)}$$
$$X_1, X_2 \geq 0 \quad \text{(nonnegative variables)}$$

Step 1: Show the constraint inequalities graphically and identify the feasible region. First, plot the linear constraint equations (Figure 3.9) by following the intercept method described in Chapter 1. For example, consider the budget constraint equation $X_1 + X_2 = 5$ and set $X_2 = 0$. The equation reduces to $X_1 + (0) = 5$ and thus $X_1 = 5$, resulting in the point (5,0) (point D). Similarly, with $X_1 = 0$, we solve for X_2 from $(0) + X_2 = 5$ and thus $X_2 = 5$. Then (0,5) is another point (point C). Connecting these two points, we obtain the linear budget line. Following the same procedure, we can readily find two points, (3.5, 0) (point A) and (0,7) (point E) on the risk constraint. And by connecting these two points, we obtain the linear risk equation. After drawing the straight lines corresponding to the constraint boundaries, we identify the region representing the intersection of the constraint solution sets. That is, this intersection lies below the constraint lines since it satisfies the direction of the constraint inequalities and is shaded as OABC. It describes the region of admissible solutions or the feasible region, and is shown in Figure 3.9.

Step 2: Find the corner (extreme) points of the feasible region. From Figure 3.9 we see that the only unknown corner point is B, where the budget constraint and risk constraint lines intersect. Obviously, point B is obtained by solving the two constraint equations simultaneously. That is,

denote the budget constraint equation as "B": $X_1 + X_2 = 5$ (B);
likewise, denote the risk constraint equation as "R": $2X_1 + X_2 = 7$ (R).

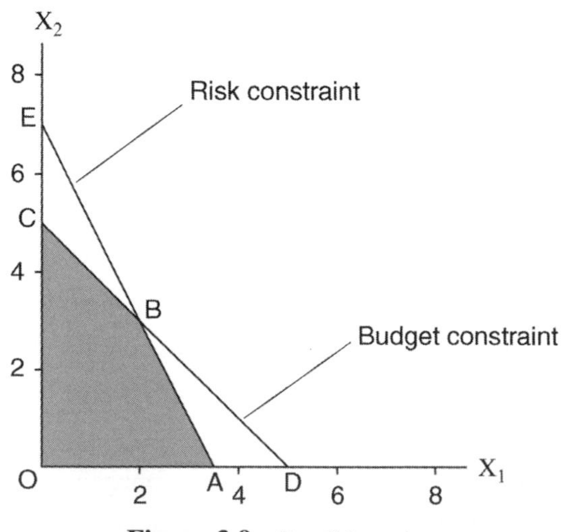

Figure 3.9 Feasible region.

Table 3.4 Objective Function Values

Extreme point (X_1, X_2)	$Z = 1.5X_1 + X_2$
$O(0, 0)$	$1.5(0) + (0) = 0$
$A(3.5, 0)$	$1.5(3.5) + (0) = 5.25$
$B(2, 3)$	$1.5(2) + (3) = 6$
$C(0, 5)$	$1.5(0) + 5 = 5$

Subtracting (B) from (R), we get $X_1 = 2$; and substituting this value into equation B, we get, $2 + X_2 = 5$ or $X_2 = 5 - 2 = 3$. Thus, the corner point B has coordinates (2,3).

Step 3: Evaluate the objective function at each of the corner (extreme) points of the feasible region (following the Extreme Point theorem stated earlier) and identify the optimum corner point corresponding to the maximum value of the objective function. These evaluations appear in Table 3.4.

It is clear from this table that the investor maximizes the total return on the portfolio by buying 2(000's) of shares of Mutual Fund A and 3(000's) shares of Mutual Fund B. The maximum return on this portfolio is \$6(000's). The extreme point B, where the portfolio return is maximized, is called the "optimal point." The values of the slack variables, X_3 and X_4, are given in Table 3.5.

Table 3.5 Slack Variables

Constraint	Left Side of the Constraint	Right Side of the Constraint	Slack
Budget	$2 + 3 = 5$	5	$X_3 = 0$
Risk	$2(2) + (3) = 7$	7	$X_4 = 0$

We may summarize the solution to this linear programming problem as $X_1 = 2$ or 2000 shares of Fund A, $X_2 = 3$ or 3000 shares of Fund B, $X_3 = 0$ (all \$5000 invested), $X_4 = 0$ (risk tolerance limit is reached), and the maximum return $Z = 6$ or \$6000.

Note: Both approaches gave identical solutions, as expected.

EXAMPLE 3.6

Solve the following linear programming problem and compute the slack values corresponding to the constraints at the optimal solution.

$$\text{Maximize } Z = X_1 + 1.5X_2 \text{ s.t.,}$$
$$X_1 + 2X_2 \leq 5 \text{ (constraint 1),}$$
$$2X_1 + X_2 \leq 4 \text{ (constraint 2),}$$
$$X_1, X_2 \geq 0 \text{ (nonnegativity conditions).}$$

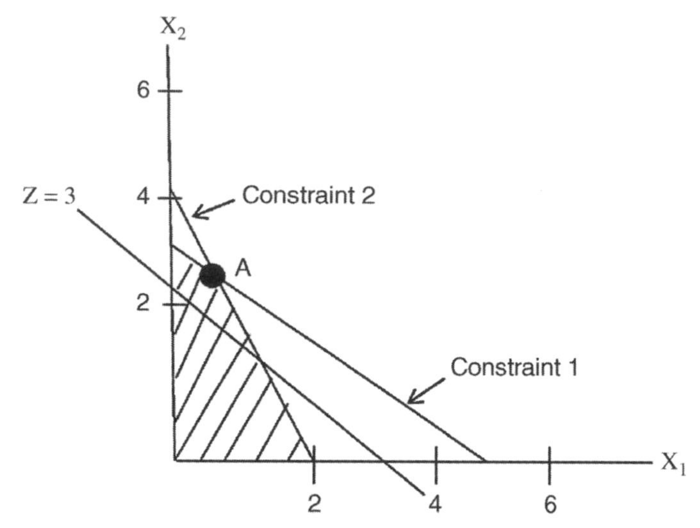

Figure 3.10 Feasible region.

SOLUTION BY GRAPHICAL METHOD—FIRST APPROACH: As a first step, let us determine the feasible region by intersecting the solution sets of the constraint system (Figure 3.10). Next, insert a reference objective function into Figure 3.10 and slide it upward (or downward) parallel to itself to determine where tangency to the feasible region occurs. For $Z = X_1 + 1.5X_2 = 3$, we see that this line is too low. Sliding it upward results in its tangency to the feasible region at extreme point A. At this extreme point, both structural constraints are binding or hold as equalities. Hence, we must solve the constraint equality system

$$X_1 + 2X_2 = 5,$$
$$2X_1 + X_2 = 4,$$

simultaneously so as to obtain the coordinates of extreme point A as (1,2). At this extreme point, the optimal objective function value is $Z(A) = 1 + 1.5(2) = 4$. With both constraints binding at this extreme point, we should anticipate that the values of their respective slack variables, X_3 and X_4, are zero, that is,

$$(\text{constraint } 1)X_3 = 5 - X_1 - 2X_2 = 5 - 1 - 2(2) = 0,$$
$$(\text{constraint } 2)X_4 = 4 - 2X_1 - X_2 = 4 - 2(1) - 2 = 0.$$

In sum, the optimal solution to this problem is $X_1 = 1$, $X_2 = 2$, $X_3 = X_4 = 0$ (no slack), and $Z = 4$.

SOLUTION BY GRAPHICAL METHOD—SECOND APPROACH:

$$\text{Maximize } Z = X_1 + 1.5X_2 \text{ s.t.,}$$
$$X_1 + 2X_2 \leq 5 \text{ (constraint 1)},$$
$$2X_1 + X_2 \leq 4 \text{ (constraint 2)},$$
$$X_1, X_2 \geq 0 \text{ (nonnegativity conditions)}.$$

Step 1: Show the constraint inequalities graphically and identify the feasible region.

As required, the nonnegativity conditions restrict the solution sets of the structural constraints to the first quadrant of the graph. We may plot the linear constraint equations graphically by following the intercept method of Chapter 1. For example, consider the constraint 1 equation $X_1 + 2X_2 = 5$ and set $X_2 = 0$. This equation reduces to $X_1 + 2(0) = 5$ and, thus,

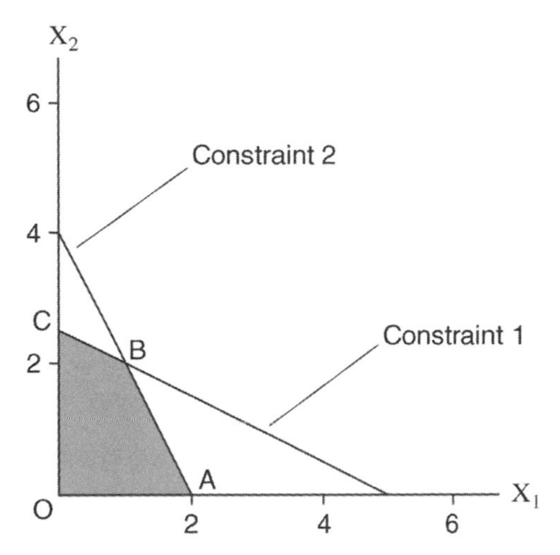

Figure 3.11 Feasible region.

$X_1 = 5$, resulting in the point (5,0) (Figure 3.11). Similarly, with $X_1 = 0$, we can solve for X_2 from $(0) + 2X_2 = 5$ so that $X_2 = 2.5$.Then (0,2.5) is another point. Connecting these two points in Figure 3.11, we obtain the constraint 1 boundary line. Applying the same procedure to constraint 2, we find two points (2,0) and (0,4) on the constraint 2 line and by connecting these two points, we get the constraint 2 boundary line. After drawing the straight lines corresponding to the boundaries of the constraints, we can easily identify the region representing the satisfaction of each of the constraints individually. We next identify the region that constitutes the intersection of the sets of admissible solutions for these two "≤" structural constraints. This intersection obviously satisfies the direction of the inequalities and is shaded as OABC. It describes the feasible region and it is shown in Figure 3.11.

Step 2: Find the corner (extreme) points of the feasible region. The only unknown corner point is point B. Obviously, it is obtained by solving the two constraint equations simultaneously.

To this end,

$$2 \times (\text{constraint 1}) \text{ yields } 2X_1 + 4X_2 = 10,$$
$$\text{constraint 2 is } 2X_1 + X_2 = 4.$$

Subtracting the second equation from the first, we obtain $3X_2 = 6$ or $X_2 = 2$. Substituting this value into the second equation yields $2X_1 + 2 = 4$ or $X_1 = 1$.

Therefore, the intersection point B has coordinates (1,2).

Step 3: Evaluate the objective function at each one of the corner points and identify the optimum extreme point corresponding to the maximum value.

It is clear from Table 3.6 that the maximum value of the objective function occurs at point B: (1,2). Thus, point B is optimal and the corresponding maximum value of the objective function is 4.

Table 3.6 Objective Function Values

Extreme Point	$Z = X_1 + 1.5X_2$
O.(0, 0)	$0 + 1.5(0) = 0$
A.(2, 0)	$2 + 1.5(0) = 2$
B.(1, 2)	$1 + 1.5(2) = 4$
C.(0, 2.5)	$0 + 1.5(2.5) = 3.75$

Table 3.7 Slack Variables

Constraint	Left Side of the Constraint	Right Side of the Constraint	Slack Value
1	$1 + 2(2) = 5$	5	$X_3 = 0$
2	$2(1) + 2 = 4$	4	$X_4 = 0$

At the optimal point B: (1,2), Table 3.7 provides the values of the slack variables (X_3 and X_4) corresponding to the two constraints.

Note: Both approaches gave identical solutions, as required. ■

3.4.2 Linear Programming: Minimization

EXAMPLE 3.7

A manufacturing company is planning to introduce a new product into the market and decides to advertise on TV and in newspapers in order to reach at least 2 million potential customers. The effectiveness of the media campaign and the requirements specified as per the company policy are given in Table 3.8.

Table 3.8 Effectiveness of the Media and the Requirements of the Company

Medium	$ Cost/Unit (000's)	Audience Reached/ Unit (000's)	Minimum Units Required
TV commercials (X_1)	3	50	20
Newspaper ads (X_2)	2	20	50

Formulate a linear programming model so that the company can minimize the cost of reaching the target audience. Determine the optimum number of TV commercials and newspaper ads the company has to buy in order to achieve its goal. Find the values of the surplus variables (defined below).

SOLUTION BY GRAPHICAL METHOD: FIRST APPROACH: The objective is to minimize the total cost of publicity, $Z = 3X_1 + 2X_2$, where X_1 is the number of TV commercials and X_2 is the number of newspaper ads.

The audience constraint is given by $50X_1 + 20X_2 \geq 2000$. (This is the result of dividing both sides of the inequality $50,000X_1 + 20,000X_2 \geq 2,000,000$ by 1000.); the TV commercials constraint is $X_1 \geq 20$ and the newspaper ads constraint is $X_2 \geq 50$.

Thus, the complete linear programming model is specified as follows:

$$\text{Minimize } Z = 3X_1 + 2X_2 \text{ s.t.,}$$
$$50X_1 + 20X_2 \geq 2000 \text{ (audience constraint)},$$
$$X_1 \geq 20 \text{ (TV commercials constraint)},$$
$$X_2 \geq 50 \text{ (newspaper ads constraint)},$$
$$X_1, X_2 \geq 0 \text{ (nonnegativity conditions)}.$$

Next, let us specify the feasible region by intersecting the solution sets of the constraint system. (The (shaded) solution set for the audience constraint taken by itself is shown in

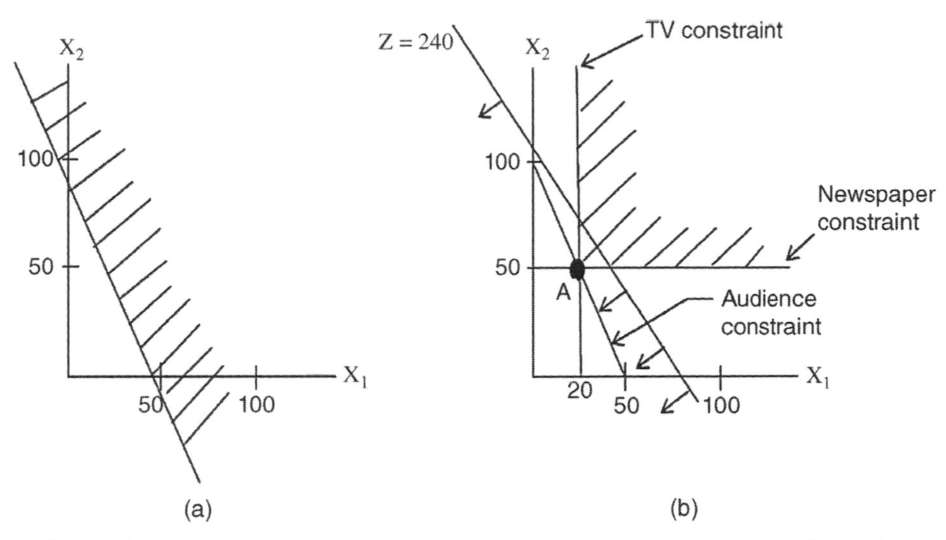

Figure 3.12 Solution sets. (a) Solution set for $50X_1 + 20X_2 \geq 2000$. (b) Feasible region.

Figure 3.12a.) This results in the shaded area depicted in Figure 3.12b. Suppose our reference objective function value is $Z = 240$. A plot of the implied objective function $3X_1 + 2X_2 = 240$ is shown in Figure 3.12b. Since this objective function is clearly not minimal, let us slide it downward over the feasible region until it is just tangent to the same. This occurs at extreme point A, where the TV and newspaper constraint boundary lines intersect. Hence, the coordinates of A are (20, 50). Then $Z(A) = 3(20) + 2(50) = 160$ represents the minimum publicity cost associated with 20 TV spots and 50 newspaper ads.

Since at extreme point A, all of the structural constraints are binding (they hold as strict equalities), we should expect that the associated *surplus variables*, X_3, X_4, and X_5, respectively, are all zero valued. That is, let us write the surplus variables in terms of the "≥" structural constraints as

$$\left.\begin{aligned} 50X_1 + 20X_2 - X_3 &= 2000 \\ X_1 - X_4 &= 20, \\ X_2 - X_5 &= 50, \end{aligned}\right\} \quad \text{or} \quad \left\{\begin{aligned} X_3 &= 50X_1 + 20X_2 - 2000, \\ X_4 &= X_1 - 20, \\ X_5 &= X_2 - 50. \end{aligned}\right.$$

(Clearly, a surplus variable accounts for the excess of the left-hand side over the right-hand side in a "≥" structural constraint.)

Then,

$$\begin{aligned} &\text{Media constraint:} && X_3 = 50(20) + 20(50) - 2000 = 0, \\ &\text{TV constraint:} && X_4 = 20 - 20 = 0, \\ &\text{Newspaper constraint:} && X_5 = 50 - 50 = 0. \end{aligned}$$

To summarize, the optimal solution to the media selection problem is $X_1 = 20$ TV commercials, $X_2 = 50$ newspaper ads, $X_3 = 0$ surplus (minimum number of TV ads attained), $X_4 = 0$ surplus (minimum number of newspaper ads attained), $X_5 = 0$ surplus (2 million audience reached and no more), and minimum total cost $Z = 160$ ($160,000). Let us make a slight adjustment to this problem. Specifically, let us replace the original objective function by $Z = 5X_1 + 2X_2$ and the original newspaper ads constraint by $X_2 \geq 30$. Hence, our revised problem has the following structure:

$$\text{Minimize } Z = 5X_1 + 2X_2 \text{ s.t.,}$$
$$50X_1 + 20X_2 \geq 2000,$$
$$X_1 \geq 20,$$
$$X_2 \geq 30,$$
$$X_1, X_2 \geq 0.$$

The new feasible region is shown in Figure 3.13.

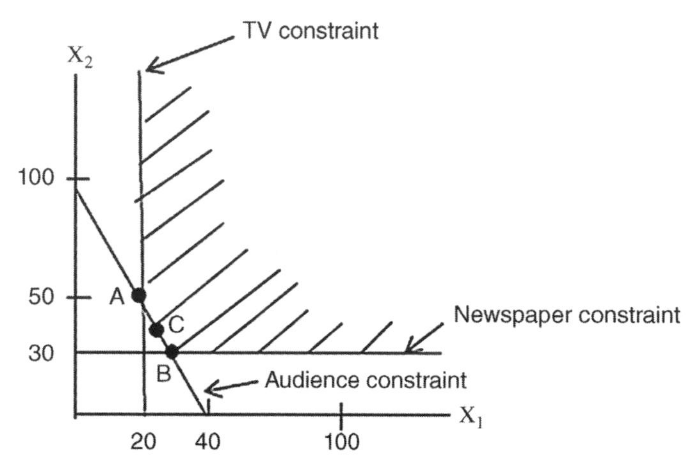

Figure 3.13 Multiple optimal solutions.

It now has two extreme points: A and B. It is readily demonstrated that the minimal objective function is tangent to the feasible region all along the edge from A to B. Hence, we have two minimal extreme point solutions and an infinity of nonextreme point solutions all along the edge between A and B. At A, the audience and TV spot constraints are binding; at B, the audience and newspaper ad constraints are binding. And at the nonextreme point C, only the audience constraint is binding. Clearly, Z(A) = Z(B) = Z(C) for any C between A and B. How does management determine which optimal solution to implement?

Could we have predicted that the case of multiple optimal solutions would emerge from this adjusted problem? The answer is *yes*! Specifically, if the objective function coefficients (5 and 2) are a multiple of the coefficients on the variables in a particular structural constraint, then the instance of multiple optimal solutions occurs. Here, the multiple is $\frac{1}{10}$ since, from the audience constraint, $5X_1 + 2X_2 = \frac{1}{10}(50X_1 + 20X_2)$.

SOLUTION BY GRAPHICAL METHOD: SECOND APPROACH

$$\text{Minimize } Z = 3X_1 + 2X_2 \text{ s.t.,}$$
$$50X_1 + 20X_2 \geq 2000,$$
$$X_1 \geq 20,$$
$$X_2 \geq 50,$$

and nonnegativity of the variables $X_1, X_2 \geq 0$.

Step 1: Show the constraint inequalities graphically and identify the feasible region. Notice that the nonnegative variables restrict the solutions of the structural constraints to the first quadrant of Figure 3.14. The feasible region is the shaded area within this figure.

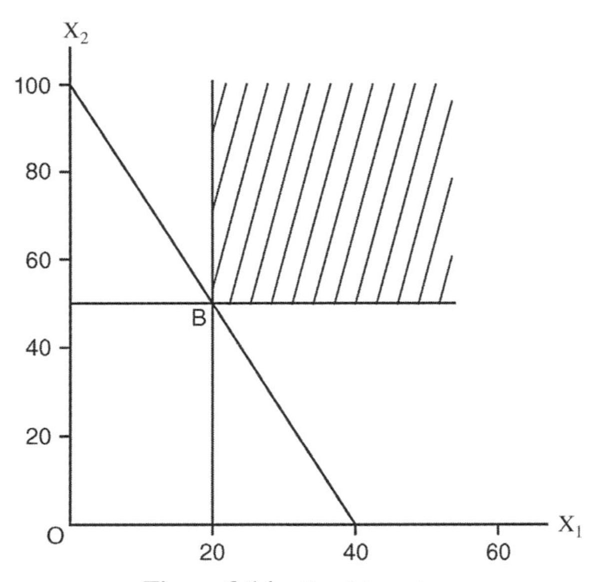

Figure 3.14 Feasible region.

Step 2: We observe that the feasible region has only one extreme point B, where $X_1 = 20$ and $X_2 = 50$ lines intersect. This point B occurs where all the three structural constraint lines intersect and thus can be obtained by solving either the first and third constraints simultaneously, or by solving first and second structural constraints simultaneously. This process results in only one extreme point B: (20,50).

Step 3: Evaluating the objective function Z at the above extreme point gives $Z = 3(20) + 2(50) = 160$ thousands of dollars. This is the minimum cost.

The surplus values of the constraints at the optimum point are given in Table 3.9.

Table 3.9 Surplus Variables

Extreme Point	Left Side of the Constraint	Right Side of the Constraint	Surplus Value
$B: (20, 50)$	$50(20) + 20(50) = 2000$	2000	$X_3 = 0$
$B: (20, 50)$	$20 + 0(50) = 20$	20	$X_4 = 0$
$B: (20, 50)$	$0(20) + 50 = 50$	50	$X_5 = 0$

To summarize, the optimal solution to the media selection problem is $X_1 = 20$ TV commercials, $X_2 = 50$ newspaper ads, $X_3 = 0$ surplus (minimum number of TV ads attained), $X_4 = 0$ surplus (minimum number of newspaper ads attained), $X_5 = 0$ surplus (2 million audience size reached), and minimum total cost $Z = 160$ (\$160,000).

Consider the following modification of this model:

$$\begin{aligned} \text{Minimize } Z &= 5\,X_1 + 2\,X_2 \text{ s.t.,} \\ 50\,X_1 + 20\,X_2 &\geq 2000, \\ X_1 &\geq 20, \\ X_2 &\geq 30, \\ X_1, X_2 &\geq 0. \end{aligned}$$

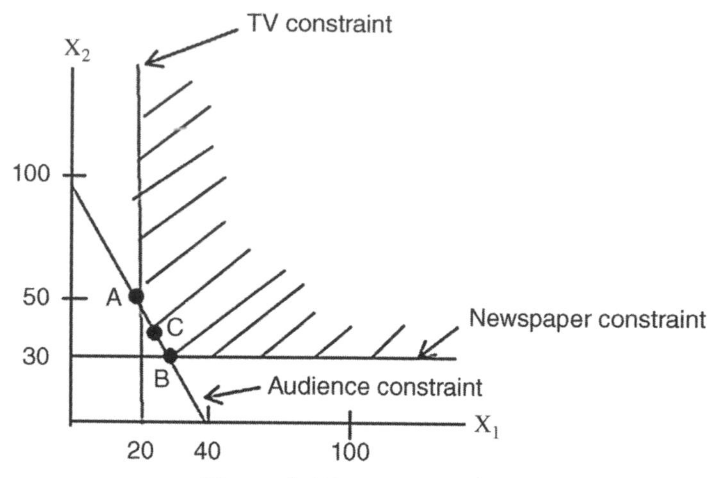

Figure 3.15 Feasible region.

Step 1: To exhibit the constraint equations graphically, first insert $X_2 = 0$ into the minimum audience constraint and solve for X_1. This results in the point (40,0). Similarly, setting $X_1 = 0$ and solving for X_2 yields the point (0,100). It is evident that the second constraint equation is a vertical line parallel to the X_2-axis at $X_1 = 20$. Likewise, the third constraint equation is a horizontal line parallel to the X_1-axis at $X_2 = 30$. Taking into consideration that we have "≥" appearing in all the three structural constraints, the shaded region in Figure 3.15, corresponding to the intersection of the constraint solution sets, defines the feasible region.

Step 2: The corner point A: (20,50) is obtained by solving the first and second constraint equations simultaneously. Similarly, corner point B: (28,30) is obtained by solving constraint equations one and three simultaneously.

Step 3: We evaluate the objective function at each one of the corner (extreme) points of the feasible region (by virtue of the extreme point theorem stated earlier) and identify the optimum corner point corresponding to the minimum value of the objective function. The evaluation is given in Table 3.10.

It is evident from Table 3.10 that this is a case of multiple optimal solutions since points A and B give the same minimum value for total cost Z = 200,000 dollars. As explained in the first approach, any point between A and B, say, for example, point C: (24,40) also gives the same minimum value for total cost $Z = 5(24) + 2(40) = 200$. Here C can be obtained as $(28 + 20)/2$, $(30 + 50)/2$, the mid-point between A and B.

A brief comment on the surplus values: Management has a choice in selecting the optimal solution. For example, if point A is selected, then there is zero surplus in the TV

Table 3.10 Objective Function Values

Extreme Point (X_1, X_2)	$Z = 5X_1 + 2X_2$
A(20, 50)	$5(20) + 2(50) = 200$
B(28, 30)	$5(28) + 2(30) = 200$

constraint, 20 units of surplus in the newspaper ads constraint, and 0 surplus in the audience constraint. What are the surplus values at point B? At point C? ■

EXAMPLE 3.8

$$\begin{aligned} &\text{Minimize } Z = 2X_1 + 3X_2 \text{ s.t.,} \\ &X_1 + 2X_2 \geq 5, \\ &X_1 + X_2 \geq 3, \\ &X_1, X_2 \geq 0. \end{aligned}$$

Identify the feasible region and find the optimal solution. What is the minimum value of the objective function? Find the values of the surplus variables.

SOLUTION BY GRAPHICAL METHOD: FIRST APPROACH The feasible region is shown in Figure 3.16. Suppose we choose $Z = 12$ in order to get a reference objective function.

A plot of $2X_1 + 3X_2 = 12$ reveals that we are not at a minimum of Z. Sliding this reference equation downward over the feasible region results in a tangency to the same at extreme point A, where both structural constraints are binding.

Solving the structural constraint equalities

$$\begin{aligned} X_1 + 2X_2 &= 5, \\ X_1 + X_2 &= 3, \end{aligned}$$

simultaneously yields the coordinates of A, namely, $X_1 = 1$ and $X_2 = 2$. At this extreme point $Z(A) = 2(1) + 3(2) = 8$, the minimum of Z. Since both structural constraints are binding at A, we anticipate that the surplus variables, X_3 and X_4, each equal zero, that is,

$$\begin{aligned} X_3 &= X_1 + 2X_2 - 5, \\ X_4 &= X_1 + X_2 - 3, \end{aligned}$$

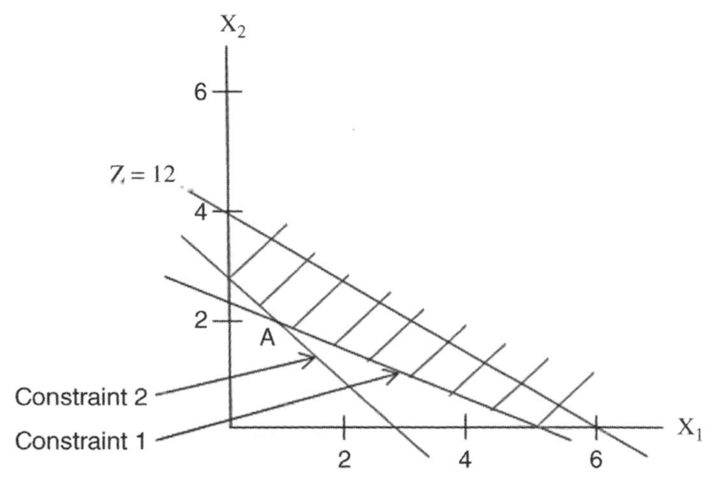

Figure 3.16 Feasible region.

or, at (1,2),

$$X_3 = 1 + 2(2) - 5 = 0,$$
$$X_4 = 1 + 2 - 3 = 0.$$

Our full solution to this linear program is thus

$$X_1 = 1, \quad X_2 = 2, \quad X_3 = X_4 = 0 \text{ (no surplus)}, \quad \text{and} \quad Z = 8.$$

SOLUTION BY GRAPHICAL METHOD: SECOND APPROACH

$$\text{Minimize } Z = 2X_1 + 3X_2 \text{ s.t.,}$$
$$X_1 + 2X_2 \geq 5,$$
$$X_1 + X_2 \geq 3,$$
$$X_1, X_2 \geq 0.$$

Step 1: Show the constraint inequalities graphically and identify the feasible region.

Notice that the nonnegative variables restrict the solutions of the constraints to the first quadrant of Figure 3.17. We first exhibit the linear constraint equations graphically by following the intercept method described in Chapter 1. For example, if we consider the first constraint equation and set $X_2 = 0$, then we can solve for X_1. We thus obtain $X_1 = 5$ or point (5,0). Similarly, with $X_1 = 0$, we can solve for X_2 so as to obtain a second point (0,2.5). Connecting these two points in Figure 3.17, we obtain the first constraint boundary line. Following this same procedure, we can also find two points (3,0) and (0,3) on the second constraint line and, by connecting them, we obtain the boundary line for the second constraint. After drawing the lines corresponding to the equality of all the constraints, we can easily identify the region representing the "inequality" of each of the constraints. This is obtained by intersecting the solution sets of the constraint system, and this intersection (common area) defines the feasible region that satisfies the direction of both of the inequalities. It constitutes the shaded portion of Figure 3.17.

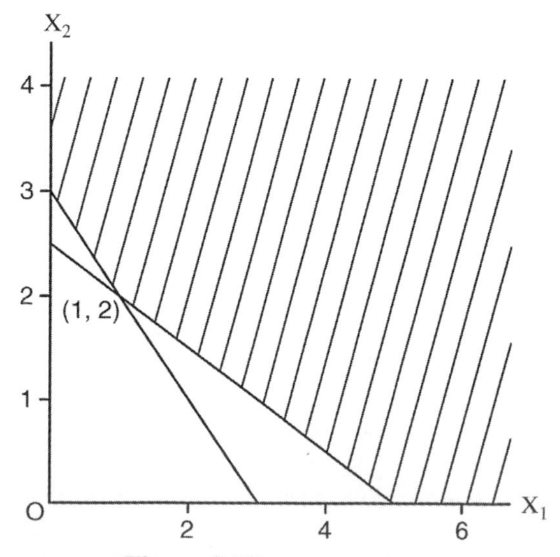

Figure 3.17 Feasible region.

Step 2: From the above feasible region, we find that there are three extreme points: (0,3), (1,2), and (5,0). The point with coordinates (1,2) is obtained by solving both constraint equations simultaneously.

Step 3: We evaluate the objective function at each one of the extreme points and identify the optimal extreme point corresponding to the minimum value of the objective function.

Table 3.11 Objective Function Values

Extreme point	$2X_1 + 3X_2$
(5,0)	$2(5) + 3(0) = 10$
(1,2)	$2(1) + 3(2) = 8$
(0,3)	$2(0) + 3(3) = 9$

From Table 3.11 we see that the extreme point (1,2) is optimal since it corresponds to the minimum value of the objective function $Z = 8$.

The surplus values are next calculated by evaluating the constraints at the optimal point, (1,2) (see Table 3.12).

Table 3.12 Surplus Variables

Constraint	Left Side of the Constraint	Right Side of the Constraint	Surplus Value
First	$1 + 2(2) = 5$	5	$X_3 = 0$
Second	$1 + 2 = 3$	3	$X_4 = 0$

■

3.5 EXCEL APPLICATIONS

EXAMPLE 3.9

Find the answers to Example 3.3 relating to the economic order quantity (EOQ) using Excel:

SOLUTION: Open the Excel spreadsheet, select an appropriate cell, and enter the formula in the formula bar as indicated below:

A	B	Formula
ANS a	1000	→ =sqrt(2*50*10000/1)
ANS b	500	→ =B1/2
ANS c	10	→ =10000/B1
ANS d	36	→ =360/B3
ANS e	138.8889	→ =5*10000/360
ANS f	500	→ =500*1
ANS g	500	→ =50*B3
ANS h	1000	→ =B6+B7

Note: ANS means answers: If you find cell addresses in the formulas, you are highlighting those cells. ■

EXAMPLE 3.10

Find the answers to Example 3.4 relating to the economic order quantity (EOQ) using Excel:

SOLUTION: Open the Excel spread sheet, select an appropriate cell, and enter the formula in the formula bar as indicated below:

A	B	Formula
ANS a	2828.43	→ =sqrt(2*50000*100/1.25)
ANS b	1414.21	→ =B1/2
ANS c	17.68	→ =50000/B1
ANS d	14.14	→ =250/B3
ANS e	800.00	→ =4*50000/250
ANS f	1767.77	→ =B2*1.25
ANS g	1767.77	→ =50*B3
ANS h	3535.53	→ =B6+B7

Note: ANS means answers: If you find cell addresses in the formulas, you are highlighting those cells. ■

EXAMPLE 3.11

Find the solution to the problem in Example 3.5 using the Solver program of Excel. Also, find slack values.

SOLUTION:

(i) Formulation of the mathematical model of the linear programming problem.

Let X_1 represent the number of shares of fund A (in thousands) and let X_2 depict the number of shares of fund B (also in thousands). Our objective is to maximize total return $Z = 1.5X_1 + X_2$ (*the objective function*) subject to the *constraint system*:

$$X_1 + X_2 \leq 5 \text{ (budget constraint)},$$
$$2X_1 + X_2 \leq 7 \text{ (risk constraint)},$$
$$X_1, X_2 \geq 0 \text{ (nonnegativity conditions)}.$$

(ii) Set up a spreadsheet of the linear programming mathematical model in Excel as given below:

1. Set up the columns as decision variables, X1 (fund A shares), X2 (fund B shares), in that order, in columns A, B, and C.

2. Set up the rows as trial values, objective function Z, budget constraint and risk constraint, in that order, in rows 2–5.

3. Show trial values 1 and 1 (arbitrary) in cell (B2) and cell (C2).
4. Show the coefficients of the objective function Z of 1.5 and 1 in cell (B3) and (C3), in that order.
5. In cell (D3) type, =Sumproduct (highlight trial values (1,1), highlight coefficients 1.5 and 1) and hit ENTER key. By doing this, you get the number 2.5 in cell (D3). *Note:* Sumproduct formula is an array formula.
6. For budget constraint in row 4, after showing the coefficient 1 of X1 in cell B4 and 1of X2 in cell C4, go to cell (D4) and type =SUMPRODUCT (highlight trial values, highlight budget constraint coefficients) and hit ENTER key. You will see 2 in cell D4. This is the effect of the trial values on the left-hand side (here after LHS) of the budget constraint. In cell E4 and E5, you type the right-hand side (here after RHS) information of the budget constraint.
7. Repeat 6 above for the risk constraint in cell (A5) and see 3 in cell D5. Click on any empty cell. Then the setup table will be as given below:

	A	B	C	D	E	F
1	Decision variables	X1	X2			
2	Trial Values	1	1			
3	Objective function, Z	1.5	1	2.5		
4	Budget Constraint	1	1	2	≤	5
5	Risk Constraint	2	1	3	≤	7

Access "Solver" Program of Excel using instructions as given in Example 1.32 Part B.

8. In the "Solver Parameters" dialogue box, highlight cell D3 in "set Objective" window. Click in the circle for "Max." Highlight the trial values cells, B2 and C2, in the window "By Changing Variable Cells" and click in the "Subject to the Constraints" window and then click on "Add" button on the RHS to get "Add Constraint" "dialogue box." Highlight cell D4 where you have SUMPRODUCT formula in "Cell Reference" window. In the next window, choose the inequality ≤. In the next window, enter 5 (or highlight cell F4), which is the RHS of the budget constraint from the mathematical model. Click on "Add" button and enter information about risk constraint similarly and click on "OK" as there are no more constraints. You will get back to "Solver Parameters" dialogue box. Verify whether the constraints entered are correct. Check the box against "Make Unconstrained Variables NonNegative." This is where the nonnegativity constraints of the variables in the mathematical model are taken care of. In the window of "Select a Solving Method," choose "Simplex LP" and click "Solve" button to get "Solver Results" dialogue box. Click OK to keep the solver solution that is given below.

	A	B	C	D	E	F
1	Decision variables	X1	X2			
2	Trial Values	2	3			
3	Objective function, Z	1.5	1	6		
4	Budget Constraint	1	1	5	≤	5
5	Risk Constraint	2	1	7	≤	7

Notice how the trial values have changed to the optimal values of X1 = 2 (in cell B2) and X2 = 3 (in cell C2) and maximum value the objective function, Z = 6 (in cell D3).

Also notice that the whole budget of 5 (in thousands of dollars) in cell D4 and the risk index of 7 in cell D5 are used completely, leaving zero slack values. ■

EXAMPLE 3.12

$$\text{Minimize } Z = 2X_1 + 3X_2 \text{ s.t.,}$$
$$X_1 + 2X_2 \geq 5,$$
$$X_1 + X_2 \geq 3,$$
$$X_1, X_2 \geq 0.$$

Find the optimal solution. What is the minimum value of the objective function? Find the values of the surplus variables. Use the "Solver" Program in Excel.

SOLUTION: Note that the linear programming mathematical model is given.

Set up a spreadsheet of the linear programming mathematical model in Excel as shown below:

1. Set up the columns as decision variables X_1 and X_2, in that order, in columns A, B, and C.

2. Set up the rows as trial values, objective function Z, constraint 1 and constraint 2, in that order, in rows 2–5.

3. Show trial values 1 and 1 (arbitrary) in cell (B2) and cell (C2).

4. Show the coefficients of the objective function Z, of 2 and 3, in cell (B3) and cell (C3), in that order.

5. In cell (D3), type =SUMPRODUCT (highlight trial values (1,1), highlight coefficients 2 and 3) and hit ENTER key. By doing this, you get the number 5 in cell (D3). *Note:* SUMPRODUCT formula is an array formula.

6. For constraint 1 in row 4, after showing the coefficient 1 of X1 in cell B4 and 2 of X2 in cell C4, go to cell (D4) and type =SUMPRODUCT (highlight trial values (1,1), highlight constraint 1 coefficients 1 and 2) and hit ENTER key. You will see 3 in cell D4 that reflects the LHS of constraint 1 if the trial values are true. In cell E4, enter ≥ and 5 in cell F4, which indicate the RHS of constraint 1.

7. Repeat 6 above for the constraint 2 in cell (A5) and see 3 in cell D5. Click on any empty cell. Then the setup table will be as shown below:

	A	B	C	D	E	F
1	Decision Variables	X1	X2			
2	Trial Values	1	1			
3	Objective function,Z	2	3	5		
4	Constraint 1	1	2	3	≥	5
5	Constraint 2	1	1	2	≥	3

Access the "Solver" program of Excel using instructions as given in Example 1.32 Part B.

8. In the "Solver parameters" dialogue box, highlight cell D3 in "set Objective" window. Click in the circle for "Min." Highlight the trial values cells, B2 and C2, in the window "By Changing Variable Cells" and click in the "Subject to the Constraints" window and then click on "Add" button on the right side to get "Add Constraint" dialogue box. Highlight cell D4 where you have "SUMPRODUCT" formula in "Cell Reference" window. In the next window, choose the inequality ≥. In the next window, enter 3 that is the RHS of constraint 1from the mathematical model. Click on "Add" button again and enter information about constraint 2 similarly and click on "OK" as there are no more constraints. You will get back to "Solver Parameters" dialogue box. Verify whether the constraints entered are correct. Check the box against "Make Unconstrained Variables Nonnegative." This is where the non-negativity constraints of the variables in the model are taken care of. In the window of "Select a Solving Method," choose "Simplex LP" and click "Solve" button to get "Solver Results" dialogue box. Click OK to keep the solver solution that is given below.

	A	B	C	D	E	F
1	Decision Variables	X1	X2			
2	Trial Values	1	2			
3	Objective function,Z	2	3	8		
4	Constraint 1	1	2	5	≥	5
5	Constraint 2	1	1	3	≥	3

Notice the trial values of the variables have changed to $X_1=1$ and $X_2=2$, which are optimal, and the minimum value of the objective function, Z, is 8 (cell D3). The surplus value of the constraint 1 is 5(cell D4) – 5 (RHS value of constraint 1) = 0. Similarly, the surplus value of the constraint 2 is 3 (cell D5) – 3 (RHS value of constraint 2) = 0. ■

CHAPTER 3 REVIEW

You should be able to:

1. State the decision rule that allows the firm to find the profit maximizing level of output. (*Hint:* Think in "marginal" terms.)
2. Distinguish between first- and second-order conditions for an extreme value.
3. Distinguish between holding cost, carrying cost, and ordering cost.
4. Specify the role of the EOQ model.
5. Specify what happens to annual total cost when annual holding and annual ordering costs are equal.
6. Distinguish between lead time and cycle time.
7. Describe the individual components of a linear programming problem.
8. Distinguish between the solution set of a given constraint and the feasible region for a linear program.
9. State the significance of the Extreme Point theorem.
10. Distinguish between slack and surplus variables.

Key Terms and Concepts:

binding constraint, 108
carrying cost, 99
closed set, 107
constraint capacities, 106
constraint system, 106
convex set, 107
cycle time, 99
EOQ model, 99-100
Existence theorem, 107
extreme point, 107
Extreme Point theorem, 107
extreme value, 95
feasible region, 107
first-order condition, 95
holding cost, 99
lead time, 101
linear programming model, 105
nonnegativity conditions, 106
objective function, 106
ordering cost, 99
profit-maximizing rule, 91
second-order condition, 95
slack variable, 109
solution set, 106-7
strictly bounded set, 107
structural constraint, 106
surplus variable, 115
total inventory cost, 99-100

CHAPTER 3 EXERCISES

1. Given the demand (price) and cost functions $P = 2000 - 40Q$ and $C(Q) = 3000 + 400Q$, respectively, find the following, using the calculus approach:
   - **a.** Profit-maximizing output
   - **b.** Profit-maximizing price
   - **c.** Maximum total profit
   - **d.** Revenue-maximizing output
2. Given the demand (price) and cost functions $P = 300 - Q$ and $C(Q) = 4000 + 100Q$, respectively, find the following, using the calculus approach:
   - **a.** Profit-maximizing output
   - **b.** Profit-maximizing price
   - **c.** Maximum total profit
   - **d.** Revenue maximizing output
3. We noted earlier in this chapter that if a function $y = f(x)$ has a maximum or minimum at $x = x_0$, then $f'(x_0) = 0$. If $f' = 0$ is linear in x, then x_0 is readily determined by setting $f' = 0$ and solving for the value of x that makes it vanish. But what if $f' = 0$ is nonlinear in x and cannot be easily factored? For example, what if it is, say, quadratic and of the form $0.2x^2 - 1.5x + 1 = 0$? To solve for the x values (*roots*) that satisfy this equation, let us employ the *quadratic formula*. To this end, suppose we have a general quadratic equation of the form

$$ax^2 + bx + c = 0, \quad a \neq 0.$$

   Then it can be shown that

$$x = \frac{-b \pm \sqrt{b^2 - 4ac}}{2a}.$$

So given $0.2x^2 - 1.5x + 1 = 0$ (here a = 0.2, b = −1.5, and c = 1),

$$x = \frac{1.5 \pm \sqrt{(1.5^2) - 4(0.2)(1)}}{2(0.2)} = \frac{1.5 \pm 1.204}{0.4}$$

so x = 6.76 or x = 0.74.

Use the quadratic formula to find the roots of

a. $3x^2 + 5x - 7 = 0$

b. $-2x^2 + 3x + 5 = 0$

c. $3x^2 - x - 2 = 0$

4. A *point of inflection* is a point where a curve crosses over its tangent line and changes the direction of its concavity from upward to downward (Figure 3.18a) or vice versa (Figure 3.18b). How do we go about finding a point of inflection?

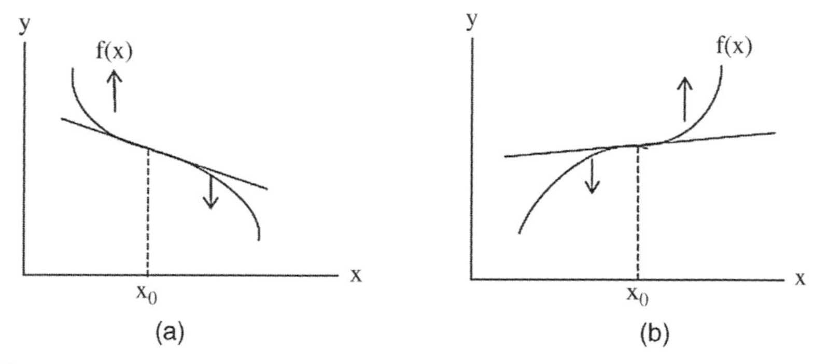

Figure 3.18 (a) Concavity changes from upward to downward. (b) Concavity changes from downward to upward.

A glance at Figure 3.18b reveals that, as we increase the value of x and move along the curve from left to right, the slope or first derivative f′ decreases, reaches a minimum at x_0, and then increases, that is, a point of inflection must be a point where f′ attains an extremum (a maximum or minimum). Now, we know that a first-order condition for f′ to attain a minimum at x_0 is that its derivative or $f''(x_0) = 0$. Moreover, the second-order condition for f′ to attain a minimum at x_0 is that $f'''(x_0) \neq 0$. In sum, if $f''(x_0) = 0$ and $f'''(x_0) \neq 0$, then f has a point of inflection at x_0. And if $f'(x_0) = f''(x_0) = 0$, and $f'''(x_0) \neq 0$, then f has a *horizontal point of inflection* at x_0 (the line tangent to f at x_0 is horizontal).

Given the preceding discussion, determine the following:

a. If $y = f(x) = x^3 - 6x^2 + 20$ has a point of inflection at x = 2.

b. If $y = f(x) = 5 - (x - 3)^3$ has a horizontal point of inflection at x = 3.

5. a. Given the total cost function $C(Q) = 0.01Q^3 - 3Q^2 + 4Q + 10$, determine the output level Q at which the marginal cost curve has a minimum. Does this output level correspond to the one for which the total cost function has a point of inflection? Verify your answer.

b. Given C(Q) above, find the average cost (AC) function. (*Note:* AC(Q) = C(Q)/Q.) At what Q value does average cost attain a minimum?

c. Suppose the firm faces a constant price of P = \$10 for its product. Find the profit-maximizing level of output. Now, let the firm face a demand (inverse) curve of the form P = 10 − 3Q with the same total cost structure. What is the new profit-maximizing level of output? At what level of output is total revenue (TR) at a maximum?

6. A fast-food restaurant sells a soft drink product that has a constant annual demand rate of 4800 cases. The cost of a case is estimated to be $4.00. The unit cost of holding is 20% of the unit cost of the product. Ordering costs are $40.00 per order irrespective of the size of the order. If there are 360 working days per year and 5 days of lead time, find the following:

   a. EOQ

   b. Number of orders per year

   c. Cycle time in days

   d. Reorder point

   e. Annual holding cost

   f. Annual ordering cost

   g. Total annual cost without the purchase cost

7. Suppose Q stands for order quantity and the annual demand for the product is 10,000 units. Assume that the cost of placing an order, independent of the size of the order, is $50. The cost of holding per unit is 50% of the unit cost of $2. Use this information to answer the following:

   a. Show that the total annual cost of maintaining the inventory is

$$Y(Q) = \frac{Q}{2} + \frac{500,000}{Q}$$

   b. Find the economic order quantity, Q^*, and the total cost corresponding to that value of Q^*.

   c. Also find the total of annual ordering and holding costs if orders are placed for 1500 units.

   d. Compare your results for (b) and (c) and comment.

8. A computer company purchases a part used in the manufacture of personal computers directly from a supplier. It requires a total of 16,000 parts per year for its production operations. Assume that the ordering costs are $50 per order and the cost of the part is $25. Further, assume that the holding cost per unit is 4% of the unit cost of the part. There are 360 working days per year and the lead time is 5 days. Answer the following questions:

   a. What is the EOQ for this part?

   b. What is the reorder point?

   c. What is the cycle time?

   d. What are the total annual holding and ordering costs associated with your recommended EOQ?

9. A firm estimates the annual demand for one of its products to be 12,000 units. Each unit costs the firm $25.00. The inventory holding cost per unit is 4% of the unit cost. The fixed cost of placing an order is $60.00. The firm operates 360 days of the year. Currently, the firm orders 1000 per month. Determine if the firm should order more or less frequently, and specify how many units will minimize its annual inventory maintenance cost.

10. An Insurance company is interested in determining the desired cash balance on hand to meet the operating expenses in a particular time period, say, 1 year. The company's estimated annual demand for cash is $16 million. The opportunity cost of holding cash is 4% per annum. The cost per transaction for withdrawing cash is $200. Find the transaction balance that minimizes the total cost of maintaining that balance and the number of transactions required during the year.

11. Production to maximize profit:

Assume an electronics company in China is planning to produce two components used in the production of a major household appliance. Based on market experience, the estimated profits on the two components are assumed to be $11 and $15, respectively. The table below represents per unit labor and cash requirements and their availabilities.

Input Requirements and Availabilities

Product	Labor hours	Cash ($)
Component 1	2	4
Component 2	3	5
Availability	12	22

Formulate a linear programming model designed to maximize the total profit. Also, present the graphical solution in detail. Identify slack variables in each of the structural constraints.

12. Maximize $Z = X_1 + 2X_2$ s. t.,

$3X_1 + 4X_2 \leq 28,$

$X_1 + 3X_2 \leq 12,$

$2X_1 + X_2 \leq 6,$

$X_1, X_2 \geq 0.$

a. Solve graphically for the optimal solution.

b. Find slack, if any, in each of the constraints.

13. A dietician suggests a diet consisting of two food items to a particular patient in a hospital. The diet is designed to satisfy basic nutritional requirements at minimum cost. The two food items cost $10.00/lb and $12.00/lb, respectively. Food 1 contains 5% protein and 60% fat. Food 2 contains 30% protein and 40% fat. The patient needs a minimum of 2 lb of protein and at least 4.8 lb of fat per week. Formulate as a linear programming model and find the food mix that minimizes the total cost of the diet.

14. Minimize $Z = 2X_1 + X_2$ s.t.,

$X_1 + X_2 \geq 6,$

$3X_1 + 2X_2 \geq 8,$

$4X_1 + X_2 \geq 7,$

$X_1, X_2 \geq 0.$

Find the optimal solution using the graphical method. Compute the optimal values of the surplus variables.

Excel Applications

15. Find solutions to Exercise 6 above using Excel.

16. Find solutions to Exercise 8 above using Excel

17. **a.** Given the following information, find:

- **i.** the profit-maximizing output, Q^*.
- **ii.** the corresponding price,
- **iii.** the maximum profit, and
- **iv.** the output Q that maximizes revenue. Use Excel and the trial and error method.

$P(Q) = 2000 - 40Q$ and $C(Q) = 3000 + 400Q$. Use the trial values $Q = 15–25$.

(*Hint:* Using Excel, you can find the solution values by trial and error. Refer to the examples in the textbook, Examples 3.1 and 3.2, and develop a similar table.)

b. Repeat Exercise 17a above with the following functions: $P = 300 - Q$ and $C(Q) = 4000 + 100Q$ in the range of $Q = 95–105$ with increments of one unit. Present the table showing detailed computations of profit, MR, and MC using Excel.

18. For Exercise 11 above, use the Solver program of Excel to find

a. optimal solution,

b. maximum profit, and

c. slack values of the constraints.

19. **a.** A hospital dietician formulates an eating plan consisting of three food items. The plan purports to satisfy the requisite nutritional requirements at minimum cost. The three food items cost \$10.00/lb, \$12.00/lb and 16/lb, respectively. Food 1 contains 5% protein and 60% fat. Food 2 contains 30% protein and 40% fat. Food 3 contains 40% protein and 30% fat. The patient needs a minimum of 2 lbs of protein and at least 4.8 lbs of fat per week. Formulate as a linear programming model and find the food mix that minimizes the total cost of the diet using the Solver Program in Excel.

a. In 19a, suppose that, due to contractual obligations, the hospital has to order at least 1 lb of Food 3 per week. Find the optimal solution under this additional constraint.

20. A refinery in Southern Louisiana is in the business of producing regular and premium unleaded gasoline. Based on its experience, light and heavy crude oil have to be combined in the ratio of 1:2 and 3:2, respectively, for regular and premium gas. Market price of light crude oil is \$0.3/gallon and \$0.2/gallon for heavy crude oil. The objective is to minimize the total production cost of regular and premium gasoline. Management wants to satisfy the market demand of 6 million gallons of regular and 10 million gallons of premium gasoline per period. Formulate the problem as a linear program and obtain the optimal solution using the Solver program in Excel. *Hint:* Define the decision variables as X_{LR} = millions of gallons of light crude oil going into regular gas, and so on.

Chapter 4

What Is Business Statistics?

4.1 INTRODUCTION

In the world of business, it is common to deal with a number of experiments. Some examples of these experiments are predicting the weekly production in a manufacturing plant, the closing prices of stocks in a stock exchange, the number of building permits that will be issued by a municipality over the coming year, among others. In the vast majority of experiments, outcomes are not known with certainty. Hence, they are known as *random experiments*. There are very few instances when the outcomes are known with certainty. The systematic recording of the outcomes of a random experiment can be thought of as collecting "statistical data." The analysis of and the drawing of inferences from statistical data constitutes what may be termed the *Theory of Statistics*. In business statistics, however, the emphasis is less on theory and more on the application of statistical tools and methods.

Statistics is essential in the process of everyday decision making in business. For example, a firm has to assess the profitability of a new project before a decision is made to undertake it. Similarly, an investor should examine the annual report of a corporation before making a final decision on a stock purchase.

Generally, business statistics consists of both descriptive statistics and inferential statistics. In *descriptive statistics*, we deal with the analysis of data using tabular, graphical, and numerical methods to display data or to simply catalog facts. This approach is essential for obtaining a clear understanding of actual business data sets such as stock prices, product sales, and interest rates. However, *inferential statistics* enables us to draw conclusions or generalizations about industry, or about the underlying population at large that are based on incomplete data and limited experience. Inferential statistics deals with the selection of a sample data set and attempts to infer conclusions from the same using the well-articulated procedures of estimation and/or hypotheses testing and prediction. For example, businesses continuously depend on surveys and or polling to arrive at a meaningful understanding of some known situation. In this regard, inferential statistics involves decision making under uncertainty.

Introduction to Quantitative Methods in Business: With Applications Using Microsoft® Office Excel®, First Edition. Bharat Kolluri, Michael J. Panik, and Rao Singamsetti.

Companion website: www.wiley.com/go/Kolluri/QuantitativeMethods

4.2 DATA DESCRIPTION

4.2.1 Some Important Concepts in Statistics

Think of a *population* as the whole set of objects of interest. Examples are all taxpayers in 2005, all viewers of a TV program on ESPN, all financial mutual funds, all registered voters in presidential elections, and all trucks produced in the United States, among others.

A *sample* is a part/subset of the population. Examples are a group of taxpayers selected by the IRS, a small amount of blood taken from a patient to test for different types of cholesterol in the body, or a small number of TV sets selected in a test for manufacturing defects. A variety of techniques are available for selecting a sample to represent the whole population.

It is often necessary to obtain knowledge and information on a variety of descriptive characteristics of a population, formally known as *parameters*. These are, for example, the population mean, median, standard deviation, and proportion, to name but a few.

Numerical measures of similar characteristics, computed instead from a sample, are termed *statistics*. Examples of these are the sample mean, median, standard deviation, and proportion, along with others.

Types of variables: Variables are generally classified as qualitative or quantitative. A *qualitative variable*, by definition, refers to an attribute or a specific category. Some examples are color, gender, race, religious affiliation, occupation, political affiliation, telephone numbers, zip codes, social security numbers, and so on. These are generally expressed as percentages or proportions and described in the form of graphs, bar charts, and pie charts.

Quantitative variables are those that are expressed as numerical measurements indicating how much or how many. These can be further classified as *discrete* (expressed as a whole number) or *continuous* (can assume fractional values such as GPA, age, family incomes, and various prices)

4.2.2 Scales of Data Measurement

We now describe categorized information (data) or attributes or the assignment of numerical values. This process is called *measurement*.

1. *Nominal Scale:* The nominal scale identifies only categories. Hence, it is meaningless to attempt to perform numerical operations on nominal measures since they possess neither numerical value nor order. For example, there is no meaning to the notion of average zip code or average Social Security Number. The only summary statistics we can offer are sample size and the proportion of each category within of the sample, for example, we can determine the proportion of households with the zip code 06117 in West Hartford, CT.

2. *Ordinal Scale:* With an ordinal scale, data values can be ranked in order of importance. The difference between any two ranks is not the same over all ranks and only a well-defined hierarchy is meaningful. Percentages or proportions are also used with an ordinal scale and they are typically described in terms of pie charts, bar charts, and tables. Generally, the measurements on an ordinal scale are summarized by the median or middle of a data set. The ordinal scale has the same properties as the nominal scale with the additional feature of ranking.
3. *Interval Scale:* The data measured on an interval scale are comparable in terms of differences between magnitudes. Clearly, this is a higher level of measurement than the ordinal scale. The only limitation is that the zero level does not mean absence of the variable. For example, a temperature of zero degrees does not imply absence of heat. The interval scale possesses both ordinal and nominal scale properties with the additional feature of comparability of numerical values.
4. *Ratio Scale:* Any classification that can be described as a ratio A/B, where B $\neq 0$ and A, B are numerical values, is a ratio scale measurement. It is to be noted that all the four basic rules of arithmetic operations (addition, subtraction, multiplication, and division) apply to A/B. Ratio level data represent the highest level of measurement in the sense that they include the properties of the nominal, ordinal, and interval scales with the added property that zero is meaningful. Some of the examples of ratio scale measurement are prices, interest rates, salaries, and so on. It is to be noted that ratio scale data can be quantitative or qualitative.

Figure 4.1 depicts the types of variables and their scales of measurements.

It is to be noted that qualitative data can only be described by using pie charts and bar charts to show the relationships among the sizes (number) in the various categories. However, in business we deal almost exclusively with quantitative data.

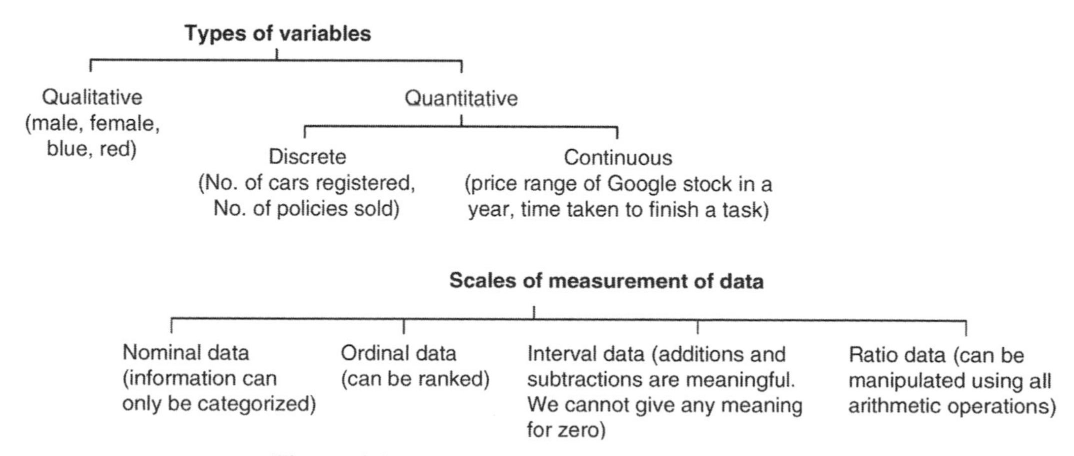

Figure 4.1 Variables and their measurement scales.

Hence, the tools developed in subsequent chapters are geared to handling quantitative data sets.

4.3 DESCRIPTIVE STATISTICS: TABULAR AND GRAPHICAL TECHNIQUES

The main purpose of this section is to present an assortment of graphical tools for describing a data set. (The numerical approach for handling data is given in subsequent sections.)

Graphical Method: In this method, we first set up a *frequency table* for the given data set. This table is used to obtain a frequency distribution, which shows the number of observations in the given data set that fall into each of several mutually exclusive and exhaustive classes. Let us see how this is accomplished.

EXAMPLE 4.1

Business Week in its June 4, 2007 issue reported the stock prices of 100 best small companies listed below (Table 4.1). *Data file:* BSCSP2007.xlsx.

For the data in Table 4.1, obtain the following:

a. Frequency distribution

b. Relative and cumulative frequency distributions

c. Histograms

d. Frequency polygon

e. Ogive

SOLUTIONS:

a. Step 1: Determine the number of classes (groups)

The number of classes chosen is arbitrary, but should not be too few or too many. Usually the number of classes chosen is anywhere between 5 and 15, depending on the size of the data set. For example, if we have a data set of 20 values, then we can arrange them into, say, five

Table 4.1 The Stock Prices of 100 Best Small Companies (in Dollars)

33	45	39	7	47	34	36	45	38	29
40	15	61	20	24	29	48	38	53	40
10	15	39	31	39	137	56	40	7	51
21	40	33	41	15	52	49	39	34	61
35	89	32	49	97	29	31	63	44	66
13	29	52	50	26	49	15	16	38	39
51	59	13	12	11	21	38	14	39	37
33	21	10	63	23	62	30	51	55	83
39	40	16	19	37	22	31	18	19	29
29	43	57	45	14	36	23	18	62	15

Table 4.2 Determining the Number of Classes

n	k	2^k
10	4	16
20	5	32
35	6	64
70	7	128

classes. Likewise, if we have a data set of 200,000 values, we may select a larger number of classes, for example, 15. There is also a mathematical formula that can be used as a guide for determining the number of classes. If k is the number of classes (an integer) and n is the number of values in the data set, then k is determined as the smallest integer satisfying $2^k \geq n$. For example, if $n = 35$, then k is 6, as is evident from Table 4.2.

In our example, the data size is n = 100 and therefore $k = 7$ since $2^7 > 100$.

Step 2: Determine the class width or class interval

The *class interval* or length of a class is obtained by dividing the range of the data set by the number of classes obtained in step 1 above and rounding off to a higher and convenient number. Following this procedure, we determine the *range* as the difference between the largest and the smallest values in the data set. In our example, range = 137 − 7 = 130. Dividing this by the number of classes, 130/7, we obtain 18.6, or rounded off to 19. It is to be noted that this approximation of class width is always toward a higher number in order to accommodate all the values in the data set. Also, note that we are assuming an equal width for all classes. It is possible to have examples with unequal width as well as with the first and/or last classes being open ended. The latter case can arise when there are no particular limits for low or high values, for example, in an age distribution, one of the classes might be "65 and over."

Step 3: Form the frequency table

We begin by selecting the *lower limit* of the first class as a convenient number equal to, or smaller than, the minimum value of the data set. Thus, in our case, since the minimum is 7, we select 6 as the lower limit of the first class. (Note that the lower limit of a class is the smallest value going into the class.) Since the length of each class is 19, the lower limit of the second class must be 6 + 19 = 25. Continuing in this fashion, we see that the lower limit of the third class is 25 + 19 = 44, the lower limit of the fourth class is 44 + 19 = 63, and so on. Since the classes must not overlap (each observation belongs to one and only one class), the *upper limit* of the first class must be 24. Thus, the upper limit of the second class must be 43 = 24 + 19; the upper limit of the third class is 43 + 19 = 62, and so on. (Clearly, the upper limit of a class is the largest value that the class contains.) This procedure yields the following seven classes, where the class from 120 to 138 accommodates the largest value in the data set:

6 – 24,
25 – 43,
44 – 62,
63 – 81,
82 – 100,
101 – 119, and
120 – 138.

Table 4.3a Frequency Table and Frequency Distribution

Serial No.	Class	Tally	Frequency
1	6–24	~~||||~~ ~~||||~~ ~~||||~~ ~~||||~~ ~~||||~~ ||||	29
2	25–43	~~||||~~ ~~||||~~ ~~||||~~ ~~||||~~ ~~||||~~ ~~||||~~ ~~||||~~ ~~||||~~	40
3	44–62	~~||||~~ ~~||||~~ ~~||||~~ ~~||||~~ ||||	24
4	63–81	|||	3
5	82–100	|||	3
6	101–119		0
7	120–138			1
Total			100

Once the classes are specified, we sort each of the 100 values of the stock prices into the classes, and then count the actual number of items in each class so as to obtain the *class frequencies*, as shown in Table 4.3a. The resulting construction (without the tally column) is known as a *frequency distribution* for grouped data.

b. Relative and cumulative frequency distributions

The relative frequency column is obtained by dividing the frequency in each class by the total number of values, n = 100; and the cumulative frequency column is obtained as shown in Table 4.3b.

c. Histograms

The frequency distribution can also be shown in graphical form. This is accomplished by showing class intervals on the horizontal axis and the frequencies on the vertical axis, with appropriate scales. It is to be noted that the class limits are sometimes replaced by *class boundaries* in order to make the histogram continuous. Class boundaries do not exist in the actual data. They are utilized to avoid gaps between the classes. (An alternative way to eliminate gaps between the classes is to exclude the upper limits in each class and to include only the lower limits. For example, in Table 4.3a, the first class could be 6 up to 25, the second one 25 up to 44, and so on.) For this frequency distribution, the class boundaries are as follows:

Table 4.3b Relative Frequency and Cumulative Frequency Table

Serial No.	Class	Frequency	Relative Frequency	Cumulative Frequency
1	6–24	29	0.29	29
2	25–43	40	0.40	29 + 40 = 69
3	44–62	24	0.24	69 + 24 = 93
4	63–81	3	0.03	93 + 3 = 96
5	82–100	3	0.03	96 + 3 = 99
6	101–119	0	0.00	99 + 0 = 99
7	120–138	1	0.01	99 + 1 = 100
Total		100	1.00	

5.5	–	24.5
24.5	–	43.5
43.5	–	62.5
62.5	–	81.5
81.5	–	100.5
100.5	–	119.5, and
119.5	–	138.5.

(Note that the length of a class can be obtained by taking the difference between its upper and lower boundaries.) Figure 4.2 shows the *frequency histogram* for our example, where it is implicitly understood that the base of each rectangle goes from its lower boundary to its upper boundary. This eliminates any gaps between the rectangles.

A similar histogram, known as a *relative frequency histogram* (*probability chart*), can also be drawn using the class intervals on the horizontal axis and the relative frequencies (shown in Table 4.3) on the vertical axis. This is given in Figure 4.3.

d. Frequency polygon

Next comes a *frequency polygon*, which is obtained by plotting the *midpoints* of the class intervals (the average of the class limits) calculated in Table 4.3c, against their frequencies and connecting the points by line segments, as in Figure 4.4. Clearly, this graph represents a polygon in form. It is simply another way to represent the frequency distribution of the data set. Two or more frequency polygons associated with similar data sets can be used in place of histograms for comparing the characteristics of the data sets.

e. Ogive

The *cumulative frequency curve* (Ogive) is obtained by plotting the upper boundaries of classes against the corresponding cumulative frequencies. This curve is useful in answering questions like, "How many stocks are there at or below a certain level of stock price?" For instance, in this example, there are 96 companies whose stock price is at or below 81.5 dollars. The cumulative frequencies are shown in Table 4.4 and the cumulative frequency curve is drawn in Figure 4.5.

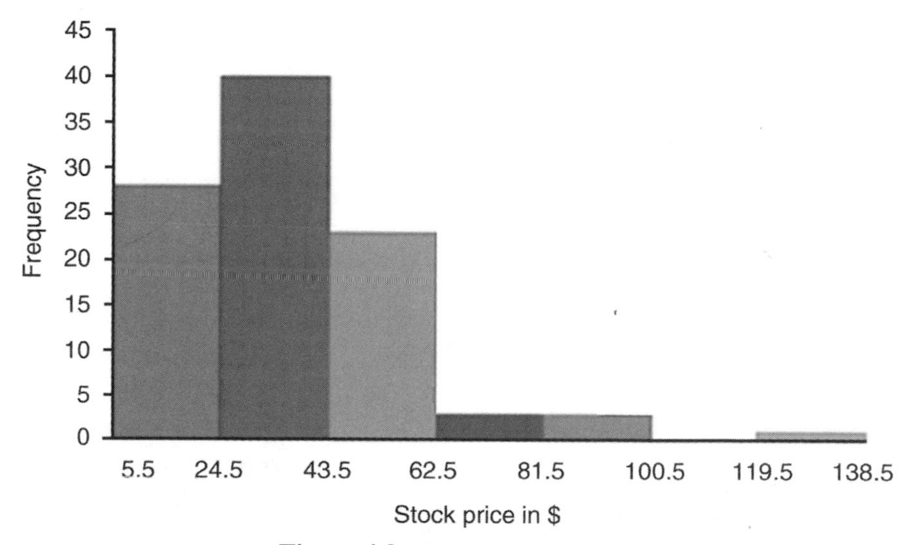

Figure 4.2 Frequency histogram.

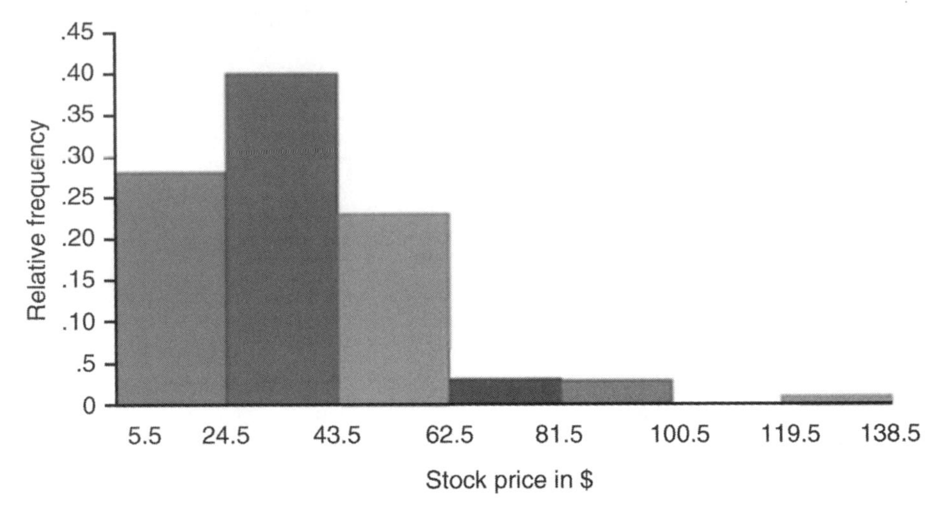

Figure 4.3 Relative frequency histogram.

Table 4.3c Frequency Table with Midpoints

Serial No.	Class	Midpoint	Frequency
1	6–24	15	29
2	25–43	34	40
3	44–62	53	24
4	63–81	72	3
5	82–100	91	3
6	101–119	110	0
7	120–138	129	1
Total			100

Figure 4.4 Frequency polygon.

Table 4.4 Cumulative Frequency Values

Upper Boundaries of Classes ($)	Cumulative Frequencies (Less Than or Equal to)
24.5	29
43.5	69
62.5	93
81.5	96
100.5	99
119.5	99
138.5	100

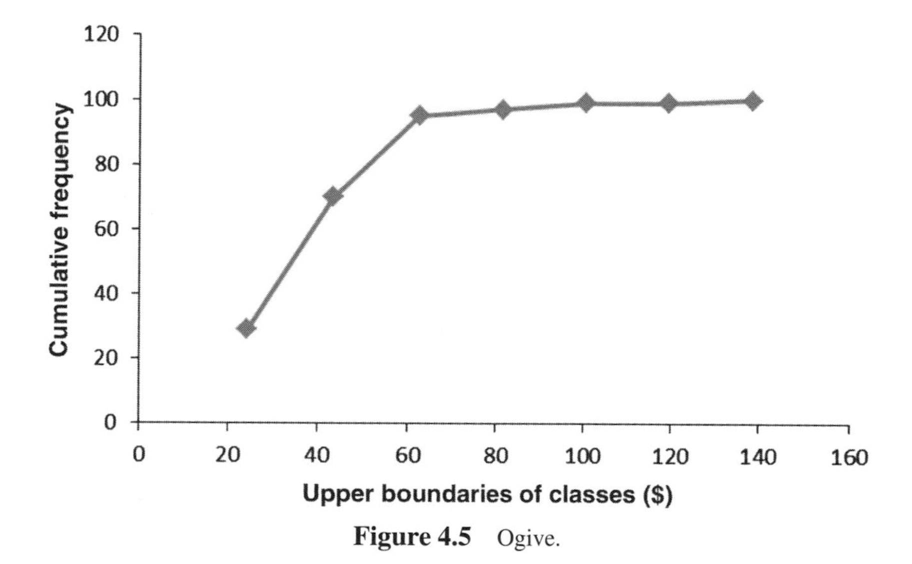

Figure 4.5 Ogive.

In a similar fashion, a graph involving the upper boundaries of the classes and the cumulative relative frequencies is known as a *cumulative relative frequency curve*. This is a useful device for finding the level of stock price at or below a given percentage of companies. ■

EXAMPLE 4.2

Based on the 2012 Statistical Abstract of the United States, the government has provided the following data on numbers of employees in manufacturing establishments in the 50 states in 2008 (Table 4.5). The highest and lowest values are highlighted. *Data file:* Employment2012.xlsx.

Using the above data set, obtain the following:

a. Frequency distribution

b. Relative and cumulative frequency distributions

c. Histograms

Table 4.5 Employment Data (in Thousands)

283	11	168	214	1560	149	211	39	372	450
13	60	727	552	223	179	260	149	65	145
330	691	336	183	308	19	102	39	87	347
34	625	613	22	829	148	178	711	60	290
37	408	851	110	44	313	269	69	500	10

Source: 2012 Statistical Abstract of the United States.

d. Frequency polygon

e. Ogive curve

SOLUTIONS:

a. Frequency distribution

To develop the frequency distribution, we follow the three steps described in Example 4.1.

Step 1: Determine the required number of classes (groups)

By using the formula $2^k \geq n$, we find that the number of classes, k, is 6 since

$$n = 50 \text{ (sample size)}, \quad 2^6 = 64, \text{and thus } 2^6 \geq 50.$$

Step 2: Determine the class width or class interval

It is obtained by dividing the range of the data set by the number of classes and rounding off to a higher convenient number. We know that the range is the difference between the highest and the lowest values in the data set, that is, $1560 - 10 = 1550$. Dividing this difference by the number of classes, we obtain $1550/6 = 258.33$ rounded off to 260.

Step 3: Form the frequency table

We begin by selecting the lower limit of the first class as a number equal to, or smaller than, the minimum value of the data set. Thus, in this case, since the minimum is 10, we select 10 as the lower limit of the first class. The lower limit of the second class is $10 + 260 = 270$, the lower limit of the third class is $270 + 260 = 530$, and so on. Since each data point belongs to one and only one class, the upper limit of the first class must be 269; the upper limit of the second class is thus 529, and so on. Continuing in this fashion yields the following six classes:

10	–	269
270	–	529
530	–	789
790	–	1049
1050	–	1309
1310	–	1569

Table 4.6a Frequency Distribution of Employment Data

Serial No.	Employment (in thousands)	Tally	Frequency
1	10–269	卌 卌 卌 卌 卌 卌	30
2	270–529	卌 卌		11
3	530–789	卌		6
4	790–1049	||	2
5	1050–1309		0
6	1310–1569			1
Total			50

These classes are all mutually exclusive or nonoverlapping. Once the classes are chosen, we sort each of the 50 values into them and count the actual number of items in each class so as to obtain the class frequencies (Table 4.6a). This table represents the desired frequency distribution, for our data in grouped form.

b. Relative and cumulative frequencies

The relative frequency column is obtained by dividing the frequency in each class by the total number of values n = 50 and the cumulative frequency column is obtained as shown in Table 4.6b.

c. Histograms

The frequency distribution can also be shown in graphical form. This is accomplished by showing continuous class intervals on the horizontal axis and the frequencies on the vertical axis, with appropriate scales. Classes can be treated as continuous by excluding the upper limits in each class and including only the lower limits, as shown below.

10 up to 270
270 up to 530
530 up to 790
790 up to1050
1050 up to 1310
1310 up to 1570

Table 4.6b Relative Frequency and Cumulative Frequency Table

Serial No.	Employment (in Thousands)	Frequency	Relative Frequency	Cumulative Frequency
1	10–269	30	0.60	30
2	270–529	11	0.22	30 + 11 = 41
3	530–789	6	0.12	41 + 6 = 47
4	790–1049	2	0.04	47 + 2 = 49
5	1050–1309	0	0.00	49 + 0 = 49
6	1310–1569	1	0.02	49 + 1 = 50
Total		50	1.00	

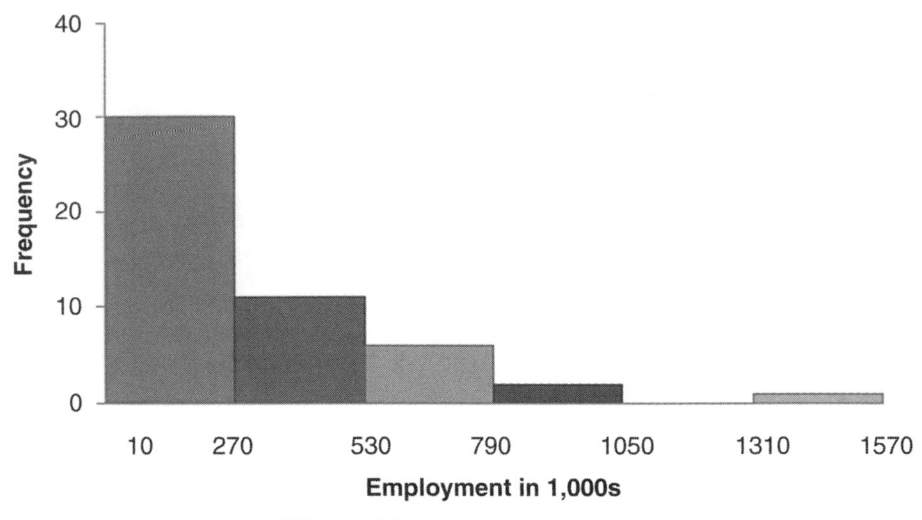

Figure 4.6 Frequency histogram.

Graphing these classes against the corresponding frequencies leads to the histogram shown in Figure 4.6.

A similar histogram, known as a relative frequency histogram (probability chart), can also be drawn using the class intervals on the horizontal axis and the relative frequencies (shown in Table 4.6b) on the vertical axis. This is shown in Figure 4.7.

d. Frequency polygon

Next comes a frequency polygon, which is obtained by plotting the midpoints of the class intervals (the average of the class limits) calculated in Table 4.6c, against their frequencies and connecting the points by line segments, as in Figure 4.8. Clearly, this graph represents a polygon in form.

e. Ogive

The cumulative frequency curve (Ogive) is obtained by plotting the upper boundary of each class against the corresponding cumulative frequency. As previously mentioned, this

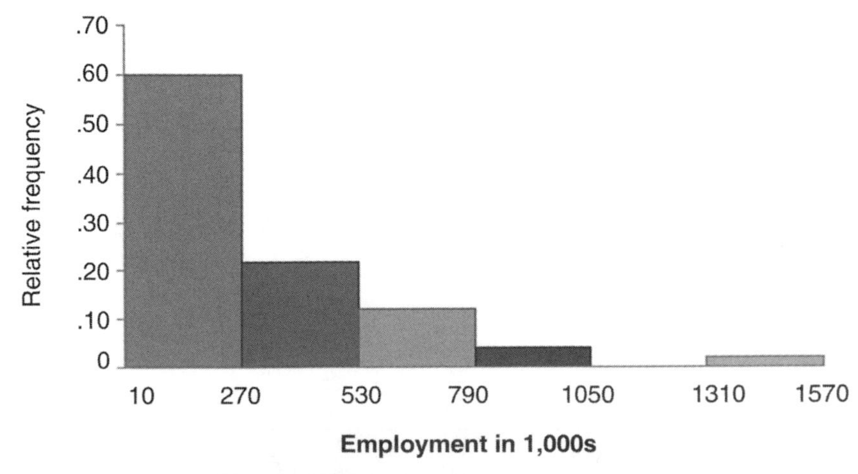

Figure 4.7 Relative frequency histogram.

Table 4.6c Frequency Table with Midpoints

Serial No.	Employment (in Thousands)	Midpoints	Frequency
1	10–270	140	30
2	270–530	400	11
3	530–790	660	6
4	790–1050	920	2
5	1050–1310	1180	0
6	1310–1570	1440	1
Total			50

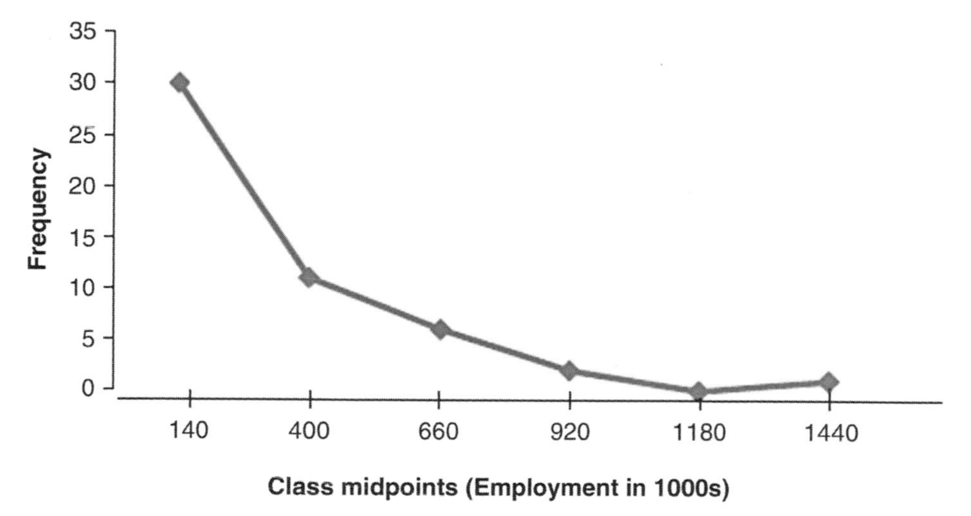

Figure 4.8 Frequency polygon.

curve is useful in answering questions like "How many states are there at or below a certain level of employment?" The cumulative frequencies are shown in Table 4.6d and the cumulative frequency curve is drawn in Figure 4.9. ■

Table 4.6d Cumulative Frequency Values

Upper Class Boundaries (Thousands)	Cumulative Frequencies (Less Than or Equal to)
270	30
530	41
790	47
1050	49
1310	49
1570	50

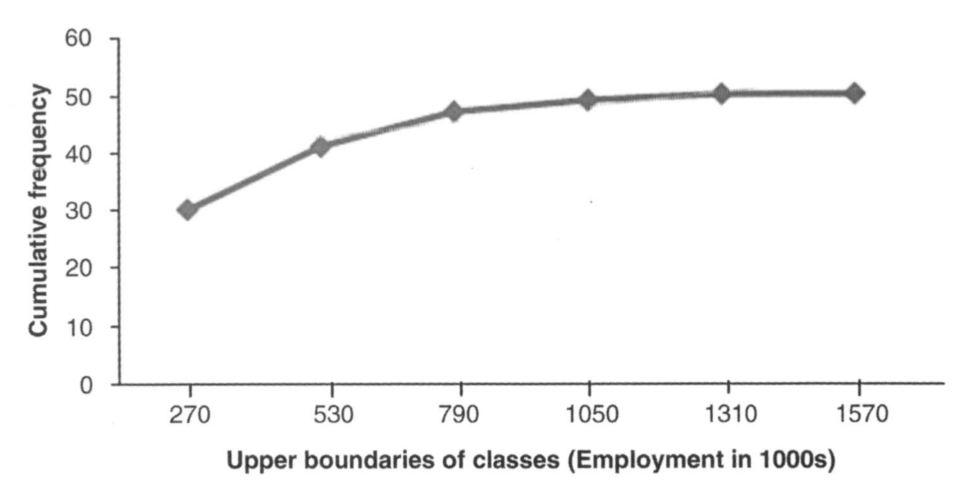

Figure 4.9 Ogive.

4.4 DESCRIPTIVE STATISTICS: NUMERICAL MEASURES OF CENTRAL TENDENCY OR LOCATION OF DATA

The tabular and graphical methods of summarizing data that were presented earlier give only an approximate way of describing a data set. A more precise description of a data set can be obtained by using "numerical measures." Such measures are of two major types: the first is a measure of central location or central tendency; and the second is a measure of variability (or measure of dispersion).

Briefly stated, *measures of location* indicate the center of gravity of the values in the data set. The data could be the *population* (the entire data set) or a *sample* (a part of the population). The numerical measures that describe a population are called *parameters*; and the measures taken at the sample level are known as *statistics*. These measures of central tendency are also called "averages."

4.4.1 Population Mean

The *population mean* (denoted by the Greek letter μ, pronounced as mu) is the sum of all values in the population divided by its size (N):

$$\mu = \frac{\sum_{i=1}^{N} X_i}{N}.$$

The subscript i, which ranges from i = 1 to N, could be dropped for convenience and if there is no ambiguity. Hence, the population mean can be rewritten more simply as

$$\mu = \frac{\sum X_i}{N}.$$

EXAMPLE 4.3

Table 4.7 shows the consumer price index for all urban consumers for the year 2014. *Data file:* Consumer Price Index.xlsx.

Find the monthly average value of the consumer price index for the year 2014. Compare your monthly average with the yearly 2014 consumer price index value of 236.43 published by the U.S. Bureau of Labor Statistics.

SOLUTION: Assuming that the data represents a population, the mean is given by

$$\mu = \frac{\sum X_i}{12},$$

where X_i is the ith monthly index and i runs from 1 to 12. Therefore,

$$\mu = \frac{X_1 + X_2 + X_3 + X_4 + X_5 + X_6 + X_7 + X_8 + X_9 + X_{10} + X_{11} + X_{12}}{12} = \frac{2837.19}{12}$$
$$= 236.43.$$

Thus, 236.43 is the monthly average of the consumer price index for all urban consumers for the year 2014, and this is identical to the 2014 annual index published by the U.S. Bureau of Labor Statistics. ■

4.4.2 Sample Mean

Let us define the *sample mean* as

$$\overline{X} = \frac{\sum_{i=1}^{n} X_i}{n} = \frac{\sum X_i}{n},$$

where n stands for the size of the sample.

Table 4.7 Consumer Price Index, 2014

Month	Consumer Price Index
January	233.9
February	234.8
March	236.3
April	237.1
May	237.9
June	238.3
July	238.3
August	237.9
September	238.0
October	237.4
November	236.2
December	234.8

Source: U.S. Bureau of Labor Statistics (1982–1984 = 100) as reported in year 2015.

EXAMPLE 4.4

The average hourly and average weekly earnings in a sample of 19 industries in the month of July 2015, as reported by the U.S. Department of Labor, are given in Table 4.8. Compute the average hourly and the weekly earnings of workers across all industries. *Data file:* Hourly and Weekly Earnings 2015.xlsx.

SOLUTION: First, the average hourly earnings of the sample of 19 industry earnings is given by dividing the sum of the earnings by 19. Thus,

$$\overline{X} = \frac{\sum_{i=1}^{19} X_i}{19} = \frac{24.55 + 25.77 + \cdots + 14.03 + 22.08}{19}$$

$$= \frac{486.34}{19} = \$25.60.$$

Thus, \$25.60 is the mean hourly earnings of workers in these 19 industries in July 2015. It is to be noted that both the population mean and the sample mean are affected by inordinately large or small extreme values in the data set. For example, if we assume that the above data set contains an extremely high value such as \$130.00 instead of, say, \$35.54, then the sample mean will be relatively high, as shown below. The new sample mean is

$$\overline{X} = \frac{486.34 - 35.54 + 130.00}{19} = \frac{580.80}{19} = \$30.57.$$

Table 4.8 U.S. Department of Labor, Bureau of Labor Statistics, Modified October 2, 2015, Seasonally Adjusted

Industry	Average Hourly Earnings	Average Weekly Earnings
Total private	\$24.55	\$846.98
Goods-producing	25.77	1,041.11
Mining and logging	30.9	1,375.05
Construction	26.81	1,045.59
Manufacturing	24.85	1,016.37
Durable goods	26.2	1,084.68
Nondurable goods	22.45	895.76
Private service-providing	24.26	807.86
Trade, transportation, and utilities	21.44	739.68
Wholesale trade	28.08	1,092.31
Retail trade	17.08	534.6
Transportation and warehousing	22.9	883.94
Utilities	35.54	1,499.79
Information	34.33	1,259.91
Financial activities	30.91	1,152.94
Professional and business services	29.39	1,063.92
Education and health services	24.77	812.46
Leisure and hospitality	14.03	367.59
Other services	22.08	702.14

Source: Bureau of Labor Statistics, October 2, 2015.

Similarly, extremely low values in the data set will pull down the mean value. These very large or small extreme values are generally known as *outliers*. Unless otherwise stated, we treat any subsequent data set as a sample.

Following the above procedure, the mean weekly earnings of the workers in the 19 industries is obtained by finding the sum of the earnings divided by 19. Thus, the mean $\overline{X} = \frac{18222.68}{19} = \$959.09$. As in the other case, this mean will also be affected in a significant way by either an extremely high value or a very low value in this sample. It is to be noted that the mean values of data sets with relatively large sizes can be easily calculated using Excel, as demonstrated in several instances in the later parts of this chapter. ■

4.4.3 Weighted Mean

The sample (or population) mean presented above assumes that all values in the data set are equally important. But quite often in the real world, especially in business, we deal with data containing values with varying levels of importance, known as weights. For example, in a typical family, a higher proportion of the family income is used to pay for housing relative to other items. Likewise, at the aggregate level, the consumer price index is essentially a weighted mean. The *weighted mean* formula is given by

$$\overline{X}_w = \frac{\Sigma X_i W_i}{\Sigma W_i},$$

where the W_i's denote the weights.

EXAMPLE 4.5

A student organized a party for his friends. He bought 2 dozen small-sized soft drinks at a price of \$0.90 per bottle, 3 dozen medium-sized bottles at a price of \$1.20 each, and 1 dozen large size bottles at a price of \$1.60 each. Find the average price per bottle of the soft drinks.

SOLUTION: The weighted average price is computed as

$$\overline{X}_w = \frac{(0.90)24 + (1.20)36 + (1.60)12}{24 + 36 + 12} = \$1.17.$$

Notes:

1. For this problem, the simple arithmetic mean price is $(0.90 + 1.20 + 1.60)/3 = \$1.23$. However, it is the wrong answer because it ignores the proportionate number of bottles of each size.

2. One has to be careful in distinguishing between the variable under consideration and its weight. The variable can be identified by the specific average required. In the example above, the variable is price because we are looking for average price.

3. The simple arithmetic mean $\overline{X}$ is a special case of the weighted mean $\overline{X}_w$ if all the weights are equal. Thus, if $W_i = 1/n$ for all i, then $\overline{X}_w = \sum X_i/n = \overline{X}$, which is a simple arithmetic mean.

4. The mean, either simple or weighted, could be a fraction and need not be one of the actual values in the data set. Generally, it should not be rounded off up or down since it may lead to wrong projections or forecasts. For example, if we assume that the birth rate in the Unites States1 is 1.2 babies per couple, for a total of 30 million couples, then the total number of newborn babies is 36 million. But if we round off the birth rate to 1.0, then the total number of newborn babies is only 30 million. ■

EXAMPLE 4.6

A student has completed five 3-credit-hour courses in the spring of 2007. He obtained an A in one course, a B in two courses, a C in one course, and a D in the fifth course. Find the grade point average for the student in the spring of 2007.

SOLUTION: The grade point average is given by

$$\overline{X}_w = \frac{4(3) + 3(6) + 2(3) + 1(3)}{3 + 6 + 3 + 3} = 2.6.$$

Thus, the student's grade-point average is 2.6. Here the weights are the credit hours relating to each grade. The variable is the grade point. ■

4.4.4 Mean of a Frequency Distribution: Grouped Data

To determine the mean of a frequency distribution for a set of grouped data, let us use the expression

$$\overline{X} = \frac{\sum_{j=1}^{K} f_j m_j}{\sum_{j=1}^{K} f_j} = \sum_{j=1}^{K} f_j m_j / n, \quad \text{and} \quad n = \sum f_j,$$

where m_j refers to the *midpoint* (*class mark*) of the jth class and K stands for the number of classes or groups. (Note that the class mark is used to depict a "representative" value from a class.) Note also that the mean for a frequency distribution is a weighted mean, where the class frequencies are the weights.

EXAMPLE 4.7

Find the average level of employment in manufacturing establishments in all the 50 states of the United States for the frequency distribution presented in Table 4.6a (reproduced below as Table 4.9a).

SOLUTION: From the Table 4.9b worksheet, the average employment level of U.S. manufacturing in thousands is calculated as

$$\overline{X} = \frac{15815}{50} = 316.3$$

Table 4.9a Frequency Distribution of Employment Data

Employment (in Thousands)	f_j
10–269	30
270–529	11
530–789	6
790–1049	2
1050–1309	0
1310–1569	1
Total	50

Table 4.9b Worksheet

Employment (in Thousands)	Class Mark m_j	f_j	$f_j m_j$
10–269	139.5	30	4,185
270–529	399.5	11	4,394.5
530–789	659.5	6	3,957
790–1,049	919.5	2	1,839
1050–1,309	1,179.5	0	0
1310–1,569	1,439.5	1	1,439.5
Total		50	15,815

Note: The mean obtained here is an approximate mean compared to the exact arithmetic mean obtained by using the raw data of the 50 states. The exact mean or simple average employment in manufacturing establishments $= (283 + 11 + 168 + \cdots + 500 + 10)/50 = 287.86$, which is different from the one obtained from the grouped data. This is due to the fact that the class midpoint is used in place of the actual values of employment within the class. ■

4.4.5 Geometric Mean

If $X_1, X_2, X_3, \ldots, X_n$ are positive values, then the nth root of the product of these n values is called the *geometric mean* and is calculated as

$$\overline{X_g} = \sqrt[n]{X_1 X_2 X_3 \ldots X_n}$$

Comment: It can be shown that the geometric mean is always less than or equal to the arithmetic mean.

The geometric mean is applied mostly in business, especially for finding the growth rate of financial assets over a period of time. More specifically, it represents the mean growth rate over a given number of periods using compounded growth in each period. When we deal with financial assets that grow in value each year at different rates, the geometric mean formula can be written as $\overline{X_R} = \sqrt[n]{(1+R_1)(1+R_2)\cdots(1+R_n)}-1$, where R_i is the growth rate in the year i.

Let us assume that $n = 2$ for purposes of illustration. Let R_1 be the growth rate in the first year and assume a principal of \$100 at the beginning of the year. Then the total proceeds at the end of the first year will be \$100 $(1+R_1)$. It is clear that $(1+R_1)$ is the factor (initial value + growth rate) by which the \$100 grows over of the first year. Note that this growth factor is independent of the initial amount of \$100. Similarly, by the end of the second year, the total amount will be $100(1+R_1)(1+R_2)$, where the new multiplying factor is $(1+R_2)$. The geometric mean of the two factors is thus $\sqrt{(1+R_1)(1+R_2)}$, which represents the mean of the initial value plus its compounded growth. Therefore, to obtain the annual growth rate over the 2-year period, we have to subtract "one" corresponding to the principal. Thus, the final formula is $\overline{X_R} = \sqrt{(1+R_1)(1+R_2)} - 1$, which represents the annual average growth rate. In general, for n periods with different growth rates,

$$\overline{X_R} = \sqrt[n]{(1+R_1)(1+R_2)\cdots(1+R_n)}-1.$$

Comparing the geometric mean presented in the first part of the $\overline{X_R}$ formula with the general formula $\overline{X_g}$, we see that $X_i = (1+R_i), i = 1, 2, \ldots, n$.

EXAMPLE 4.8

The salary for an individual increased by 10% in the first year and by 20% in the second year. Find the average annual percent increase in the salary.

SOLUTION: Note that the answer is not an arithmetic mean, or $(10\% + 20\%)/2 = 15\%$. However, it is the geometric mean since at the end of the first year, a 10% increase in salary results in 1.1 times the initial salary, assuming that the initial salary is 1. At the end of the second year, it will be 1.2 times the salary that results at the end of the first year. Now, the geometric mean is $\overline{X}_g = \sqrt{(1.1)(1.2)} = 1.14891$. Therefore, the geometric mean growth rate is $1.1489 - 1 = 0.1489$ or 14.89%, which is less than the arithmetic mean growth rate of 15%.

To continue, assume that an individual has an annual salary of \$100,000. At the end of the first year, his total salary is $\$100,000 + \$100,000(0.1) = 100,000 + 10,000 = \$110,000$. At the end of the second year, with 20% growth, his total salary becomes $110,000 + 110,000(0.2) = 110,000 + 22,000 = \$132,000$. The total increase in salary for the 2 years is thus $10,000 + 22,000 = \$32,000$. This is the same as the increase in salary for the first year, which is $100,000(0.14891) = 14,891$, plus the increase for the second year $= 114,891(0.14891) = \$17,109$. The total increase in 2 years combined is $14,891 + 17,109 = \$32,000$, which is what we obtained before, that is, \$32,000. However, if we apply the 15% arithmetic mean growth rate, the first year increase is $100,000(0.15) = \$15,000$ and the second year increase is $115,000(0.15) = \$17,250$. Thus, the total raise over 2 years is $15,000 + 17,250 = \$32,250$, which is far above the actual increase in salary for the 2 years. So, in such cases, the arithmetic mean growth rate is inappropriate. ■

EXAMPLE 4.9

The successive returns on a technology mutual fund for 4 years are 20% in the first year, 10% in the second year, 50% in the third year, and, finally, 100% in the fourth year. Find the annual average return in the last 4 years.

SOLUTION: Since we have different returns from year to year, we have to find the average return by using the geometric mean:

$$\overline{X_R} = \sqrt[4]{(1+R_1)(1+R_2)(1+R_3)(1+R_4)} - 1$$

$$= \sqrt[4]{(1+0.2)(1+0.1)(1+0.5)(1+1.0)} - 1 = \sqrt[4]{(1.2)(1.1)(1.5)(2)} - 1 = 1.4107 - 1 = 0.4107$$

or 41%, approximately.

Note:

If we apply the arithmetic mean formula, we obtain $\frac{(0.20+0.1+0.50+1.0)}{4} = \frac{1.80}{4} = 0.45$ or 45%, which is the wrong answer because if we apply this annual rate of growth for all periods, the actual amount at the end of the fourth year will be inflated. ■

4.4.6 Median

The *median* is the middle value of a data set when the actual members of the set are arranged in an increasing (or decreasing) order of magnitude. If $X_1, X_2, \ldots, X_n$ are the values in the data set arranged in increasing order of magnitude, then the median is $X_{(n+1)/2}$ if n is odd and $\left(X_{n/2} + X_{(n/2)+1}\right)/2$ if n is even.

Interpretation

The median divides the distribution into two segments, where there are as many data points below the median as there are above it.

EXAMPLE 4.10

The following data show the list prices ($) of a sample of 13 single-family homes in the greater Hartford area in the year 2010. Find the median list price ($) of these homes and interpret this value. *Data file:* Home prices.xlsx.

384900,	239900,	329900,	298700,	255000,	229900,	439900,	595000,	374900,	514900,	344500,	535000,	429750

SOLUTION: First arrange the data in increasing order of magnitude as shown here:

229900,	239900,	255000,	298700,	329900,	344500,	374900,	384900,	429750,	439900,	514900,	535000,	595000

Since the number of values is odd, the middle value, $X_{(n+1)/2} = X_{(13+1)/2} = X_7 = 374,900$, is the median price ($) of the sample of homes.

But if the data contain another value of a single-family home whose listed price is $599,900, then there is an even number of values in the data set. Then the median is calculated as the average of the middle two price values. Thus, the median is the mean of the $n/2 = 14/2 = 7$th value and the $n/2 + 1 = 14/2 + 1 = 7 + 1 = 8$th value in the arranged data set, or the median price ($) is $(374,900 + 384,900)/2 = 759,800/2 = 379,900$.

Note: The median is unaffected by extreme values in the data set. ■

4.4.7 Quantiles, Quartiles, Deciles, and Percentiles

Quantiles are data values that partition the data set into equal segments when the observations are arranged in increasing (or decreasing) order of magnitude. When the arranged data are partitioned into four equal parts, the three points of division are called quartiles (Q_1, Q_2, and Q_3). The first quartile, denoted by Q_1, indicates that 25% of the data values are below Q_1 and 75% of the data values are equal to or above Q_1. Likewise, Q_2, the second quartile, has 50% of the values below it and 50% are equal to or above Q_2. Obviously, by definition, the second quartile of the data set equals the median. Finally, the third quartile, Q_3, partitions the data in a fashion such that 75% of the values are below it and 25% are equal to or above Q_3. In a similar vein, there are nine values, known as *deciles*, which split the arranged data set into 10 equal parts.

Percentiles, which are 99 in number, split the arranged data values into 100 equal segments. The general location of the ith percentile is given by the formula

$$P_i = \frac{i(n+1)}{100}, \quad i = 1,\ 2,\ 3, \ldots, 99.$$

Here n stands for the number of values that are arranged in increasing order of magnitude.

Note that P_{25} locates the first quartile, Q_1, so that

$$P_{25} \text{ is the value of the } \frac{25}{100}(n+1) = \frac{1}{4}(n+1)\text{th term;}$$

P_{50} locates the second quartile, Q_2, or the median, and so

$$P_{50} \text{is the value of the } \frac{50}{100}(n+1) = \frac{1}{2}(n+1)\text{th term;}$$

and P_{75} locates the third quartile, Q_3, or

$$P_{75} \text{is the value of the } \frac{75}{100}(n+1) = \frac{3}{4}(n+1)\text{th term.}$$

Similarly, P_{10} locates the first decile, P_{20} locates the second decile, and P_{90} will be the ninth decile locator. If any one of these locators is not a whole number, a simple interpolation can be used to find the approximate value of the quantile.

EXAMPLE 4.11

Suppose the following are the reported P/E ratios of 10 international companies: 14, 17, 12, 22, 12, 15, 11, 12, 13, 12.

Find the 85*th* percentile and the 3*rd* quartile for this data set.

SOLUTION: Arranging the numbers in increasing order of magnitude yields 11, 12, 12, 12, 12, 13, 14, 15, 17, 22. Then,

$$P_{85}\text{is found at the position}\frac{85(10+1)}{100}=9.35\text{, that is,}$$

the location of the 85th percentile is between the 9th and 10th values. The 85th percentile is found by interpolation as follows. Since the 9th observation is 17, we add to it 0.35 of the distance between the 9th and 10th data points or $P_{85} = 17 + 0.35(22 - 17) = 17 + 0.35(5) = 17 + 1.75 = 18.75$.

The third quartile, Q_3, is located at the position $\frac{75(10+1)}{100} = \frac{3}{4(11)} = \frac{33}{4} = 8.25$. It is one fourth of the distance between the eighth and ninth data points, or $Q_3 = 15 + 0.25(17 - 15) = 15 + 0.5 = 15.5$, where 15 is the eighth observation. ■

4.4.8 Mode

The *mode* is defined as the most frequently occurring or most common value in any given data set. There may be more than one mode (or no mode) in a data set.

EXAMPLE 4.12

For the data of Example 4.11, find the mode and interpret the answer.

SOLUTION: In the data set of P/E ratios (11, 12, 12, 12, 12, 13, 14, 15, 17, 22), the value of 12 occurs most frequently (four times). Therefore, the mode here is 12. ■

EXAMPLE 4.13

The following are the temperature forecasts, high/low for 10 days, June 5–15, 2007, in the greater Hartford, Connecticut, area as reported on the Internet.

81/63	82/60	81/64	81/62	83/66	86/69	91/69	87/69	87/64	85/64

Find the modes of the high temperature as well as of the low temperature forecasts.

SOLUTION: The data values on high temperatures are 81, 81, 81, 82, 83, 85, 86, 87, 87, 91. This data set has two modes: 81 occurs three times and 87 occurs two times. Thus, this data set is termed "bimodal" and the two modal values are 81 and 87.

Forecasts on low temperatures are 60, 62, 63, 64, 64, 64, 66, 69, 69, and 69. This, again, is a bimodal distribution, with two modes: 64 (occurring three times) and 69 (occurring threc times).

Comments on the Mean, Mcdian, and Mode

1. *Comparison:* The mean, median, and mode are all equal in the case of a bell-shaped (mound-shaped) and symmetric or *normal distribution* (Figure 4.10a).

2. In a *right-skewed* or *positively skewed* distribution, mean > median > mode. This is because the mean is influenced by extremely large values (Figure 4.10b).

3. In a *left-skewed* or *negatively skewed* distribution, mean < median < mode. This is due to the fact that the mean is affected by extremely small values in the data (Figure 4.10c).

4. Applications:

a. The mean is a function of all sample values and therefore useful in statistical inference when assessing central location.
b. The median is a better measure of central location when you have a few large or small extreme values in the data set. It is used when the values are ranked, usually in the social sciences (e.g., tax cut planning, subsidized housing, financial aid, and Medicare)
c. The mode is useful in demand analysis (e.g., the shirt size that is in highest demand). The mode is a useful measure of center if, in a set of several categories, you have a predominance of a specific category (e.g., when you have a group of families with many of them having very low incomes or very high incomes). ■

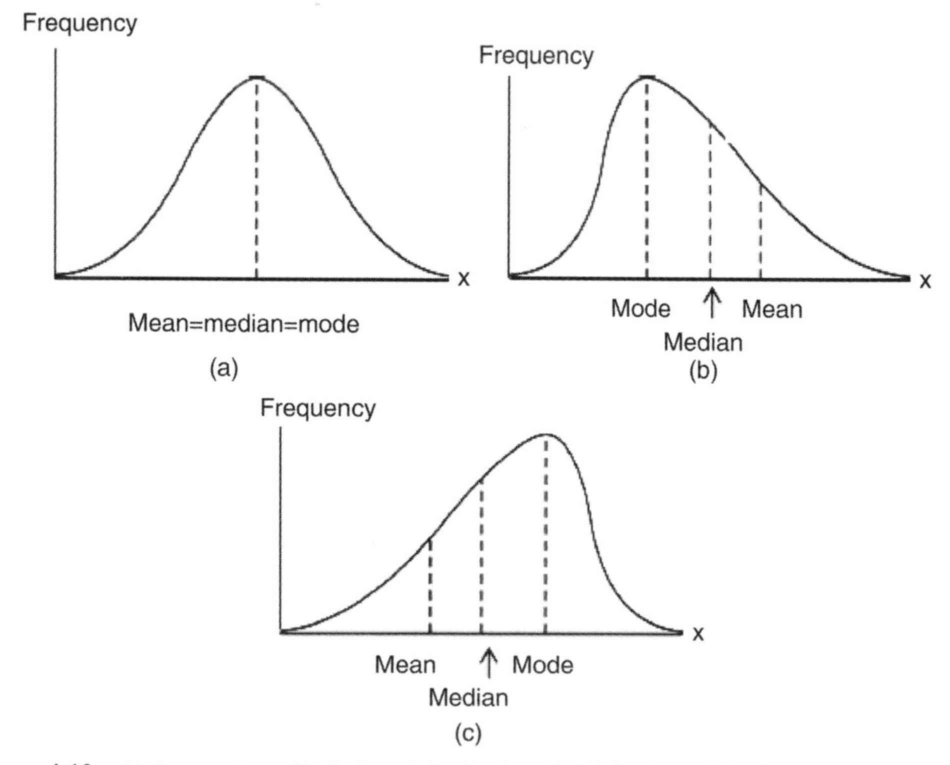

Figure 4.10 (a) Symmetric and bell-shaped distribution. (b) Right-skewed distribution. (c) Left-skewed distribution.

4.5 DESCRIPTIVE STATISTICS: MEASURES OF DISPERSION—VARIABILITY OR SPREAD

Measures of location describe the center of a given data set. In order to have a comprehensive understanding of the whole data set, we also need an indication of how the actual values of the data set vary from a chosen measure of location. This is the task of a "measure of variability." For example, to decide on whether to buy or sell a particular company stock, you need a measure of location such as mean price over a given period of time; and you also need a measure of variability (fluctuations) or the "riskiness" of the stock price over that same time period. To this end, we consider four different basic measures of variability. They are as follows:

a. Range

b. Variance

c. Standard deviation

d. Coefficient of variation

4.5.1 Range

The *range* is the difference between the maximum and minimum values in the data set.

EXAMPLE 4.14

Find the range of the numbers of employees in manufacturing establishments in the 50 states of the United States, as given earlier in Table 4.5.

SOLUTION: The maximum employment value is 1560 (in thousands), while the minimum employment value is 10 (in thousands). Therefore, the range is $1560-10=1550$ (in thousands).

It is to be noted that the range takes into account only two values, the minimum and the maximum of the data set. It is not based on all the values. In addition, we can have two data sets of different variability, but with the same means and the same ranges, thus suggesting that we need a more useful measure of variation. ■

4.5.2 Variance

The *variance* is defined as the average of the squares of the deviations of the measurements from the mean of the data set. First, we introduce the population variance. If $X_1, X_2, \ldots, X_N$ denote the population values (N is the population size), then the *population variance*, denoted by σ^2 (the square of the Greek letter σ, pronounced as *sigma*, is used), is given by

$$\sigma^2 = \sum_{i=1}^{N} (X_i - \mu)^2/N = \frac{(X_1 - \mu)^2 + (X_2 - \mu)^2 + \cdots + (X_N - \mu)^2}{N}.$$

Here, μ denotes the *population mean* and is obtained as

$$\mu = \sum_{i=1}^{N} X_i / N.$$

By expanding the squared terms and then simplifying, we get an alternative form of the variance as

$$\sigma^2 = \frac{1}{N}\sum_{i=1}^{N} X_i^2 - \mu^2.$$

This is usually described as a "short formula" in the sense that the computation involves relatively fewer steps.

EXAMPLE 4.15

Find the mean and variance for the following class size (number of students enrolled) values for five sections of an introductory quantitative methods course in the spring of 2010 in an undergraduate business program. Assume we have a population.

21	23	22	24	15

SOLUTION: The following worksheet (Table 4.10) illustrates the steps involved in the computation of the mean and variance.

Here, $\mu = 105/5 = 21$, where 5 is the population size, N.

Note that the sum of deviations from the mean is always zero, as it should be in any data set. (The positive deviations always cancel with the negative deviations, thus resulting in zero for the sum of the deviations.) To circumvent this problem, the squared deviations are considered in the variance formula.

Using the definitional or "long formula" for the variance, we obtain

$$\sigma^2 = \sum_{i=1}^{N} (X_i - \mu)^2 / N = \sum_{i=1}^{5} (X_i - 21)^2 / 5 = \frac{50}{5} = 10;$$

Table 4.10 Worksheet for Computation of Population Variance

	X	X-μ	(X-μ)^2	(X)^2
	21	0	0	441
	23	2	4	529
	22	1	1	484
	24	3	9	576
	15	−6	36	225
Total	105	0	50	2255

and using the "short formula" for the variance, we have

$$\sigma^2 = \frac{1}{N}\sum_{i=1}^{N} X_i^2 - \mu^2 = \frac{1}{5}(2255) - 21^2 = 451 - 441 = 10.$$

As can be readily seen, both formulas, the long and the short, yield the same final answer.

Sample variance (s^2)

We may calculate the *sample variance* as

$$s^2 = \frac{\sum_{i=1}^{n}\left(X_i - \overline{X}\right)^2}{n-1}.$$

Here, n is the sample size and $\overline{X}$ is the sample mean, which is given by the formula

$$\overline{X} = \sum_{i=1}^{n} X_i/n.$$

While the preceding "long formula" for s^2 is definitional, its computational version is given by the "short formula":

$$s^2 = \frac{\sum X_i^2 - \frac{\left(\sum X_i\right)^2}{n}}{n-1}.$$

Note:

In the "long" expression for the sample variance, the sum of the squared deviations from the mean is divided by $n - 1$ rather than by n, because $\sum\left(X_i - \overline{X}\right) = 0$ for any data set (a restriction on the deviations of values from their mean). Therefore, it follows that there are only $n - 1$ statistically independent deviations. One can establish that the sample variance, obtained by dividing the sum of the squared deviations from the mean by $n - 1$ rather than by n, is an "unbiased estimate" of the population variance. ■

EXAMPLE 4.16

Data on the number of individuals who accessed information about the MBA program of a business school listed on the Internet on a daily basis during the first week of July, 2010, is provided in Table 4.11a. Note that $n = 7$.

Find the variance of this sample using both the long formula and the short formula.

SOLUTION: Using the information presented in Table 4.11b, we have
sample mean $= \overline{X} = \frac{238}{7} = 34$;

Table 4.11a Information on Number of Hits

Day	Sunday	Monday	Tuesday	Wednesday	Thursday	Friday	Saturday
Number of hits	23	55	30	20	41	51	18

Table 4.11b Detailed Computations on Sample Variance

	X	X − 34	(X − 34)^2	X^2
	23	−11	121	529
	55	21	441	3025
	30	−4	16	900
	20	−14	196	400
	41	7	49	1681
	51	17	289	2601
	18	−16	256	324
Total	238	0	1368	9460

sample variance by the "long formula" is

$$s^2 = \frac{\sum_{i=1}^{n}(X_i - \overline{X})^2}{n-1} = \frac{1368}{7-1} = \frac{1368}{6} = 228;$$

and the sample variance by the "short formula" is

$$s^2 = \frac{\sum X_i^2 - \frac{(\sum X_i)^2}{n}}{n-1} = \frac{9460 - \frac{238^2}{7}}{7-1} = \frac{9460 - \frac{56644}{7}}{6} = \frac{9460 - 8092}{6} = \frac{1368}{6} = 228.$$

■

4.5.3 Standard Deviation

The *standard deviation* is defined as the positive square root of the variance.

In the case of a population, it is denoted by

$$\sigma = \sqrt{\frac{\sum (X_i - \mu)^2}{N}};$$

and in the case of a sample, its standard deviation is denoted as

$$s = \sqrt{\frac{\sum (X_i - \overline{X})^2}{n-1}}.$$

In the process of developing a measure of variability, the variance, as it is defined, reflects the measure in terms of the squared deviations of the data values from their mean. By taking the positive square root of the variance, we obtain a measure of dispersion that is expressed in the same units as the original data values.

EXAMPLE 4.17

Compute the standard deviation for the population and sample data given in Examples 4.15 and 4.16.

SOLUTION: For the population variance obtained in Example 4.15, taking the positive square root of $\sigma^2 = 10$ (students)^2 gives the population standard deviation $\sigma = \sqrt{10} = 3.162278$ students.

Similarly, for the sample variance obtained in Example 4.16, taking the positive square root of $s^2 = 228$ (hits)^2 renders the sample standard deviation $s = \sqrt{228} = 15.09967$ hits. ■

EXAMPLE 4.18

Compute the variance and the standard deviation by both the long and the short formulas for the grouped data in Example 4.7. The requisite calculations appear in Table 4.12. (*Note*: Both m_j and $(m_j - \overline{X})^2$ are "weighted" by the class frequencies.)

SOLUTION:

$$\text{Mean} = \overline{X} = \frac{\sum_{j=1}^{k} f_j m_j}{\sum_{j=1}^{k} f_j} = \frac{15815}{50} = 316.3.$$

Variance:

From the long formula,

$$s^2 = \frac{\sum_{j=1}^{k} f_j (m_j - \overline{x})^2}{n-1} = \frac{3709888}{50-1} = \frac{3709888}{49} = 75712.$$

From the short formula,

$$s^2 = \frac{\sum_{j=1}^{k} f_j m_j^2 - \left[\left(\sum_{j=1}^{k} f_j m_j\right)^2 / n\right]}{n-1} = \frac{8712173 - (15815^2/50)}{50-1} = \frac{8712173 - 5002284.5}{49}$$
$$= \frac{3709888.5}{49} = 75712.$$

Standard Deviation: From the long formula,

$$s = \sqrt{\frac{\sum_{j=1}^{k} f_j (m_j - \overline{X})^2}{n-1}} = \sqrt{\frac{3709888}{50-1}} = \sqrt{75712} = 275.16.$$

Table 4.12 Detailed Calculations

Classes	m_j	f_j	$f_j m_j$	$m_j - \overline{X}$	$(m_j - \overline{X})^2$	$f_j(m_j - \overline{X})^2$	m_j^2	$f_j m_j^2$
10–269	139.5	30	4,185	−176.8	31,258.24	937,747.2	19,460.25	583,807.5
270–529	399.5	11	4,394.5	83.2	6,922.24	76,144.64	159,600.25	1,755,602.75
530–789	659.5	6	3,957	343.2	117,786.24	706,717.44	434,940.25	2,609,641.5
790–1049	919.5	2	1,839	603.2	363,850.24	727,700.48	845,480.25	1,690,960.5
1,050–1,309	1,179.5	0	0	863.2	745,114.24	0	1,391,220.25	0
1,310–1,569	1,439.5	1	1,439.5	1,123.2	1,261,578.24	1,261,578.24	2,072,160.25	2,072,160.25
Total		50	1,5815			3,709,888		8,712,172.5

From the short formula,

$$s = \sqrt{\frac{\sum_{j=1}^{k} f_j m_j^2 - \left[\left(\sum_{j=1}^{k} f_j m_j\right)^2 / n\right]}{n-1}} = \sqrt{\frac{8712173 - (15815^2/50)}{50-1}} = \frac{\sqrt{3709888.5}}{49}$$

$$= \sqrt{75712} = 275.16.$$

Comments on the Standard Deviation

1. The standard deviation, by definition, is nonnegative. A zero standard deviation implies that all the members of the data set are equal in value.

2. A low standard deviation reflects a lower variability (or more stability) about the mean relative to a high standard deviation that shows higher variability (less stability) about the mean.

3. The variance is always expressed in squared units of the data values and the standard deviation is always stated in the same units as the data values. For example, if the data values are measured in \$, variance is in $\$^2$, and the standard deviation is in \$. ■

4.5.4 Coefficient of Variation

Quite often there is a need to compare the variability of two or more data sets. For example, comparison of the measures of variability of stocks and bonds (risk measures) cannot be done in a meaningful way by only considering the standard deviations of returns and neglecting their mean values (unless the mean values are equal). Also, standard deviations cannot be compared directly if the members of the data sets are measured in different units such as salaries (\$) and ages (years). Therefore, for comparison purposes, we need a measure of relative dispersion, which is independent of units. A popular measure of relative dispersions is known as the *coefficient of variation* (CV). It is defined as

$$CV = \left(\frac{\text{Standard deviation}}{\text{Mean}}\right) \times 100\%,$$

where the mean is positive. Clearly, CV is the standard deviation expressed as a fraction of the mean times 100%. (In the instance where one or more of the means happen to be negative, the sign of the coefficient of variation can be ignored.)

EXAMPLE 4.19a

Forbes.com 2012 reported the CEO compensation along with their ages, among other things, of the top 500 companies. Table 4.13a contains $n = 10$ data combinations of CEO total compensation, and their ages. Compare the variability of total compensation against that of the ages of the CEO's of the top 10 companies.

Table 4.13a Total Compensation of CEOs and Their Ages

Total CEO Compensation (Millions of Dollars)	Age (Years)	Total CEO Compensation (Millions of Dollars)	Age (Years)
131	53	51	57
67	72	50	55
64	55	49	59
61	67	44	61
56	59	43	60

Table 4.13b Calculation Details on Coefficients of Variation

	Total Compensation	Age
Mean	61.6 (millions dollars)	59.8 (years)
Standard deviation	25.6999 (millions dollars)	5.81 (years)
CV	42 (%)	10 (%)

SOLUTION: Table 4.13b shows the calculated values of the mean, standard deviations, and coefficients of variation for the two data sets given above.

As can be seen from Table 4.13b, total compensation shows more variability (42%) relative to age (10%). That is, the standard deviation of total compensation is 42% of its mean, whereas the standard deviation of age is only 10% of its mean. ■

EXAMPLE 4.19b

Suppose we are given a sample data as shown in Table 4.13c on total assets of the top 10 U.S. banks and thrifts, along with their total deposits. Compare the variability of total assets with the variability of total deposits of the top 10 U.S. banks and thrifts. *Data file:* Top10 US Banks2014.xlsx.

Table 4.13c Total Assets of Top 10 U.S. Banks and Thrifts and Their Deposits

Total Assets (Billions of Dollars)	Total Deposits (Billions of Dollars)	Total Assets (Billions of Dollars)	Total Deposits (Billions of Dollars)
2,527	1,335	386	265
2,124	1,112	334	226
1,883	943	300	204
1,637	1,131	280	105
391	273	275	208

Source: Wall Street Journal, December 10, 2014.

Table 4.13d Calculation Details on Coefficients of Variation

	Total Assets	Total Deposits
Mean	1,014 (billions of dollars)	580 (billions of dollars)
Standard deviation	913 (billions of dollars)	484 (billions of dollars)
CV	90 (%)	84 (%)

SOLUTION: Table 4.13d shows the calculated values of the means, standard deviations, and coefficients of variation for the two data series given above.

As can be seen in Table 4.13d, total assets show more variability (90%) relative to the variability of the total deposits (84%). That is, the standard deviation of total assets is 90% of its mean, whereas the standard deviation of total deposits is 84% of its mean. ■

EXAMPLE 4.20

Table 4.14 shows the means and standard deviations of penny stocks versus the blue chip type of expensive stocks. Compare the relative risks of these two investments incurred by an investor who hates risk.

SOLUTION: Though the two groups of values are expressed in the same units of measurement ($), their means are far apart. So, we must use the coefficient of variation to compare the variability of the two groups. The CV of penny stocks is

$$CV = \left(\frac{3.5}{12.5}\right) \times 100\% = 0.28 \times 100\% = 28\% \text{ (the standard deviation is 28\% of the mean),}$$

and the CV of blue chip stocks is

$$CV = \left(\frac{14.7}{220}\right) \times 100\% = 0.07 \times 100\% = 7\% \text{ (the standard deviation is 7\% of the mean).}$$

By ignoring the means and comparing standard deviations alone, one may conclude that penny stocks have lower variability compared to the variability of blue chip stocks. Obviously, this is a misleading conclusion. The correct way to make a risk comparison is to calculate the coefficient of variation for each and compare these relative measures of variation. Under these measures, 28% is far higher than the 7% level and, therefore, one can legitimately draw the conclusion that the variability (risk) of penny stocks is higher than that of the blue chip stocks. So, the investor who hates risk should consider blue chip stocks for investment purposes. ■

Table 4.14 Means and Standard Deviations of Stocks

	Penny Stocks	Blue Chip Stocks
Mean ($)	12.5	220
Standard deviation ($)	3.5	14.7

4.5.5 Some Important Uses of the Standard Deviation

1. *Standardization of Values*

Often it is extremely useful to standardize the actual values in a data set by using their mean and standard deviation. For example, this procedure can be used to compare data sets with different units of measurement. In addition, when dealing with distributions that are bell-shaped, convenient probability tables are available for finding standardized values. The *standardization procedure* is carried out as follows. If we let the *Z-score* Z_i denote the standardization of X_i, then

$$Z_i = \left(\frac{X_i - \mu}{\sigma}\right) \quad \text{for a population data set;}$$

$$Z_i = \left(\frac{X_i - \overline{X}}{s}\right) \quad \text{for a sample data set.}$$

Note that these Z_i-scores represent the distance of the observation X_i from the mean in terms of standard deviation units. If a Z_i-score is greater than zero, then the X_i value lies above the mean. If Z_i is less than zero, then X_i lies below the mean, for example, if $Z_i = 2$, then X_i lies two standard deviations above the mean. And if $Z_i = -0.5$, then X_i lies one half of a standard deviation below the mean. If the Z-score for X_i is zero, where does X_i lie?

EXAMPLE 4.21

Mean salary of recent business school graduates is \$31,000 with a standard deviation of \$5,000. Find the Z-scores for the salary levels of \$36,000, \$21,000, and \$42,000.

SOLUTION:

$$\text{Z-score for \$36,000} = \left(\frac{36,000 - 31,000}{5000}\right) = \left(\frac{5000}{5000}\right) = 1.$$

This means that the salary \$36,000 is located one standard deviation above the mean.

$$\text{Z-score for \$21,000} = \left(\frac{21,000 - 31,000}{5000}\right) = \left(\frac{-10,000}{5000}\right) = -2.$$

This means that the salary \$21,000 is located two standard deviations below the mean.

$$\text{Z-score for \$42,000} = \left(\frac{42,000 - 31,000}{5000}\right) = \left(\frac{11,000}{5000}\right) = 2.2.$$

Here the salary \$42,000 is located a distance of 2.2 standard deviations above the mean.

It is to be noted that, in general, a Z-score refers to how many standard deviations a particular X value is from the mean. A negative Z-score indicates that an observation lies below

Table 4.15 Chebysheff's Percentages for Different Values of K

K	$1-\frac{1}{K^2}$	%
2	$1-\frac{1}{2^2}=1-\frac{1}{4}=\frac{3}{4}=0.75$	75
3	$1-\frac{1}{3^2}=1-\frac{1}{9}=\frac{8}{9}=0.89$	89
4	$1-\frac{1}{4^2}=1-\frac{1}{16}=\frac{15}{16}=0.94$	94

the mean; and a positive Z-score indicates that a data point lies above the mean. (How do we interpret a zero Z-score?) The method of standardization is also known as the *linear transformation* of X to Z (shifting the mean of X to zero and dividing by the standard deviation). Thus, standardized values are pure numbers without any units of measurement.

2. *Chebysheff's Theorem*The standard deviation can be very useful in describing the distribution of a given data set. This is facilitated by the famous *Chebysheff's theorem.* It states that for any set of data, population or sample, and any constant $K > 1$, at least $1 - (1/(K^2))$ of the data must lie within K standard deviations of the mean. Table 4.15 illustrates this theorem for different values of K.

Thus, at least 75% of the measurements are located within two standard deviations of the mean, at least 89% of the data are located within three standard deviations of the mean, and at least 94% of the data values are located within four standard deviations of the mean. In general, the implied interval for $K > 1$ (assuming a sample of data) is $\overline{X} \pm Ks$ or $\left(\overline{X} - Ks, \overline{X} + Ks\right)$. ■

EXAMPLE 4.22

It is reported that the mean salary of recent graduates with an undergraduate degree from a business school is \$31,000 with a standard deviation of \$5000. Find the proportion of graduates with incomes between \$21,000 and \$41,000.

SOLUTION: Use the endpoints of the preceding interval to get K. That is, set $21{,}000 = 31{,}000 - K5000$ or $K = \frac{31{,}000-21{,}000}{5000} = 2$; or set $41{,}000 = 31{,}000 + K5000$ so that, again, $K = \frac{41{,}000-31{,}000}{5000} = 2$.Then, with $K = 2$, Chebysheff's theorem informs us that at least

$$1 - \frac{1}{K^2} = 1 - \frac{1}{2^2} = 1 - \frac{1}{4} = \frac{3}{4},$$

or 75% of the graduates have salaries between \$21,000 and \$41,000. What is the connection between K and a Z-score? Both represent the distance of an observation from the mean in standard deviation units. ■

EXAMPLE 4.23

For the year 2010–2011, suppose that the GMAT scores for the first-time takers in a particular state have a mean of 531 with a standard deviation of 119. (a) Find Z-scores (values of K) corresponding to the GMAT scores of 349 and 709 for two randomly selected students. (b) Find the proportion (or %) of students who scored between these two test scores.

SOLUTION: As an alternative to the solution offered in the preceding example, let us start with the calculation of the implied Z-scores.

We assume that X stands for the GMAT test score of a randomly selected student. At $X_1 = 349$,

$$Z_1 = \frac{X_1 - \overline{X}}{s} = \frac{349 - 531}{119} = \frac{-182}{119} = -1.5;$$

and at $X_2 = 709$,

$$Z_2 = \frac{X_2 - \overline{X}}{s} = \frac{709 - 531}{119} = \frac{178}{119} = 1.5.$$

Since $Z = 1.5 > 1$ (thus, the value of $K > 1$), at least

$$1 - \frac{1}{1.5^2} = 1 - \frac{1}{2.25} = 1 - 0.44 = 0.56 \text{ or } 56\%,$$

or 56% of students attained GMAT scores between 349 and 709.

Note that Chebysheff's theorem cannot be applied meaningfully if we consider two points with different (not equal) distances from the mean.

■

4.5.6 Empirical Rule

According to the *empirical rule,* if the distribution of, say, a sample data set is mound-shaped (symmetric and bell-shaped or normal), then

i. approximately 68% of the observations fall within 1 standard deviation of the mean or within the interval $(\overline{X} - 1\,s, \overline{X} + 1\,s)$;

ii. approximately 95% of the observations fall within 2 standard deviations of the mean or within the interval $(\overline{X} - 2\,s, \overline{X} + 2\,s)$; and

iii. almost all or approximately 99.7% all the observations fall within 3 standard deviations of the mean or within the interval $(\overline{X} - 3\,s, \overline{X} + 3\,s)$.

While this rule applies only to mound-shaped distributions, Chebysheff's theorem applies to any distribution. Also, note that Chebysheff's theorem has a theoretical foundation, whereas the empirical rule is based on observational experience.

However, the empirical rule can be described as a special case of Chebysheff's theorem when the distribution is mound-shaped. Remember that there is a strong restriction on K in Chebysheff's theorem, where, again, K stands for the number of standard deviations that any point X_i is away from the mean. The restriction is that K must be greater than 1. Such a restriction is not required in the empirical rule.

EXAMPLE 4.24

The weekly incomes of part-time undergraduate students are observed to be approximately distributed as bell-shaped, with a mean of \$300 and a standard deviation \$20. Find what proportion of incomes fall between

a. 280 and 320

b. 260 and 340

c. 240 and 360

SOLUTION:

a. $\overline{X} - 1s = 300 - 20 = 280$ and $\overline{X} + 1s = 300 + 20 = 320$.

Therefore, by the first part of the empirical rule, approximately 68% of the incomes fall between \$280 and \$320.

b. $\overline{X} - 2s = 300 - 2(20) = 300 - 40 = 260$ and $\overline{X} + 2s = 300 + 2(20) = 300 + 40 = 340$.

So, following the second part of the empirical rule, about 95% of the incomes are located between \$260 and \$340.

c. $\overline{X} - 3s = 300 - 3(20) = 300 - 60 = 240$ and $\overline{X} + 3s = 300 + 3(20) = 300 + 60 = 360$.

Hence, by the third part of the empirical rule, almost all (99.7%) incomes fall between \$240 and \$360.

If for some reason we are not sure that the distribution of our data is mound-shaped, we can still apply Chebysheff's theorem. That is, we can say, for example, following Chebysheff's theorem that at least 75% of the incomes fall between \$260 and \$340. ■

EXAMPLE 4.25

In Example 4.22 we used Chebysheff's theorem to find the proportion of graduates with incomes between \$21,000 and \$41,000. Suppose we now assume that the income distribution is mound-shaped. Explain how you can reconcile this solution with the result obtained by Chebysheff's theorem.

SOLUTION: Given that the said distribution is mound-shaped, we calculate

$$\overline{X} - 2s = 31,000 - 2(5000) = 31,000 - 10,000 = 21,000,$$

$$\overline{X} + 2s = 31,000 + 2(5000) = 31,000 + 10,000 = 41,000.$$

Therefore, by the empirical rule, approximately 95% of the incomes fall between \$21,000 and \$41,000. Note that this 95% is consistent with at least 75% as given by Chebysheff's theorem.

■

4.6 MEASURING SKEWNESS

As noted earlier, *skewness* relates to the shape of the distribution. In particular, it reflects a lack of symmetry in the data set. Any discussion of skewness is based on the relative positions of the mean, mode, and median of the distribution. To review:

a. If mean = median = mode, then the distribution is *symmetrical* or lacks skewness (Figure 4.10a).

b. If mean < median < mode, the distribution is *skewed to the left* (the left-hand tail is elongated) or is *negatively skewed* (Figure 4.10c).

c. If mode < median < mean, then the distribution is *skewed to the right* (the right-hand tail is elongated) or is *positively skewed* (Figure 4.10b.).

To measure the degree of skewness, we can use the *coefficient of skewness*:

$$\text{SK} = (\text{mean} - \text{mode})/\text{s}.$$

If the mode cannot be uniquely determined, use

$$\text{SK} = 3(\text{mean} - \text{median})/\text{s}.$$

Since we divide by s, these measures of skewness are independent of the units of measurement of the data. Hence, they can be used to compare the relative skewness of two or more distributions.

EXAMPLE 4.26

Find the degree of skewness of the total compensation data (in millions of dollars) for insurance company CEOs, as given below:

647	322	295	180	174	153	142	115	100	98

SOLUTION: Using the formulas provided in Table 4.16, we find that

$$\text{mean} = 222,$$

$$\text{median} = 163.5 \quad \text{(upon rearrangement of the data), and}$$

$$\text{standard deviation(s)} = 167.6.$$

Since the mode does not exist, we use

$$\text{SK} = \frac{3(\text{mean} - \text{median})}{\text{s}} = \frac{3\ (222.6 - 163.5)}{167.6} = 1.058 > 0.$$

Clearly, this distribution is positively skewed.

(Note that Table 4.16 reviews all of the key formulas for computing descriptive statistics for both ungrouped and grouped data sets.) ■

Table 4.16 Review of Key Formulas for Descriptive Statistics

Ungrouped Data		
	Sample Statistics (size = n)	Population Parameters (size = N)
Mean	$\overline{X} = \frac{\sum_{i=1}^{n} X_i}{n}$	$\mu = \frac{\sum_{i=1}^{N} X_i}{N}$
Variance	$s^2 = \frac{\sum_{i=1}^{n} (X_i - \overline{X})^2}{n-1} = \frac{\sum_{i=1}^{n} X_i^2 - \left[\left(\sum_{i=1}^{n} X_i\right)^2 / n\right]}{n-1}$	$\sigma^2 = \frac{\sum_{i=1}^{N} (X_i - \mu)^2}{N} = \frac{\sum_{i=1}^{N} X_i^2 - \left[\left(\sum_{i=1}^{N} X_i\right)^2 / N\right]}{N}$
Standard deviation	$s = \sqrt{s^2}$	$\sigma = \sqrt{\sigma^2}$
Coefficient of variation (CV)	$CV = \frac{s}{\overline{X}} \times 100$	$CV = \frac{\sigma}{\mu} \times 100$
Z	$Z = \frac{X - \overline{X}}{s}$	$Z = \frac{X - \mu}{\sigma}$
Grouped Data		
	Long formula	Short formula
Mean		$\overline{X} = \frac{\sum_{j=1}^{k} f_j m_j}{\sum_{j=1}^{k} f_j}$
Variance	$s^2 = \frac{\sum_{j=1}^{k} f_j (m_j - \overline{X})^2}{n-1}$	$s^2 = \frac{\sum_{j=1}^{k} f_j m_j^2 - \left[\left(\sum_{j=1}^{k} f_j m_j\right)^2 / n\right]}{n-1}$
Standard deviation	$s = \sqrt{\frac{\sum_{j=1}^{k} f_j (m_j - \overline{X})^2}{n-1}}$	$s = \sqrt{\frac{\sum_{j=1}^{k} f_j m_j^2 - \left[\left(\sum_{j=1}^{k} f_j m_j\right)^2 / n\right]}{n-1}}$

4.7 EXCEL APPLICATIONS

DESCRIPTIVE STATISTICS

EXAMPLE 4.27

America's Best Small Companies in Year 2014: Sales (in Millions of Dollars)

Data file: BSCS2014.xlsx.

Use Excel (Descriptive Statistics of the Data Analysis tool) to answer the following questions for the data set given below:

680	382	837	827	285	423	781	988	488	119
754	174	87	815	136	623	245	843	605	256
633	539	215	253	590	794	184	717	387	416
185	958	712	913	216	254	80	407	301	222
701	772	685	223	973	920	222	788	741	408
824	778	881	259	797	938	742	656	787	282
641	451	159	162	605	758	818	779	109	314
274	331	727	131	354	389	58	30	477	510
676	57	63	743	455	386	576	489	505	904
99	511	774	94	203	420	545	555	834	175

Source: Forbes.com

a. Find the mean, median, and the mode for the data set.

b. Based on the values of the mean, median, and mode, can you comment on the shape of the distribution? More specifically, does it look bell-shaped, skewed to the right, or skewed to the left?

c. Find the variance, standard deviation, and the coefficient of variation for the given data set.

SOLUTION: Steps for computing "Descriptive Statistics" by using the *Data Analysis* tool of Excel

Step 1: Open the file named "BSCS2014.xlsx" in Excel data sets.

Step 2: Click on *Data* on the menu bar and select *Data Analysis* on the extreme right side of the ribbon.

If you do not see "Data Analysis," go to *File* and select "Options." Click "Add-Ins" option. Click *GO*, select the first two boxes (Analysis ToolPak and Analysis ToolPak-VAB), and press "OK."

Step 3: Select "Descriptive Statistics" from the Data Analysis menu.

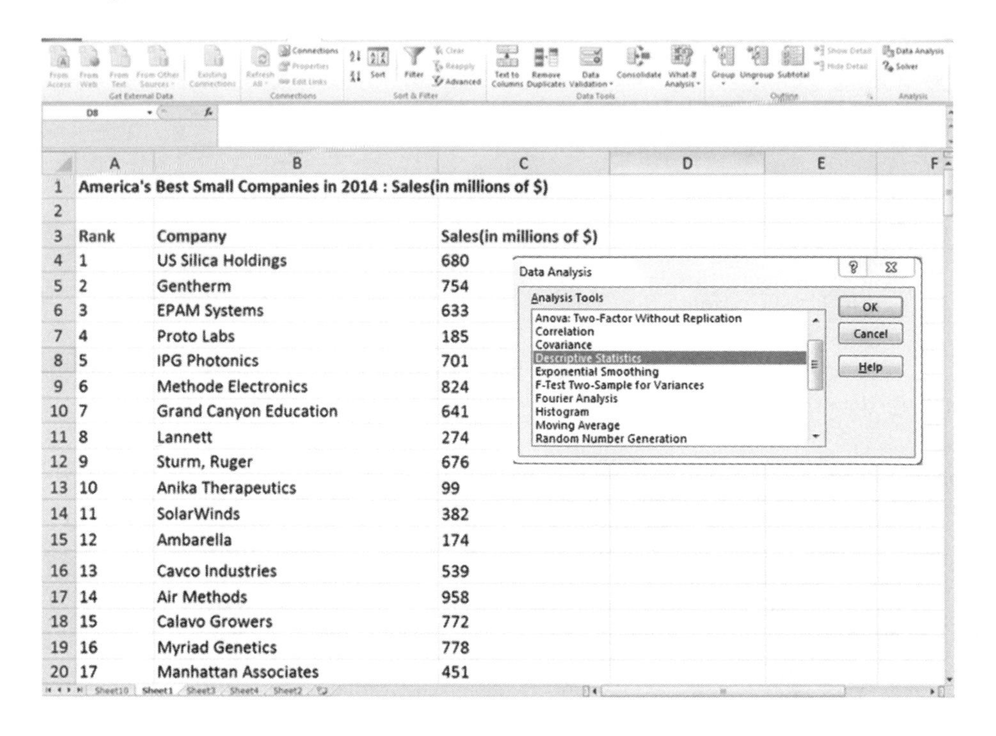
America's Best Small Companies in 2014 : Sales(in millions of $)
Rank
Company
Sales(in millions of $)
1 US Silica Holdings 680
2 Gentherm 754
3 EPAM Systems 633
4 Proto Labs 185
5 IPG Photonics 701
6 Methode Electronics 824
7 Grand Canyon Education 641
8 Lannett 274
9 Sturm, Ruger 676
10 Anika Therapeutics 99
11 SolarWinds 382
12 Ambarella 174
13 Cavco Industries 539
14 Air Methods 958
15 Calavo Growers 772
16 Myriad Genetics 778
17 Manhattan Associates 451
Data Analysis
Analysis Tools
Anova: Two-Factor Without Replication
Correlation
Covariance
Descriptive Statistics
Exponential Smoothing
F-Test Two-Sample for Variances
Fourier Analysis
Histogram
Moving Average
Random Number Generation
OK
Cancel
Help

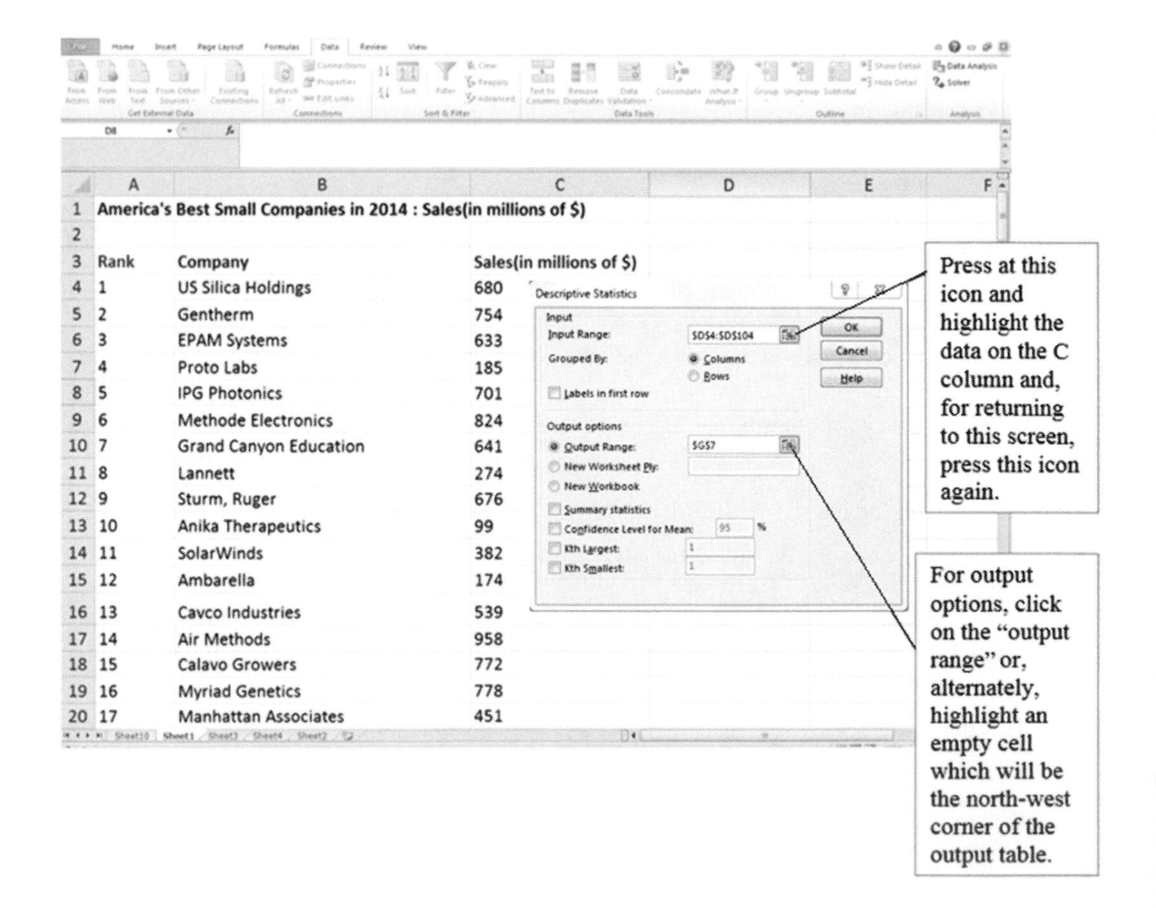
America's Best Small Companies in 2014 : Sales(in millions of $)
Rank
Company
Sales(in millions of $)
1 US Silica Holdings 680
2 Gentherm 754
3 EPAM Systems 633
4 Proto Labs 185
5 IPG Photonics 701
6 Methode Electronics 824
7 Grand Canyon Education 641
8 Lannett 274
9 Sturm, Ruger 676
10 Anika Therapeutics 99
11 SolarWinds 382
12 Ambarella 174
13 Cavco Industries 539
14 Air Methods 958
15 Calavo Growers 772
16 Myriad Genetics 778
17 Manhattan Associates 451
Descriptive Statistics
Input
Input Range:
Grouped By:
Columns
Rows
Labels in first row
Output options
Output Range:
New Worksheet Ply:
New Workbook
Summary statistics
Confidence Level for Mean:
Kth Largest:
Kth Smallest:
OK
Cancel
Help
Press at this icon and highlight the data on the C column and, for returning to this screen, press this icon again.
For output options, click on the "output range" or, alternately, highlight an empty cell which will be the north-west corner of the output table.

Press at this icon and highlight the data on the C column and, for returning to this screen, press this icon again.

Note: If you highlighted the label "Sales (in millions of dollars)" in the input range, you check the box in front of "labels in First Row" in the dialog box of "Descriptive Statistics."

An easy way of highlighting the input range is to highlight the first couple of cell values (input) in column B and keep the buttons "Shift" and "Ctrl" depressed at the same time with your left-hand fingers while you press the "down arrow" key with your right-hand finger. This action will highlight the data only in the column B.

Press **"OK"** on the screen above. You will have the descriptive statistics output table in the Excel sheet as shown below:

America's Best Small Companies in 2014 : Sales (in millions of $)				
Rank	Company	Sales (in millions of $)		
1	US Silica Holdings	680	*Sales(in millions of $)*	
2	Gentherm	754		
3	EPAM Systems	633	**Mean**	**498.42**
4	ProtoLabs	185	Standard Error	27.59245315
5	IPG Photonics	701	**Median**	**497**
6	Methode Electronics	824	**Mode**	**605**
7	Grand Canyon Education	641	Standard Deviation	275.9245315
8	Lannett	274	Sample Variance	76134.34707
9	Sturm, Ruger	676	Kurtosis	−1.312901521
10	Anika Therapeutics	99	Skewness	−0.01473354
11	Solar Winds	382	Range	958
12	Ambarella	174	Minimum	30
13	CavcoIndustries	539	Maximum	988
14	Air Methods	958	Sum	49842
15	Calavo Growers	772	Count	100
16	Myriad Genetics	778		
17	Manhattan Associates	451		

The mean, median, and mode are highlighted in bold above.

a. Based on the relative values of the mean, median, and mode, and also on the value of the coefficient of skewness, we conclude that the distribution of small companies' sales seems to be left-skewed or long left-tailed.

b. The sample variance of the data set from the output above is 76,134.34707 and the sample standard deviation is 275.9245315.

The coefficient of variation is computed as ((standard deviation)/mean) × 100% = (275.9245315/498.42) × 100% = 55.359%. ■

EXAMPLE 4.28

Find the mean and standard deviation for the data in Example 4.27 using the "Function" key.

SOLUTION: Some Summary Statistics by Using the Function Key f_x

You will find below a set of instructions (with illustrations) on how to compute the mean, variance, and standard deviation by using the function key f_x.

Steps for computing "Mean"

Open the Excel *Data file:* BSCS2014.xlsx.

Click on *function* f_x

Select "Statistical" as *Function Category* and "AVERAGE" as *Function Name*. Hit the OK button.

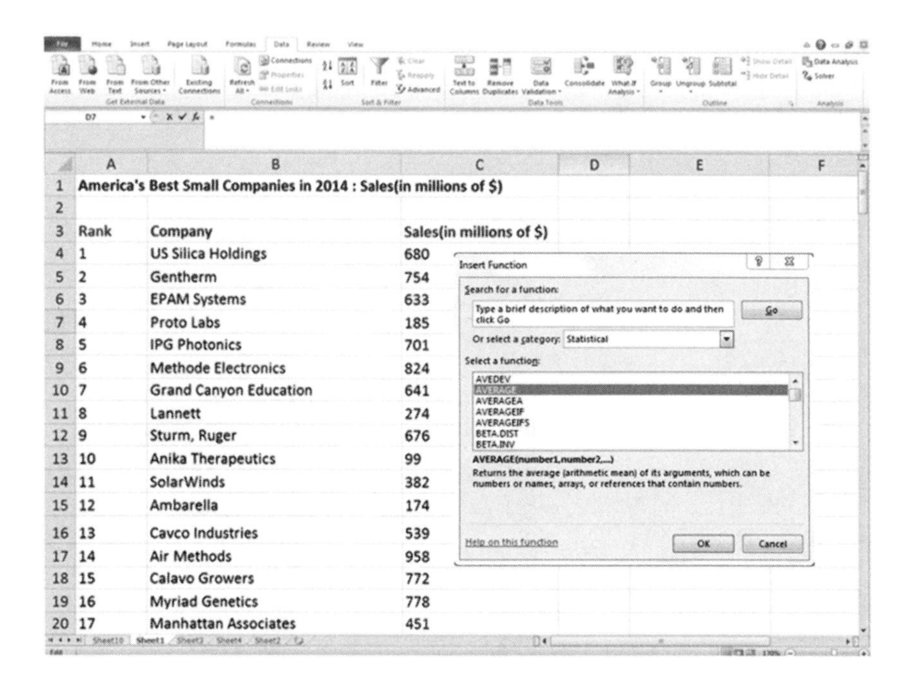

America's Best Small Companies in 2014 : Sales(in millions of $)

Rank	Company	Sales(in millions of $)
1	US Silica Holdings	680
2	Gentherm	754
3	EPAM Systems	633
4	Proto Labs	185
5	IPG Photonics	701
6	Methode Electronics	824
7	Grand Canyon Education	641
8	Lannett	274
9	Sturm, Ruger	676
10	Anika Therapeutics	99
11	SolarWinds	382
12	Ambarella	174
13	Cavco Industries	539
14	Air Methods	958
15	Calavo Growers	772
16	Myriad Genetics	778
17	Manhattan Associates	451

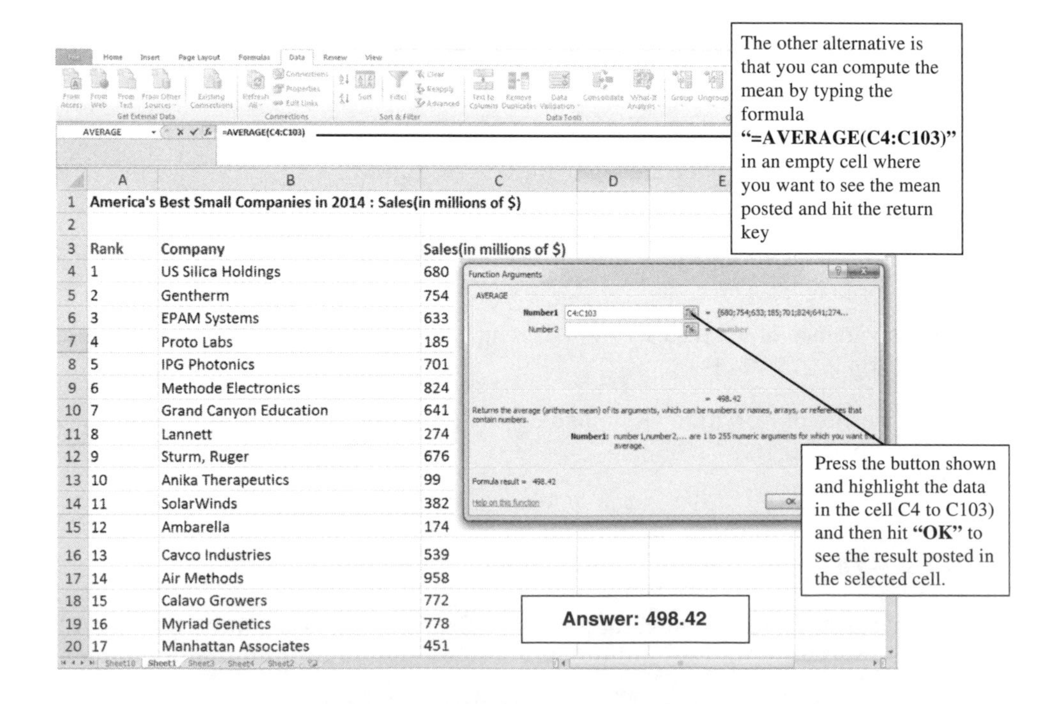

America's Best Small Companies in 2014 : Sales(in millions of $)

Rank	Company	Sales(in millions of $)
1	US Silica Holdings	680
2	Gentherm	754
3	EPAM Systems	633
4	Proto Labs	185
5	IPG Photonics	701
6	Methode Electronics	824
7	Grand Canyon Education	641
8	Lannett	274
9	Sturm, Ruger	676
10	Anika Therapeutics	99
11	SolarWinds	382
12	Ambarella	174
13	Cavco Industries	539
14	Air Methods	958
15	Calavo Growers	772
16	Myriad Genetics	778
17	Manhattan Associates	451

Steps for computing "***Standard Deviation***"

1. Click on *function* f_x

2. Select "Statistical" as *Function Category* and "STDEV.P" (population) or "STDEV. S" (samples) as *Function Name*. Hit the OK button.

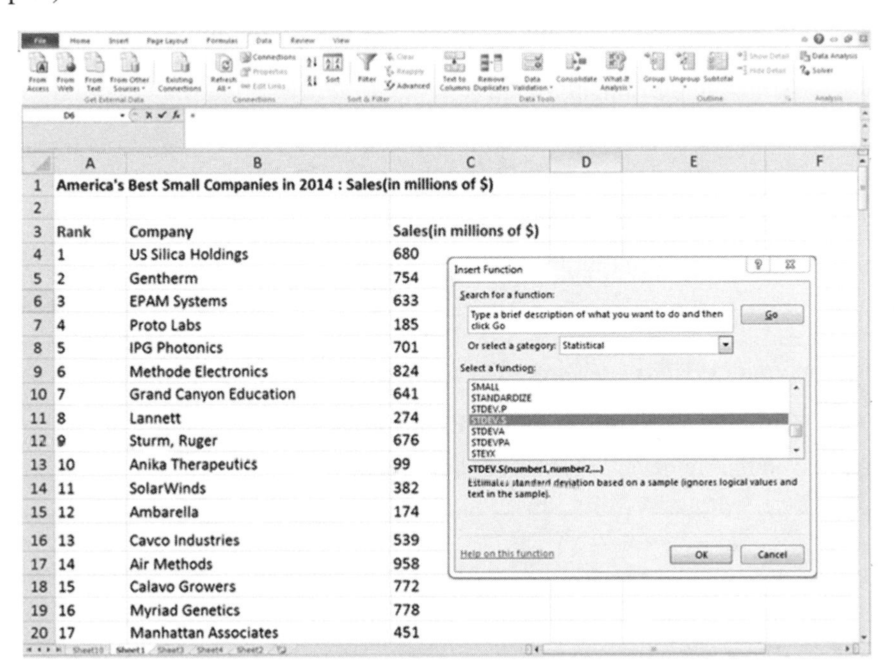

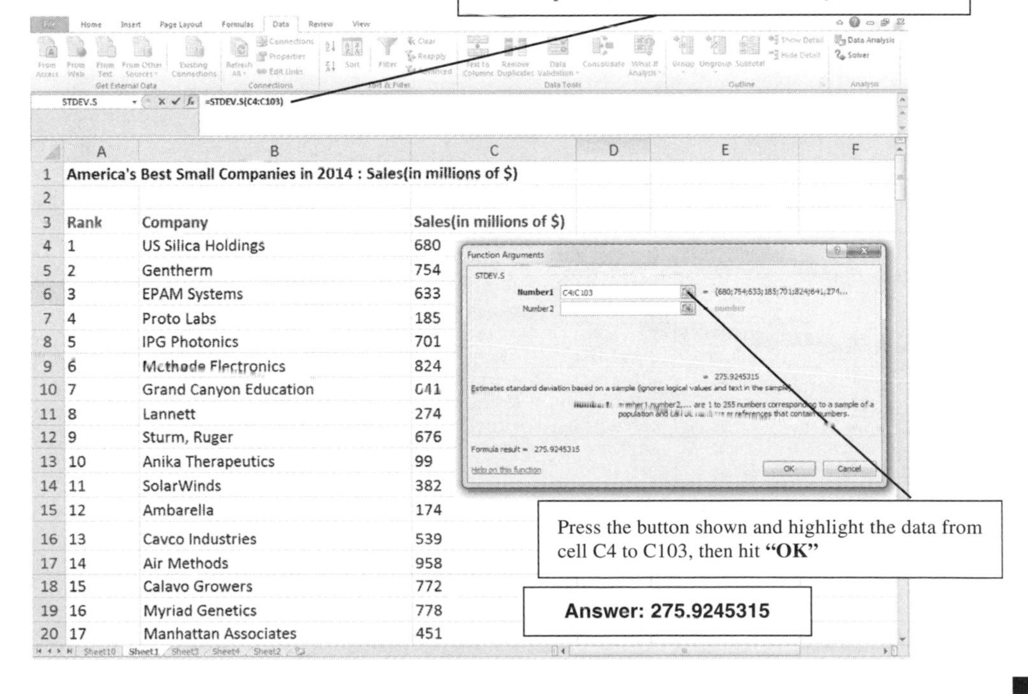

■

EXAMPLE 4.29

Use Excel to calculate the variance and standard deviation of the preceding data set. *Data file:* BSCS2014.xlsx of Example 4.27 by both long and short formulas.

SOLUTION: Use the long and the short formulas to calculate the standard deviation.

To this end,

Variance of the sample (long formula) is

$$s^2 = \frac{\sum_{n}^{i=1} (X_i - \overline{X})^2}{n-1}.$$

Variance of the sample (short formula) is

$$s^2 = \frac{\sum_{n}^{i=1} X_i^2 - \frac{\left(\sum X_i\right)^2}{n}}{n-1}.$$

Steps for computing "standard deviation" by using long and short formulas in Excel.

1. Open the data file: BSCS2014.xlsx in Excel data sets.
2. Create columns in new worksheet as shown below.

X_i	$(X_i - \overline{X})$	$(X_i - \overline{X})^2$	X_i^2
.	.	.	.
.	.	.	.
.	.	.	.

3. We begin by entering the formulas into the cells. A formula in Excel always begins with an "equal" sign (=). Instead of typing the values in the cells, we will refer to the cell address for the values.

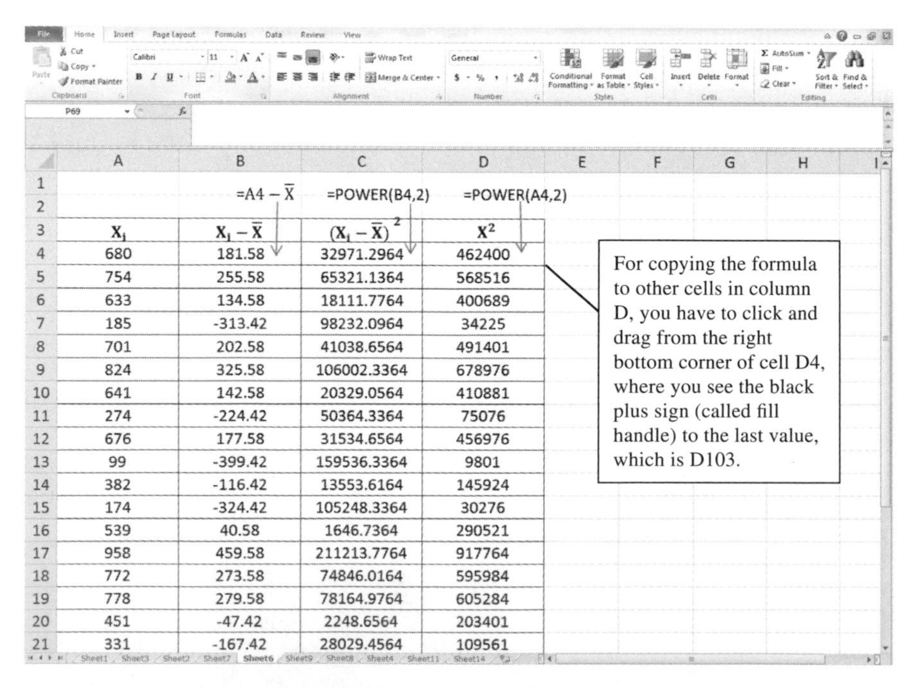

	A	B	C	D
1		=A4 − X̄	=POWER(B4,2)	=POWER(A4,2)
2				
3	X_i	$X_i - \overline{X}$	$(X_i - \overline{X})^2$	X^2
4	680	181.58	32971.2964	462400
5	754	255.58	65321.1364	568516
6	633	134.58	18111.7764	400689
7	185	-313.42	98232.0964	34225
8	701	202.58	41038.6564	491401
9	824	325.58	106002.3364	678976
10	641	142.58	20329.0564	410881
11	274	-224.42	50364.3364	75076
12	676	177.58	31534.6564	456976
13	99	-399.42	159536.3364	9801
14	382	-116.42	13553.6164	145924
15	174	-324.42	105248.3364	30276
16	539	40.58	1646.7364	290521
17	958	459.58	211213.7764	917764
18	772	273.58	74846.0164	595984
19	778	279.58	78164.9764	605284
20	451	-47.42	2248.6564	203401
21	331	-167.42	28029.4564	109561

A125

	A	B	C	D	E	F
90	109	-389.42	151647.9364	11881		
91	477	-21.42	458.8164	227529		
92	505	6.58	43.2964	255025		
93	834	335.58	112613.9364	695556		
94	119	-379.42	143959.5364	14161		
95	256	-242.42	58767.4564	65536		
96	416	-82.42	6793.0564	173056		
97	222	-276.42	76408.0164	49284		
98	408	-90.42	8175.7764	166464		
99	282	-216.42	46837.6164	79524		
100	314	-184.42	34010.7364	98596		
101	510	11.58	134.0964	260100		
102	904	405.58	164495.1364	817216		
103	175	-323.42	104600.4964	30625		
104	∑X = 49842		7537300.36	32379550		
105	X̄= 498.42					
106	$(\sum X)^2$=2484224964					
107						
108						
109						
110						

= power(A104,2) = A104/50 = sum(A4: A103) $\sum(X_i - \bar{X})^2$= sum(C4:C103) $\sum X^2$= sum(D4:D103)

Sheet1 Sheet3 Sheet2 Sheet7 Sheet6 Sheet9 Sheet8 Sheet4 Sheet11 Sheet14

Variance:
From the long formula,

$$s^2 = \frac{\sum_{i=1}^{n}(X_i - \overline{X})^2}{n-1} = \frac{7,537,300.36}{100-1} = \frac{7,537,300.36}{99} = 76,134.35;$$

and from the short formula,

$$s^2 = \frac{\sum_{i=1}^{n} X_i^2 - \left[\left(\sum x_i\right)/n\right]^2}{n-1} = \frac{32,379,550 - \frac{2,484,224,964}{100}}{100-1} = \frac{32,379,550 - 24,842,249.64}{99}$$
$$= \frac{753,7301}{99} = 76,134.35.$$

Standard Deviation:
From the long formula,

$$s = \sqrt{\frac{\sum_{i=1}^{n}(X_i - \overline{X})^2}{n-1}} = \sqrt{\frac{7,537,300.36}{100-1}} = \sqrt{76,134.35} = 275.92$$

and from the short formula,

$$s = \sqrt{\frac{\sum_{i=1}^{n} X_i^2 - \left[\left(\sum x_i\right)/n\right]^2}{n-1}} = \sqrt{\frac{32,379,550 - \frac{2,484,224,964}{100}}{100-1}}$$
$$= \sqrt{\frac{32,379,550 - 24,842,249.64}{99}} = \sqrt{76,134.35} = 275.92.$$

Note:
The mean and standard deviation (computed from the sample data in the data file: BSCS2014.xlsx) in the preceding three examples (Examples 4.27,4.28, and 4.29) are all identical. ■

EXAMPLE 4.30

Using Excel, find the first quartile (25th percentile), median (50th percentile), third quartile (75th percentile), and 80th percentile of the data in the *Data file:* BSCS2014.xlsx.

SOLUTION:

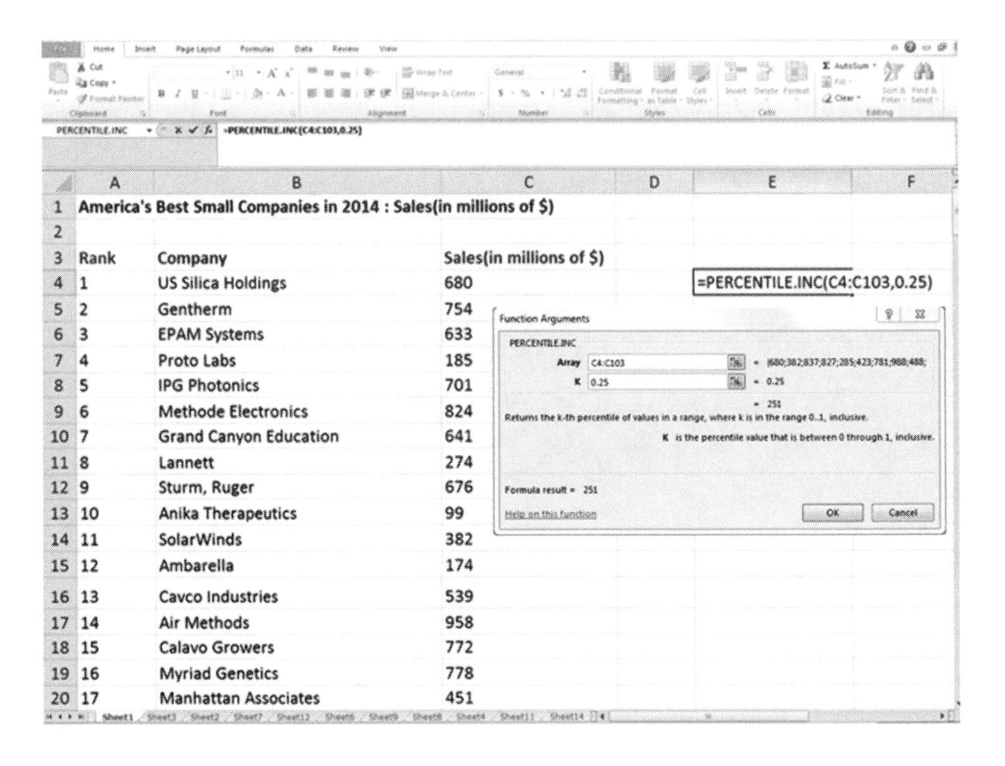

=PERCENTILE.INC(C4:C103,0.25)

America's Best Small Companies in 2014 : Sales(in millions of $)

Rank	Company	Sales(in millions of $)
1	US Silica Holdings	680
2	Gentherm	754
3	EPAM Systems	633
4	Proto Labs	185
5	IPG Photonics	701
6	Methode Electronics	824
7	Grand Canyon Education	641
8	Lannett	274
9	Sturm, Ruger	676
10	Anika Therapeutics	99
11	SolarWinds	382
12	Ambarella	174
13	Cavco Industries	539
14	Air Methods	958
15	Calavo Growers	772
16	Myriad Genetics	778
17	Manhattan Associates	451

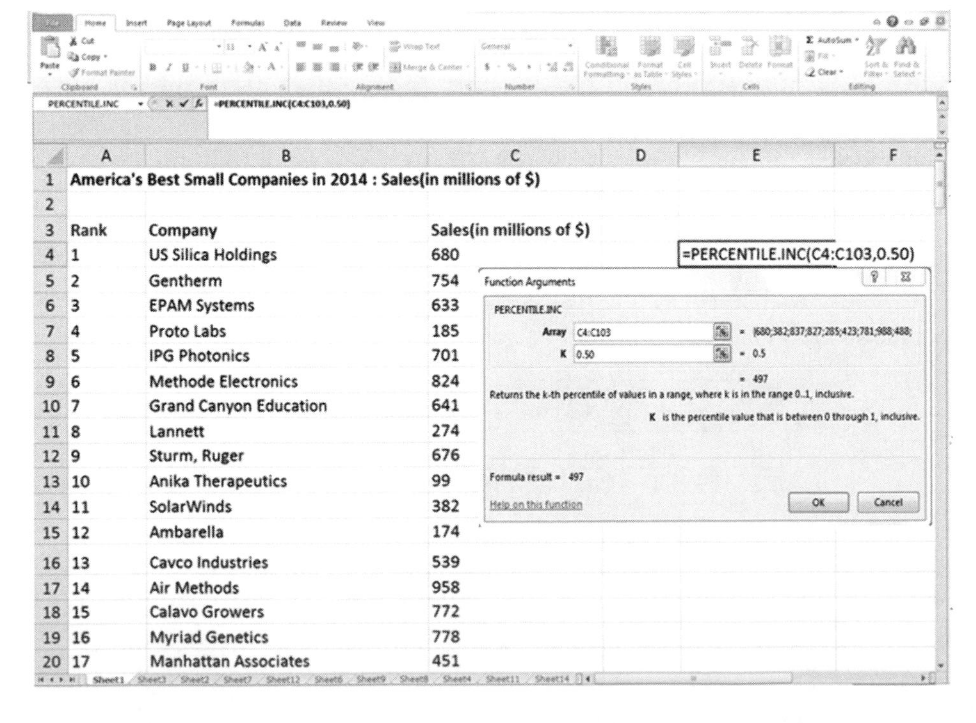

=PERCENTILE.INC(C4:C103,0.50)

America's Best Small Companies in 2014 : Sales(in millions of $)

Rank	Company	Sales(in millions of $)
1	US Silica Holdings	680
2	Gentherm	754
3	EPAM Systems	633
4	Proto Labs	185
5	IPG Photonics	701
6	Methode Electronics	824
7	Grand Canyon Education	641
8	Lannett	274
9	Sturm, Ruger	676
10	Anika Therapeutics	99
11	SolarWinds	382
12	Ambarella	174
13	Cavco Industries	539
14	Air Methods	958
15	Calavo Growers	772
16	Myriad Genetics	778
17	Manhattan Associates	451

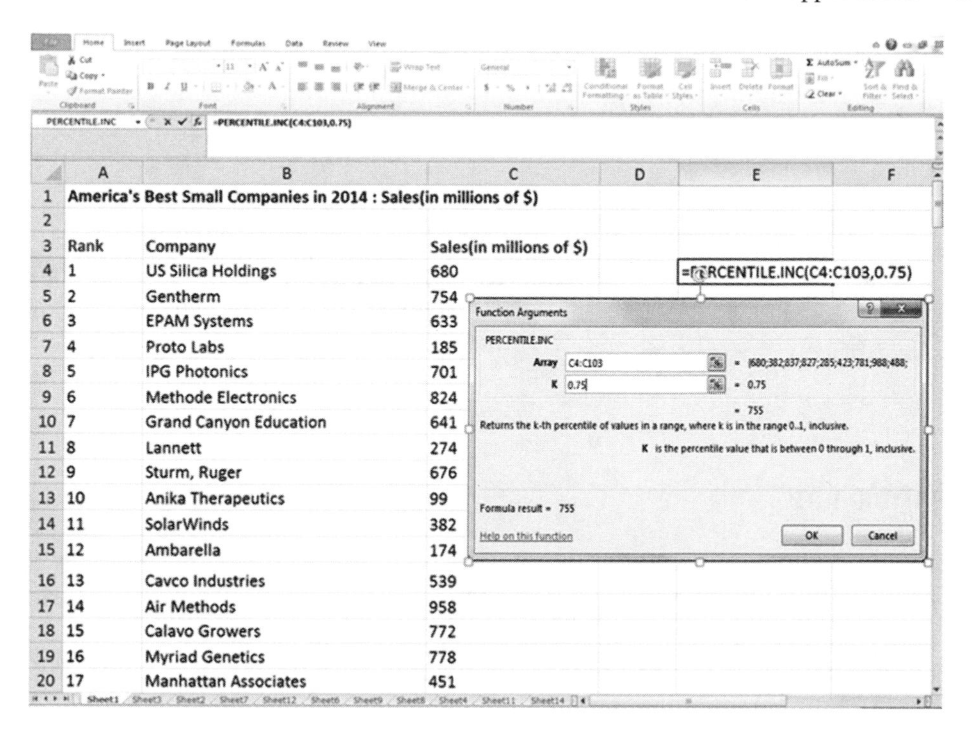
=PERCENTILE.INC(C4:C103,0.75)
America's Best Small Companies in 2014 : Sales(in millions of $)
Rank
Company
Sales(in millions of $)
1 US Silica Holdings 680
2 Gentherm 754
3 EPAM Systems 633
4 Proto Labs 185
5 IPG Photonics 701
6 Methode Electronics 824
7 Grand Canyon Education 641
8 Lannett 274
9 Sturm, Ruger 676
10 Anika Therapeutics 99
11 SolarWinds 382
12 Ambarella 174
13 Cavco Industries 539
14 Air Methods 958
15 Calavo Growers 772
16 Myriad Genetics 778
17 Manhattan Associates 451
Function Arguments
PERCENTILE.INC
Array C4:C103 = {680;382;837;827;285;423;781;988;488;
K 0.75 = 0.75
= 755
Returns the k-th percentile of values in a range, where k is in the range 0..1, inclusive.
K is the percentile value that is between 0 through 1, inclusive.
Formula result = 755
Help on this function
OK
Cancel

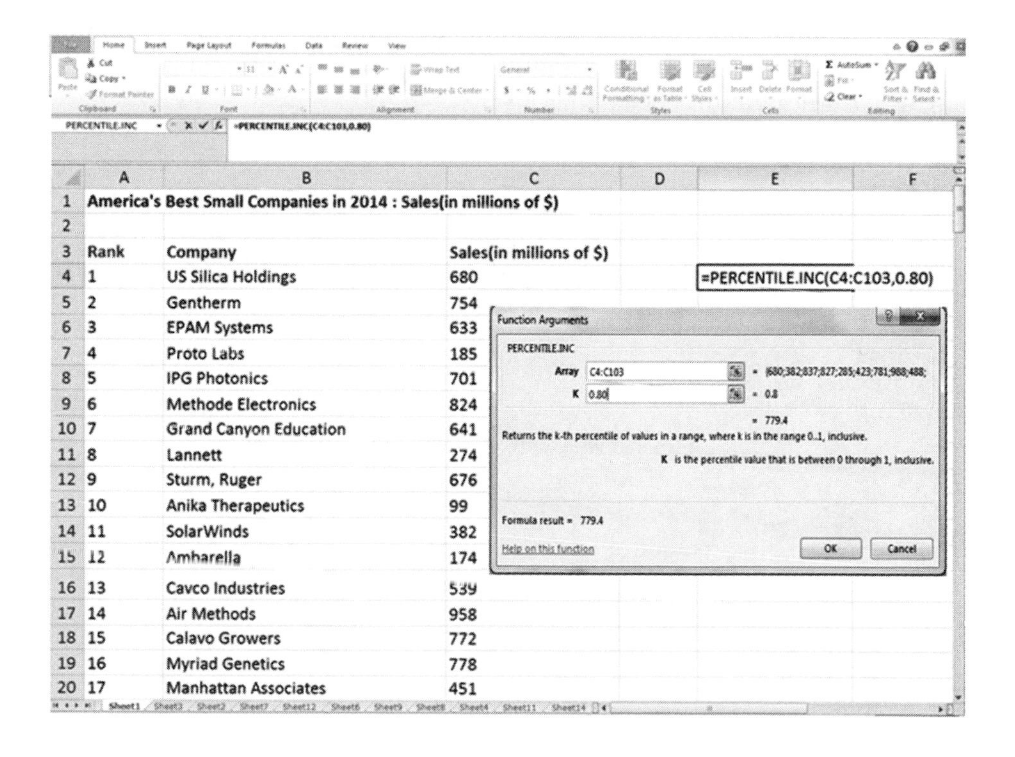
=PERCENTILE.INC(C4:C103,0.80)
America's Best Small Companies in 2014 : Sales(in millions of $)
Rank
Company
Sales(in millions of $)
1 US Silica Holdings 680
2 Gentherm 754
3 EPAM Systems 633
4 Proto Labs 185
5 IPG Photonics 701
6 Methode Electronics 824
7 Grand Canyon Education 641
8 Lannett 274
9 Sturm, Ruger 676
10 Anika Therapeutics 99
11 SolarWinds 382
12 Ambarella 174
13 Cavco Industries 539
14 Air Methods 958
15 Calavo Growers 772
16 Myriad Genetics 778
17 Manhattan Associates 451
Function Arguments
PERCENTILE.INC
Array C4:C103 = {680;382;837;827;285;423;781;988;488;
K 0.80 = 0.8
= 779.4
Returns the k-th percentile of values in a range, where k is in the range 0..1, inclusive.
K is the percentile value that is between 0 through 1, inclusive.
Formula result = 779.4
Help on this function
OK
Cancel

The answers are summarized below:

Percentile	Formula	Answer
First quartile (25th percentile)	=PERCENTILE.INC(C4:C103,0.25)	251
Median (50th percentile)	=PERCENTILE.INC(C4:C103,0.50)	497
Third quartile (75th percentile)	=PERCENTILE.INC(C4:C103,0.75)	755
80th percentile	=PERCENTILE.INC(C4:C103,0.80)	779.4

■

EXAMPLE 4.31

Using Excel, find the 40th percentile, 70th percentile, and 3rd quartile (75th percentile) for the following data (*Data file:* Employment2012.xlsx):

Employment Data (in thousands)

283	11	168	214	1560	149	211	39	372	450
13	60	727	552	223	179	260	149	65	145
330	691	336	183	308	19	102	39	87	347
34	625	613	22	829	148	178	711	60	290
37	408	851	110	44	313	269	69	500	10

Source: 2012 Statistical Abstract of the United States.

SOLUTION:

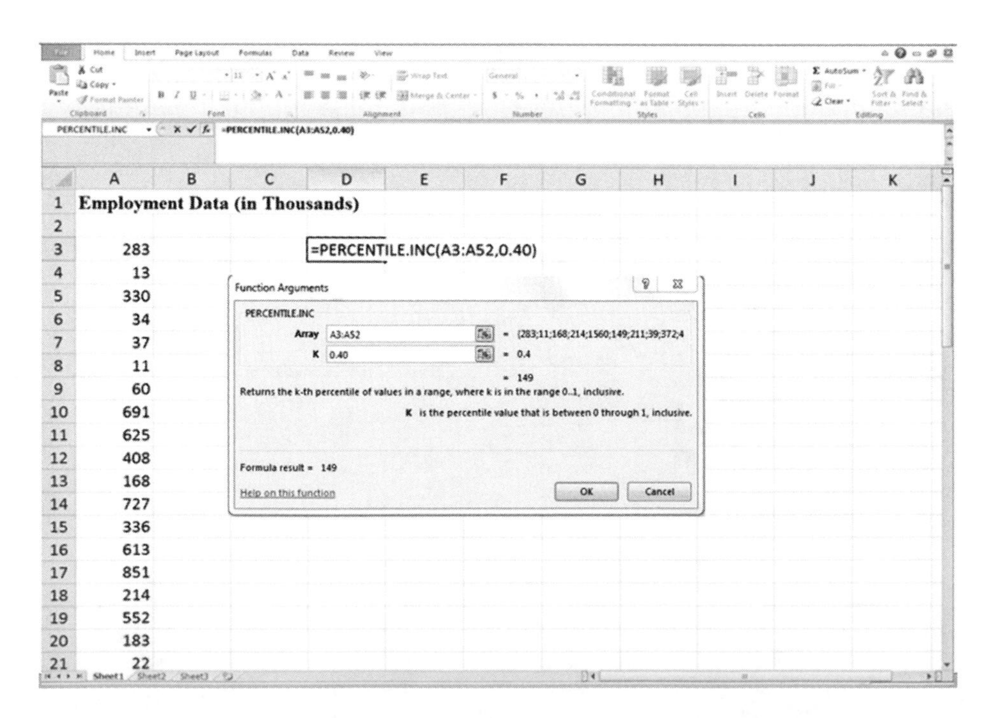

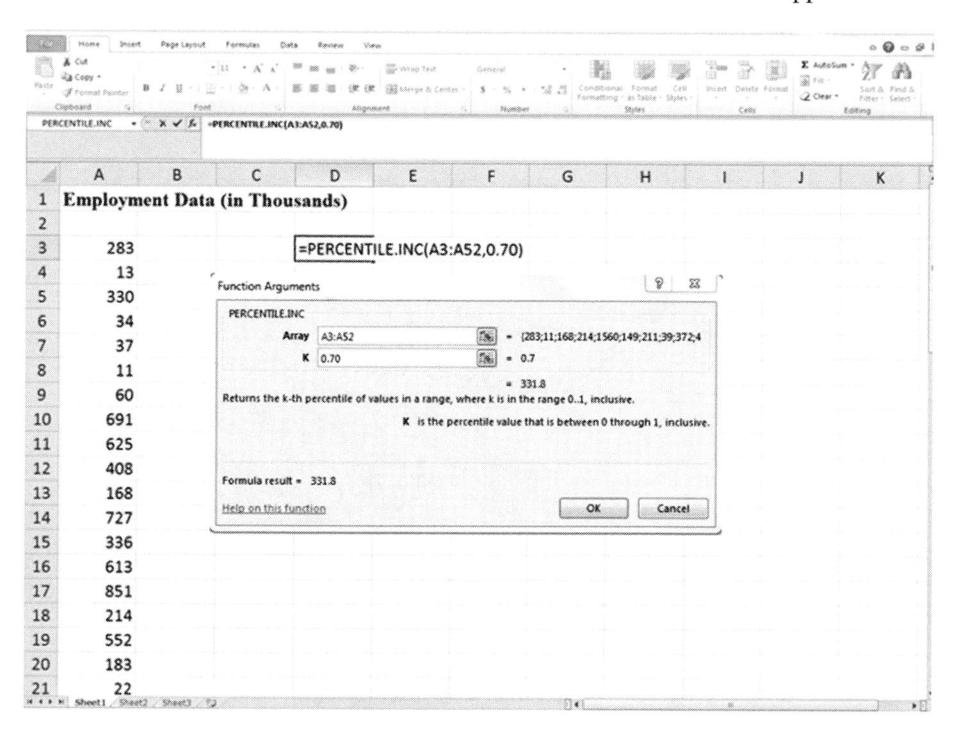
Employment Data (in Thousands)
283
13
330
34
37
11
60
691
625
408
168
727
336
613
851
214
552
183
22
=PERCENTILE.INC(A3:A52,0.70)
Function Arguments
PERCENTILE.INC
Array A3:A52 = {283;11;168;214;1560;149;211;39;372;4
K 0.70 = 0.7
= 331.8
Returns the k-th percentile of values in a range, where k is in the range 0..1, inclusive.
K is the percentile value that is between 0 through 1, inclusive.
Formula result = 331.8
Help on this function
OK
Cancel

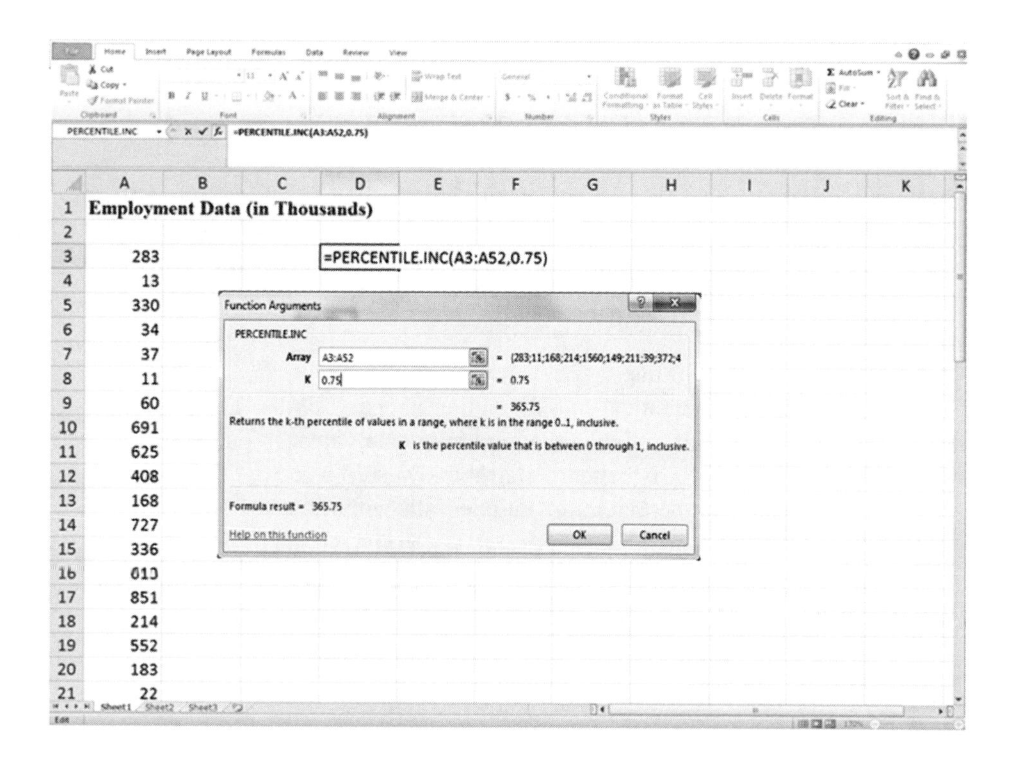
Employment Data (in Thousands)
283
13
330
34
37
11
60
691
625
408
168
727
336
613
851
214
552
183
22
=PERCENTILE.INC(A3:A52,0.75)
Function Arguments
PERCENTILE.INC
Array A3:A52 = {283;11;168;214;1560;149;211;39;372;4
K 0.75 = 0.75
= 365.75
Returns the k-th percentile of values in a range, where k is in the range 0..1, inclusive.
K is the percentile value that is between 0 through 1, inclusive.
Formula result = 365.75
Help on this function
OK
Cancel

The answers are summarized below:

Percentile	Formula	Answer
40th percentile	=PERCENTILE.INC(A3:A52,0.40)	149
70th percentile	=PERCENTILE.INC(A3:A52,0.70)	331.8
Third quartile (75th percentile)	=PERCENTILE.INC(A3:A52,0.75)	365.75

■

EXAMPLE 4.32

Use Excel to calculate the following for the data in data file: BSCS2014.xlsx.

- Form the frequency distribution (group the data)
- Histograms

SOLUTION: *Method 1:* Creating a frequency table and histogram by the standard procedure

1. Open the data file: BSCS2014.xlsx in Excel Spreadsheet.
2. Decide on the number of classes k:
   To determine the number of classes, follow the "2 to the *k* rule." This rule suggests that we select the smallest number k for the number of classes such that 2^k is greater than or equal to the number of observations n. In our example, we have 100 observations.
   Therefore, we know 2^6 equals to 64 and it is less than 100; and 2^7 equals to 128 and it is greater than 100. Hence, we select 7 as the number of classes (k).
3. Determine the class interval or width as follows:
   Approximate class width = (largest data value−smallest data value)/number of classes.
   We can readily observe the largest and smallest data values from the descriptive statistics table (output) as given under "Minimum" and "Maximum." Use these values in the preceding formula and find the class width by rounding off the result to the nearest whole number. In this case, the class width $= \frac{988-30}{7} = 136.85$ (round off to 137).
4. Set the individual class limits.
   There is no specific mathematical rule for setting the class limits. The lower limit of the first class should be close to but less than the minimum data value. In our example, the minimum number is 30. Therefore, it is better to start the lower limit at 30. Obviously, it is a trial and error process. Be sure that the maximum number falls within the last class.
5. Enter the class limits in the worksheet from cell (D5) as shown.
6. Select the cells E6:E12 in order to insert the class frequencies. (You can also select cells in a different column to house the frequencies.)
7. Type the following formula in the formula bar:
   =FREQUENCY (C4:C103, {167,305,443,581,719,857,995})
   Here "C4:C103" refers to the input range;
   "167,305,443,581,719,857,995" refer to the upper limits of each class interval.
8. Press CTRL + SHIFT + ENTER at the same time and the array formula will be copied into each of the cells from E6 to E12. The numbers in the cells are the frequencies of each class.

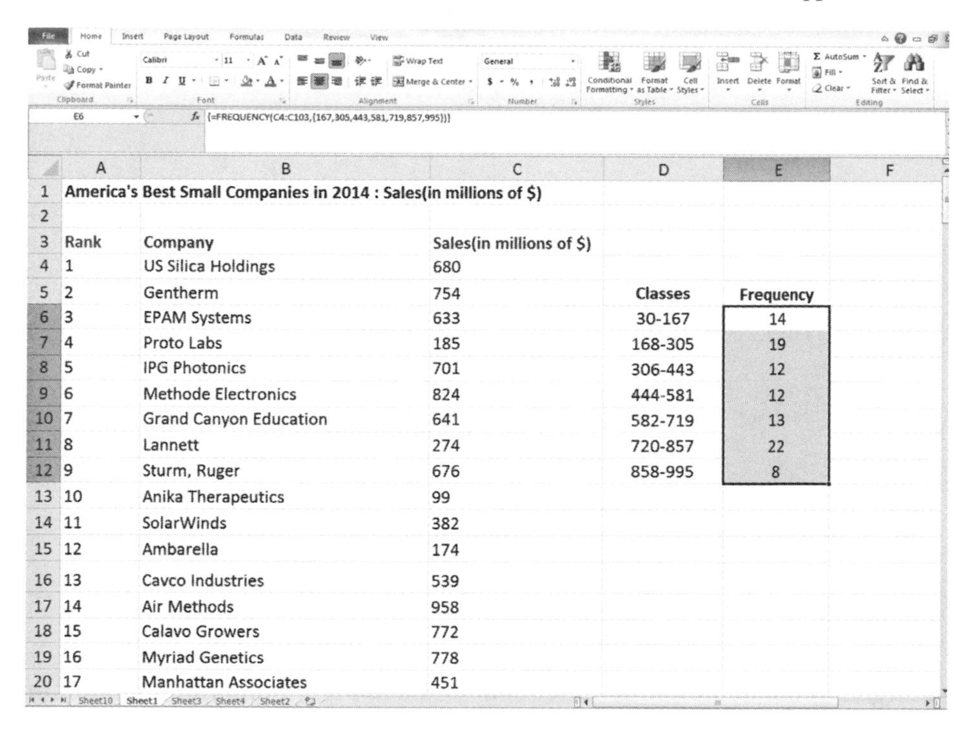

E6 fx {=FREQUENCY(C4:C103,{167,305,443,581,719,857,995})}

	A	B	C	D	E
1	**America's Best Small Companies in 2014 : Sales(in millions of $)**				
2					
3	**Rank**	**Company**	**Sales(in millions of $)**		
4	1	US Silica Holdings	680		
5	2	Gentherm	754	**Classes**	**Frequency**
6	3	EPAM Systems	633	30-167	14
7	4	Proto Labs	185	168-305	19
8	5	IPG Photonics	701	306-443	12
9	6	Methode Electronics	824	444-581	12
10	7	Grand Canyon Education	641	582-719	13
11	8	Lannett	274	720-857	22
12	9	Sturm, Ruger	676	858-995	8
13	10	Anika Therapeutics	99		
14	11	SolarWinds	382		
15	12	Ambarella	174		
16	13	Cavco Industries	539		
17	14	Air Methods	958		
18	15	Calavo Growers	772		
19	16	Myriad Genetics	778		
20	17	Manhattan Associates	451		

Using Excel's chart wizard tool to create the histogram graph

1. Select all the classes and their frequencies (D6:E12)
2. Click insert and select the chart button on the toolbar ( )
3. Select "Column" as chart type and the first chart subtype as shown below.

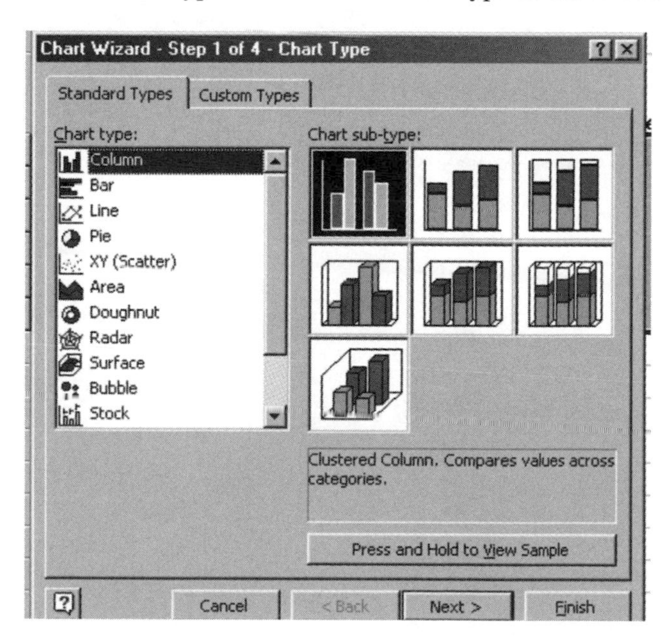

4. Adjust the chart to get the desired histogram as shown below:

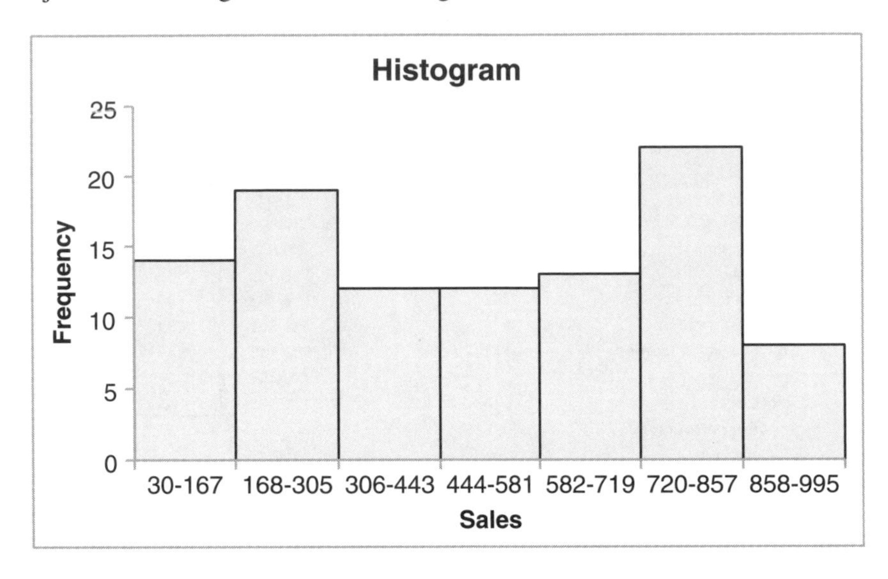

Method 2: To form a frequency table and histogram using the data analysis tool of Excel

1. Repeat steps 1–4 as in method 1 above.

2. Enter the upper limits, also known as bins, of the classes in a column of your choice.

3. Click "Data," and then click "Data Analysis."

4. Click "Histogram" in the Data Analysis list and click OK.

5. Follow the procedure given in the diagram below including highlighting the bins typed in step 2 above in the Bin Range box.

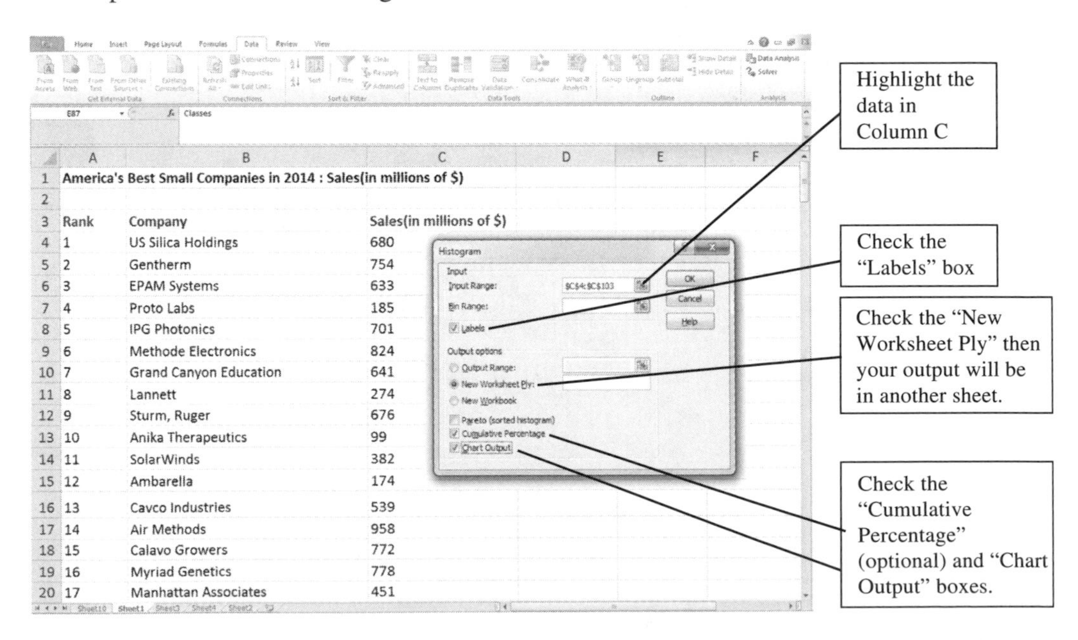

6. Click OK. You will see the histogram and the frequency table in a new worksheet.

7. Format the histogram and frequency table by resizing. The results are given below.

Bin	Frequency
167	14
305	19
443	12
581	12
719	13
857	22
995	8

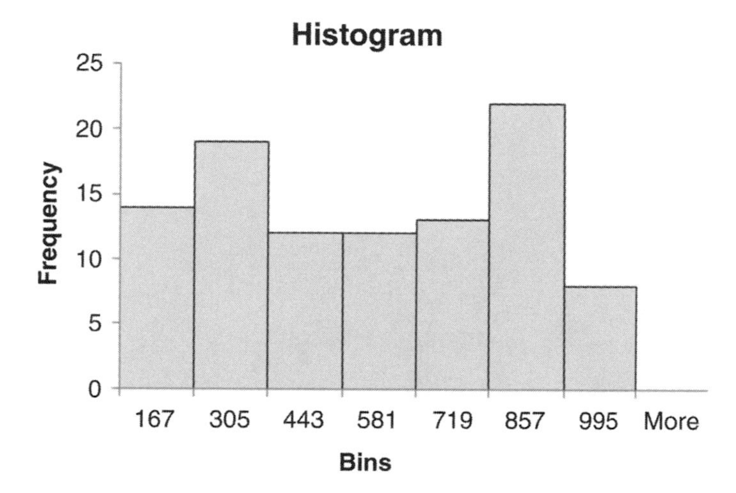

■

EXAMPLE 4.33

Referring to the Example 4.32 results, find the mean for the grouped data or frequency distribution obtained by method 1. Explain why the mean value for the grouped data is different from the mean value for the ungrouped data, 498.42, obtained under Descriptive Statistics of Data Analysis in Example 4.27. Which one of these is a more accurate measure of the mean and why?

SOLUTION: Mean for the grouped data is computed below:

Classes	f_j	m_j	$f_j\ m_j$
30–167	14	98.5	1,379
168–305	19	236.5	4,493.5
306–443	12	374.5	4,494
444–581	12	512.5	6,150
582–719	13	650.5	8,456.5
720–857	22	788.5	17,347
858–995	8	926.5	7,412
Total	**100**		**49,732**

Mean = (49,732/100) = 497.32

Note: In the table above, m_j refers to the midpoint of the jth class, and f_j refers to the class frequency of the jth class.

The mean sales value 498.42 determined from the ungrouped data is more accurate than the mean sale level of 497.32 computed from the grouped data because some accuracy is lost when the data values in each interval are represented by the midpoint of that class interval. ■

EXAMPLE 4.34

Find the variance and standard deviation by both the long and short formulas for the frequency table generated in Example 4.32, using Excel.

SOLUTION:

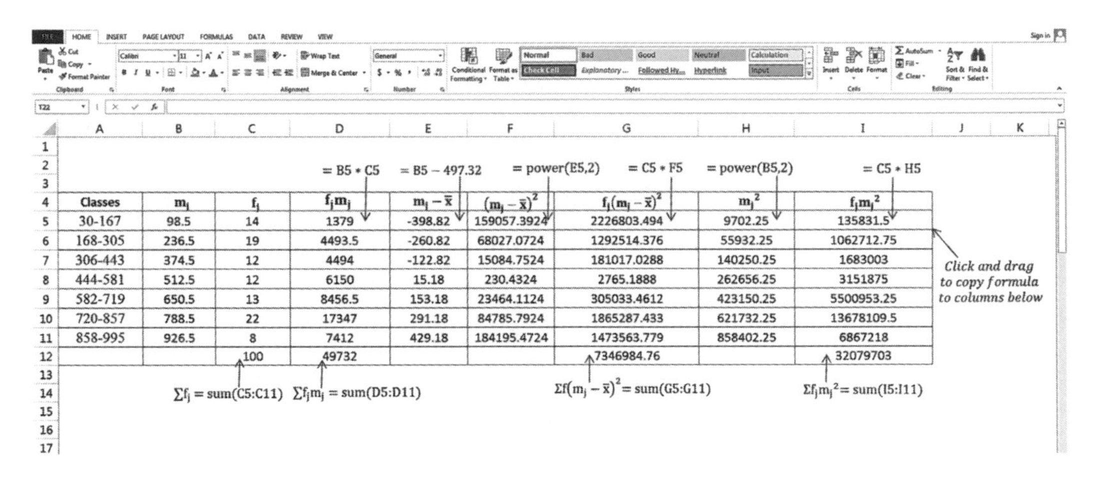

	= B5 * C5	= B5 − 497.32	= power(E5,2)	= C5 * F5	= power(B5,2)	= C5 * H5

Classes	m_j	f_j	$f_j m_j$	$m_j - \bar{x}$	$(m_j - \bar{x})^2$	$f_j(m_j - \bar{x})^2$	m_j^2	$f_j m_j^2$
30-167	98.5	14	1379	-398.82	159057.3924	2226803.494	9702.25	135831.5
168-305	236.5	19	4493.5	-260.82	68027.0724	1292514.376	55932.25	1062712.75
306-443	374.5	12	4494	-122.82	15084.7524	181017.0288	140250.25	1683003
444-581	512.5	12	6150	15.18	230.4324	2765.1888	262656.25	3151875
582-719	650.5	13	8456.5	153.18	23464.1124	305033.4612	423150.25	5500953.25
720-857	788.5	22	17347	291.18	84785.7924	1865287.433	621732.25	13678109.5
858-995	926.5	8	7412	429.18	184195.4724	1473563.779	858402.25	6867218
		100	49732			7346984.76		32079703

We first find

$$\overline{X} = \frac{\sum f_j m_j}{\sum f_j} = \frac{49,732}{100} = 497.32.$$

Then,

Variance.

From the long formula

$$s^2 = \frac{\sum f_j(m_j - \overline{X})^2}{n-1} = \frac{7,346,984.76}{100-1} = \frac{7,346,984.76}{99} = 74,211.97,$$

and from the short formula

$$s^2 = \frac{\sum f_j m_j^2 - \left[\left(\sum_{j=1}^{k} f_j m_j\right)^2 / n\right]}{n-1} = \frac{32,079,703 - \left(49,732^2/100\right)}{100-1}$$

$$= \frac{32,079,703 - 24,732,718.24}{99} = \frac{7,346,984.76}{99} = 74,211.97.$$

Standard deviation

From the long formula

$$s = \sqrt{\frac{\sum f_j(m_j - \overline{X})^2}{n-1}} = \sqrt{\frac{7,346,984.76}{100-1}} = \sqrt{74,211.97} = 272.42,$$

and from the short formula

$$s=\sqrt{\frac{\sum f_j m_j-\left[\left(\sum_{j=1}^{k} f_j m_j\right)^2/n\right]}{n-1}}=\sqrt{\frac{32,079,703-\left(49,732^2/100\right)}{100-1}}=\sqrt{\frac{7,346,984.76}{99}}$$

$$=\sqrt{74,211.97}=272.42.$$

■

EXAMPLE 4.35

Find the variance and standard deviation for the grouped data pertaining to the level of employment in manufacturing establishments in all the 50 states of the United States (Table 4.6a) by using the long and short formulas.

Detailed calculations:

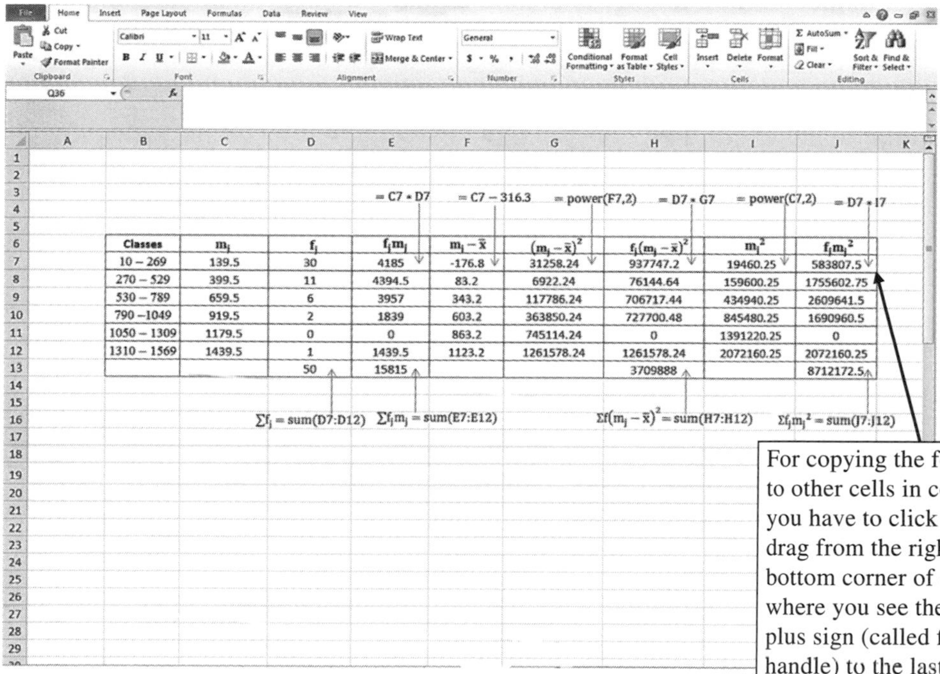

Formulas (row 3): E7 = C7 * D7; F7 = C7 − 316.3; G7 = power(F7,2); H7 = D7 * G7; I7 = power(C7,2); J7 = D7 * I7

Classes	m_j	f_j	$f_j m_j$	$m_j-\bar{x}$	$(m_j-\bar{x})^2$	$f_j(m_j-\bar{x})^2$	m_j^2	$f_j m_j^2$
10 – 269	139.5	30	4185	-176.8	31258.24	937747.2	19460.25	583807.5
270 – 529	399.5	11	4394.5	83.2	6922.24	76144.64	159600.25	1755602.75
530 – 789	659.5	6	3957	343.2	117786.24	706717.44	434940.25	2609641.5
790 –1049	919.5	2	1839	603.2	363850.24	727700.48	845480.25	1690960.5
1050 – 1309	1179.5	0	0	863.2	745114.24	0	1391220.25	0
1310 – 1569	1439.5	1	1439.5	1123.2	1261578.24	1261578.24	2072160.25	2072160.25
		50	15815			3709888		8712172.5

$\sum f_j$ = sum(D7:D12) $\sum f_j m_j$ = sum(E7:E12) $\sum f(m_j-\bar{x})^2$ = sum(H7:H12) $\sum f_j m_j^2$ = sum(J7:J12)

For copying the formula to other cells in column J, you have to click and drag from the right bottom corner of cell J7, where you see the black plus sign (called fill handle) to the last value, which is J12.

The following are the long and short formulae used to calculate the variance and the standard deviation. Given that the mean is

$$\bar{X}=\frac{\sum_{j=1}^{k} f_j m_j}{\sum_{j=1}^{k} f_j}=\frac{15,815}{50}=316.3,$$

we can next find
Variance
From the long formula

$$s^2 = \frac{\sum_{j=1}^{k} f_j(m_j - \overline{X})^2}{n-1} = \frac{3,709,888}{50-1} = \frac{3,709,888}{49} = 75,712,$$

and from the short formula

$$s^2 = \frac{\sum_{j=1}^{k} f_j m_j^2 - \left[\left(\sum_{j=1}^{k} f_j m_j\right)^2/n\right]}{n-1} = \frac{8,712,173 - (15,815^2/50)}{50-1}$$
$$= \frac{8,712,173 - 5,002,284.5}{49} = \frac{3,709,888.5}{49} = 75,712.$$

Standard deviation:
From the long formula

$$s = \sqrt{\frac{\sum_{j=1}^{k} f_j(m_j - \overline{X})^2}{n-1}} = \sqrt{\frac{3,709,888}{50-1}} = \sqrt{75,712} = 275.16,$$

and from the short formula

$$s = \sqrt{\frac{\sum_{j=1}^{k} f_j m_j^2 - \left[\left(\sum_{j=1}^{k} f_j m_j\right)^2/n\right]}{n-1}} = \sqrt{\frac{8,712,173 - (15,815^2/50)}{50-1}} = \sqrt{\frac{3,709,888.5}{49}}$$
$$= \sqrt{75,712} = 275.16.$$

■

CHAPTER 4 REVIEW

You should be able to:

1. Distinguish between descriptive and inferential statistics.
2. Distinguish between a sample and a population,
3. Give an example of a "statistic."
4. Distinguish between quantitative and qualitative data as well as between discrete and continuous data.
5. Differentiate between the various types or scales of data.
6. Distinguish between absolute and relative frequency distributions; and between histograms and frequency polygons.
7. Distinguish between a class mark and a class boundary.
8. Distinguish between a parameter and a statistic.

9. Discuss the distinction between a measure of location and a measure of dispersion.

10. Distinguish between a simple arithmetic mean, a weighted mean, and a geometric mean.

11. Name the three measures of central location and explain their usage.

12. Distinguish between quartiles, deciles, and percentiles.

13. Differentiate between distributions that are symmetrical and those that are skewed.

14. Distinguish between a measure of absolute variation and a measure of relative variation.

15. Calculate and interpret a Z-score.

16. Discuss the role of Chebysheff's theorem. How does it relate to the Empirical Rule?

Key Terms and Concepts:

Chebysheff's theorem, 164
class boundary, 136
class frequency, 136
class interval, 135
coefficient of skewness, 167
coefficient of variation, 160
continuous data, 132
cumulative frequency distributions, 139
cumulative relative frequencies, 139
decile, 152
descriptive statistics, 155
discrete data, 132
Empirical Rule, 165
frequency distribution, 136
frequency histogram, 137
frequency polygon, 137
geometric mean, 149-150
inferential statistics, 131
interval scale, 133
lower class limit, 135
median, 151
midpoint, 137
mode, 153
nominal scale, 132
normal distribution, 154
ordinal scale, 133
parameter, 144
percentile, 152
population, 132
population mean, 144
population standard deviation, 158
population variance, 155
qualitative data, 132
quantitative data, 132
quartile, 152
range, 155
ratio scale, 133
relative frequency histogram, 137
sample, 132
sample mean, 145
sample standard deviation, 158
sample variance, 157
skewed left, 154
skewed right, 154
statistics, 132
symmetrical, 167
upper class limit, 135
weighted mean, 147
Z-score, 163

EXERCISES

1. The following data set represents merchandise trade as a percentage of GDP of 100 countries in the year 2014. *Data file:* Merchandise.xlsx.

40	57	25	54	33	51	102	42	101	173
75	66	92	20	81	115	50	30	41	145
34	53	55	58	42	31	49	67	37	159
61	43	53	33	56	45	70	57	71	70
42	44	50	53	32	90	51	99	157	61
38	40	34	67	77	47	47	33	87	56
40	78	100	56	80	97	51	102	136	77
50	63	96	92	61	65	102	91	44	145
103	46	28	46	29	74	74	40	45	79
62	81	43	69	61	49	41	111	57	40

Source: http://data.worldbank.org/2015.

(Merchandise trade as a share of GDP is the sum of merchandise exports and imports divided by the value of GDP, all in current U.S. dollars.)

For the above data, obtain the following:

a. Frequency distribution

b. Relative and cumulative frequency distributions

c. Histograms

d. Frequency polygon

e. Ogive curve

2. The following data set describes the number of FDIC insured financial institutions in the 50 states of the United States in the year 2006. *Data file:* FDIC.xlsx.

144	54	9	326	165	73	53	248	83	115
6	27	18	198	136	224	186	34	191	79
40	6	607	156	404	29	100	216	615	65
130	247	146	29	91	24	92	14	57	276
272	268	360	87	336	117	239	83	14	37

Source: Statistical abstract of the United States, 2006.

For the above data, obtain the following:

a. Frequency distribution

b. Relative and cumulative frequency distributions

c. Histograms

d. Frequency polygon

e. Ogive curve

3. The following data set represents new home mortgage yields in the United States from 1963 to 2014.
Data file: Mrtgyld.xlsx.

5.89	5.83	5.81	6.25	6.46	6.97	7.81	8.45
7.74	7.60	7.96	8.92	9.0	9.0	9.02	9.56
10.78	12.66	14.70	15.14	12.57	12.38	11.55	10.17
9.31	9.19	10.13	10.05	9.32	8.24	7.20	7.49
7.87	7.8	7.71	7.07	7.04	7.52	7.0	6.43
5.80	5.77	5.94	6.63	6.41	6.05	5.14	4.80
4.56	3.69	4.00	4.22				

Source: Economic Report of the President, 2015, Table B-17.

For the above data, obtain the following:

a. Frequency distribution

b. Relative and cumulative frequency distributions

c. Histograms

d. Frequency polygon

e. Ogive curve

4. An investor wants to start a small business with a total capital of $10 million. This investor is able to raise an equity (stock) capital of $6 million at an estimated cost of 10%. The remaining balance of $4 million is obtained as debt (bond) at a 5% rate. Find the cost of capital (average cost of total capital).

5. An undergraduate student needs a total financial commitment of $45,000 for 1 year of his college education. The student received the following financial aid package. A Pell grant of $5000 at a 3% rate, a federal loan of $18,000 at a 7% rate, family contribution of $15,000 at the rate of 11%, a grant (0% interest rate) of $4000, and the balance of $3000 is financed from individual earnings (0% interest). Find the average cost of obtaining this aid package of $45,000 for the year, ignoring opportunity cost.

6. Find the average stock price of 100 small companies from the frequency distribution given below:

Stock Price ($)	Frequency
5–24	28
24–43	40
43–62	23
62–81	5
81–100	3
100–119	0
119–138	1

7. Find the average growth in GDP over a period of 3 years if the GDP grew at the rate of 4% in the first year, 3% in the second year, and 5% in the third year.

8. Between 2005 and 2014, the annual average U.S. inflation rates, as reported in the Bureau of Labor Statistics (BLS), are given below. *Data file:* Inflation.xlsx.

Year	2005	2006	2007	2008	2009	2010	2011	2012	2013	2014
Inflation	3.4	3.2	2.8	3.8	−0.4	1.6	3.2	2.1	1.5	1.6

Find the average rate of inflation from 2005 to 2014.

During the 1970s, the annual average U.S. inflation rates, as reported in the Business Statistics (BLS), are given below. *Data file:* Inflation.xlsx.

Year	1974	1975	1976	1977	1978	1979	1980	1981	1982	1983
Inflation	11.0	9.1	5.8	6.5	7.6	11.3	13.5	10.3	6.2	3.2

Find the average rate of inflation during the 1970s.

9. It is reported that the total medical care cost in the United States over the period of 10 years from 1996 to 2005 grew at 42%. Find the annual average growth rate (medical care cost inflation) during this 10-year period.

Hint: Consider $(1 + R)^{10} = 1.42$ and find R.

10. The following data set refers to the annual cost ($) of attending 4-year private business schools, as reported by the forbes.com 2012 list of best business schools:

44,766	43,380	45,250	43,935	42,990	43,436	42,860	42,500	41,950
42,000	41,822	41,670	39,600					

Find the 30th percentile and second quartile. What is the median value?

11. Suppose that the mean GPA of a sample of 280 MBA students is 3.57, with a standard deviation of 0.23. Find the proportion of students with GPAs ranging from 3.30 to 3.84.

12. In a survey to determine the factors contributing to the value added of having an MBA degree, it is reported that the mean and the standard deviation of the "competency" variable are, respectively, 4.96 and 0.34. Find the following:

a. The proportion of students that fell within the competence range of 4.32–5.60.

b. A range of the "competency" variable within which 70% of the values are guaranteed to fall.

c. Assuming the distribution of the "competency" variable is mound-shaped, find the proportion of students that lie within a "competence" range of 4.32–5.60. Comment on the results you obtained here, and in item (a) above.

13. The Bureau of Labor Statistics reported that the mean price per gallon of unleaded regular gas for the year 2011 was $3.53 per gallon, with a standard deviation of $0.25 per gallon. Assuming that the distribution of gas prices is mound-shaped, find a gas price per gallon range centered around the mean within which 95% of the prices would fall.

14. The hourly earnings of workers in nine geographical areas of the United States in July 2010 are as follows. Compute the mean for this sample.

Geographical Areas	Hourly Earnings ($)
New England	24.55
Middle Atlantic	24.28
East North Central	20.51
West North Central	17.09
South Atlantic	19.97
East South Central	14.66
West South Central	16.36
Mountain	20.71
Pacific	24.11

Source: Bureau of Labor Statistics.

15. A sample of P/E (price/earnings) ratios of 10 international companies appears below.

15,	7,	26,	16,	11,	9,	5,	14,	11,	28

Find the 85th percentile and third quartile for this data set.

16. The accompanying sample data set refers to the annual cost ($) of attending a 4-year private business school in 2006:

32,364	31,540	32,300	39,794	32,540	31,656	32,042	31,650	31,180	32,008	30,950	32,140	34,850

Find the 30th percentile and second quartile. What is the median value?

17. The following are the temperature forecasts, high/low, for 10 days from June 5 to June 15, 2012, in the greater Hartford area of Connecticut, as reported on weather.com.

63/52	72/53	75/56	76/59	80/64	85/65	85/67	88/67	83/63	80/59

Find the modes of the high temperature values as well as of the low temperature values of the forecasts.

18. Suppose the GMAT scores for the first time takers in the United States have a mean of 529, with a standard deviation of 120. (a) Find Z-scores corresponding to the GMAT scores of 349 and 709 obtained for two randomly selected students. (b) Find the proportion (or %) of students who scored between these two test scores.

EXCEL APPLICATIONS

19. The accompanying sample data set shows the external debt of countries in the year 2014 in millions of dollars. *Data file:* External debt.xlsx.

Find the median debt of the above countries and interpret its value.

20. The following data set shows the list prices ($) of single-family luxury homes in the greater Hartford area in year 2012. Find the median list price ($) of these luxuries homes and interpret its value. *Data file:* Luxury homes.xlsx.

21. Find a measure of the degree of skewness for the employment data of Table 4.5. Comment on the nature of skewness. *Data file:* Employment2012.xlsx.

22. Find a measure of the degree of skewness for the data on stock prices of the 100 best small companies. Comment on the shape of the data set. *Data file:* BSCSP2007.xlsx.

23. Using Excel, find the 40th percentile, median, and third quartile of the following data on average SAT scores of students in undergraduate B-schools, as reported by the May 8, 2006 issue of *Business Week* magazine. Interpret the values obtained. *Data file:* SAT scores.xlsx.

24. Find the modes for both male and female unemployment data given in the data set. *Data file:* UnemploymentMF2013.xlsx.

25. Compare the relative variability of exports and imports of the United States from 1986 to 2011 as shown in the table below. *Data file:* Exports and Imports.xlsx.

Year	Exports	Imports
1986	320.3	452.9
1987	363.8	508.7
1988	443.9	554
⋮	⋮	⋮
2009	1583.00	1974.60
2010	1839.00	2356.70
2011	2087.60	2665.80

26. Compare the variability in the weekly earnings (in $) of males and females as shown in the table below. *Data file:* Male Female earnings.xlsx.

Year	Men	Women
1979	291	182
1980	312	201
1981	339	219
⋮	⋮	⋮
2009	819	657
2010	824	669
2011	832	684

Source: Current Population Survey, Bureau of Labor Statistics and Bureau of Census. Dollars, not seasonally adjusted.

27. Use Excel (Data Analysis tool) to answer the following questions for the data on the unemployment rate for both males and females. *Data file:* UnemploymentMF2013.xlsx.

a. Find the mean, median, and the mode for the data set.

b. Based on the values of the mean, median, and mode, can you comment on the shape of the distribution? More specifically, does it look like bell-shaped, skewed to the right, or skewed to the left?

c. Find the variance, standard deviation, and the coefficient of variation for the given data set.

28. Find the mean and standard deviation for the unemployment rate data for both males and females using the "Function" key of Excel. *Data file:* UnemploymentMF2013.xlsx.

Chapter 5

Probability and Applications

5.1 INTRODUCTION

In business situations, more often than not, the outcomes of actions are not known with certainty. Probability calculations let you estimate the chance or likelihood of occurrence of these outcomes. In extreme cases, the probability of a certain event is 1 or 100%, and 0 if the event does not exist or is impossible. But, in general, the probability of any outcome falls between 0 and 1. In this regard, the more likely an event happens to be, the closer is its probability to 1. The following are the situations where the concept of probability or likelihood is essential:

i. The probability of college tuition increasing by more than 3% next year is important for a student in making a decision to secure the necessary finances.

ii. Most businesses would like to know the probability or chance of facing higher inflation and higher interest rates in order to make suitable business plans.

iii. Any local government is interested in finding the probability of a heavy snowfall in the upcoming winter so that it can plan accordingly in order to put forward a suitable budget.

These few examples illustrate the need to evaluate the probability or likelihood of uncertain events. In addition, probability is central to the concepts of sampling distributions and statistical inference introduced in more advanced courses in statistics. Finally, probability is the foundation for prediction and forecasting in the world of uncertain events.

In this chapter, we first introduce some basic concepts that are essential to the understanding of the concept of probability. In addition, various classifications of probability and the different approaches of determining probability are also provided.

Introduction to Quantitative Methods in Business: With Applications Using Microsoft® Office Excel®, First Edition. Bharat Kolluri, Michael J. Panik, and Rao Singamsetti.

Companion website: www.wiley.com/go/Kolluri/QuantitativeMethods

5.2 SOME USEFUL DEFINITIONS

Random Experiment: A *random experiment* is a process that has an unknown outcome or outcomes that are known only after the process is completed. For example, recording the daily production of a manufacturing plant, predicting a town's tax revenue for the next fiscal year, the findings of a physical examination, rolling a die and observing the score on the face that appears, and finding whether a newborn baby is a boy or a girl are all random experiments.

Event: The outcome of a random experiment is an *event*.

Sample Point: The graphical representation of an event is a *sample point* (or collection of sample points), and the set of all sample points forms the *sample space* (*S*).

Figures 5.1 and 5.2 depict two common sample spaces.

An event may be simple or compound. That is, an event that cannot be subdivided further is known as a *simple event*. The graphical version of a simple event is shown above as a single sample point (e.g., point H in Figure 5.1a). A combination of more than one simple event forms a *compound event* (event C in Figure 5.2).

EXAMPLE 5.1

Define simple and some compound events associated with the rolling of a single six-sided die (Figure 5.1b).

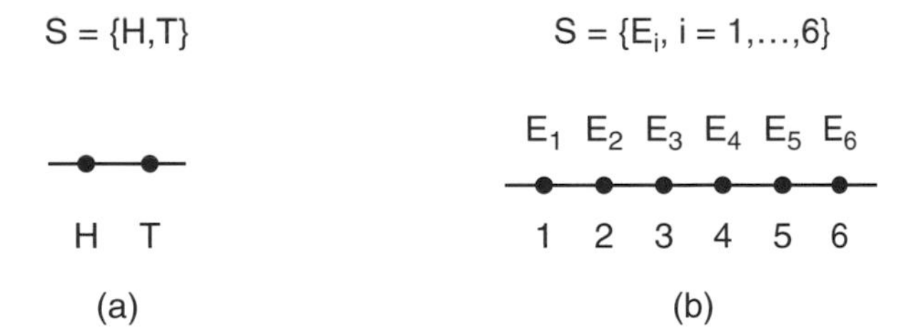

Figure 5.1 (a) Sample space related to flipping a coin. (b) Sample space related to rolling a six-sided die.

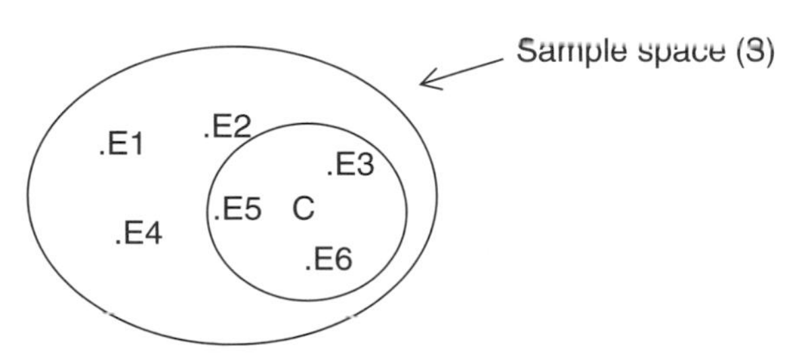

Figure 5.2 A compound event is a collection of simple events.

SOLUTION: Any single outcome point E_i for i = 1, 2, 3, 4, 5, 6 is a simple event. Define

A: observe an odd number: E_1, E_3, E_5

B: observe a number less than 4: E_1, E_2, E_3

Clearly, A and B are compound events. ■

5.3 PROBABILITY SOURCES

5.3.1 Objective Probability

If in a random experiment there are N_A outcomes favorable to event A out of N possible equally likely simple events, then the *probability of event A*, denoted by $P(A) = \frac{N_A}{N}$, is known as *objective or classical probability*. For example, if you toss a fair coin once, the probability of heads showing up is 1/2, where 1 refers to the number of heads and 2 refers to the total number of possible simple events within the same space. Note that heads and tails are equally likely to occur on any single flip of a fair coin.

Another type of objective probability is known as the *relative frequency* interpretation of probability. If a random experiment is repeated a large number of times and the outcomes favorable to event A are observed n_A times, then $P(A) = \frac{n_A}{n}$, where n is the number of times the experiment is repeated. For example, the probability of obtaining heads when tossing a coin once is 1/2 according to the classical definition of probability. However, if we toss a coin n = 10 times and observe heads $n_A = 7$ times, then the relative frequency of heads showing is $n_A/n = 7/10$. Note that the number of times a coin was tossed here is not very large. So, why is there a divergence between the classical probability of 1/2 and the relative frequency value, which is 7/10? The answer is that if this experiment of tossing the coin is done a very large number of times, then the frequency ratio n_A/n will tend to the limit 1/2, which is the classical probability value.

5.3.2 Subjective Probability

This is the probability we assign to the outcome of an experiment based on experience, when the experiment cannot be repeated a large number of times. For example, the probability of landing on the moon, the probability of a 9/11-type event, the probability of a market crash like the one in 1929, or the probability of a tsunami.

For all practical purposes, mostly in business situations, we use the relative frequency definition of probability. Note that, by definition, probability is always a number between 0 and 1, both inclusive. That is, for some event A,

$$0 \leq P(A) \leq 1.$$

Thus, the probability of the sample space is 1: P (S) = 1. Here S can be viewed as the *certain event* or the event that will definitely happen.

Some other examples of probability are the following:

- Probability of a sunny day in Arizona State during the month of June is 1 or 100%.
- Probability of the sun rising in the east is 1.
- Probability of a sunny day in Alaska during the month of January is 0.
- Probability of an *impossible event* (depicted as the null or empty set ϕ) is 0: $P(\phi) = 0$.
- Probability of flying in a commercial passenger plane to Moscow, Russia, from JFK, New York, in an hour is 0.
- Probability of obtaining tails on the toss of a fair coin is 1/2 or 50%.

EXAMPLE 5.2

Let us verify that the probability of tossing a coin twice (or tossing two coins simultaneously) and observing two heads is 1/4.

SOLUTION: This answer emerges because there are four possible outcomes in the sample space: {HH, HT, TH, TT}. Here H refers to heads and T refers to tails. The first letter in each of the four pairs is the outcome obtained on the first toss, and the second letter is the outcome resulting on the second toss. Therefore, the probability of observing two heads equals 1/4 using the classical probability rule. Can you draw the sample space S? ■

EXAMPLE 5.3

Find the probability of rolling an odd score if you roll a single six-sided die once.

SOLUTION: There are three favorable outcomes of an odd score when you roll the die once. These are E_1, E_3, and E_5 out of the six possible outcomes. Therefore, the probability of observing an odd score when you roll the die once is $3/6 = 1/2$. ■

EXAMPLE 5.4

In a deck of 52 playing cards, find the probability of getting (a) a spade and (b) a king.

SOLUTION: There are 13 spades in a deck of 52 playing cards. Hence, the probability of getting a spade is $13/52 = 1/4$. Similarly, there are four kings in the deck of 52 cards.

So, the probability of getting a king is $4/52 = 1/13$. ■

5.4 SOME USEFUL DEFINITIONS INVOLVING SETS OF EVENTS IN THE SAMPLE SPACE

These set concepts from Chapter 1 are presented here again in the context of probability for the reader's convenience. To this end, let A and B be events in the sample space S.

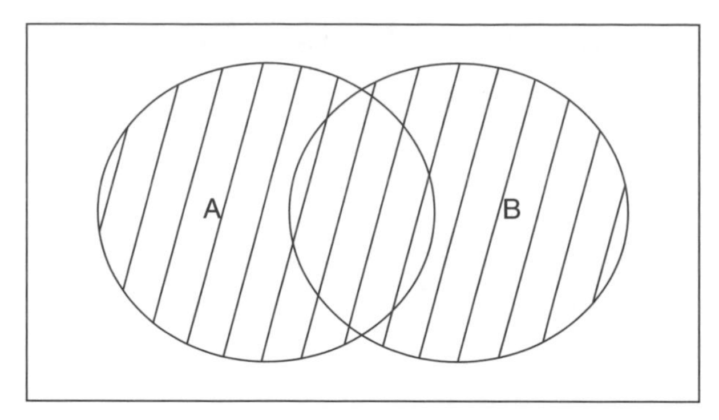

Figure 5.3 Venn diagram related to the union of sets A and B.

The *union* of A and B, denoted by A∪B, is the set of all events that are in at least one of A, B (either in A or in B or in both). This relationship between sets A and B can be illustrated by a Venn diagram (Figure 5.3). The graphical presentation of the entire sample space is simply a rectangle.

The *intersection* of A and B, denoted by A∩B (also written as AB), is the set of all events that are common to A and B (events that are in both A and B simultaneously) and is shown in Figure 5.4.

EXAMPLE 5.5

In the case of rolling a six-sided die, let A stand for an even-numbered score (E_2, E_4, E_6); and let B stand for a number greater than 3 (E_4, E_5, E_6). Find A∪B and A∩B.

SOLUTION: $A \cup B = \{E_2, E_4, E_5, E_6\}$. This is shown in Figure 5.5.
$A \cap B = \{E_4, E_6\}$.This is shown in Figure 5.6. ■

EXAMPLE 5.6

In the MBA program at a Business School, 60 students signed up to take the managerial statistics course, 70 students signed up to take the managerial economics course, and 40 students signed up for both courses. Find the total number of students that signed up for at least one of the two courses.

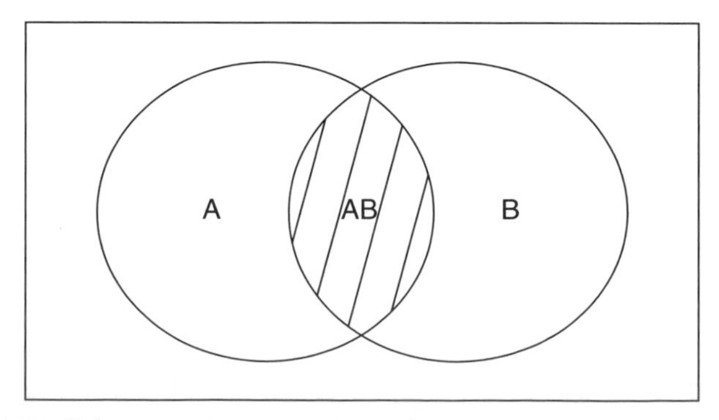

Figure 5.4 Venn diagram related to the intersection of sets A and B.

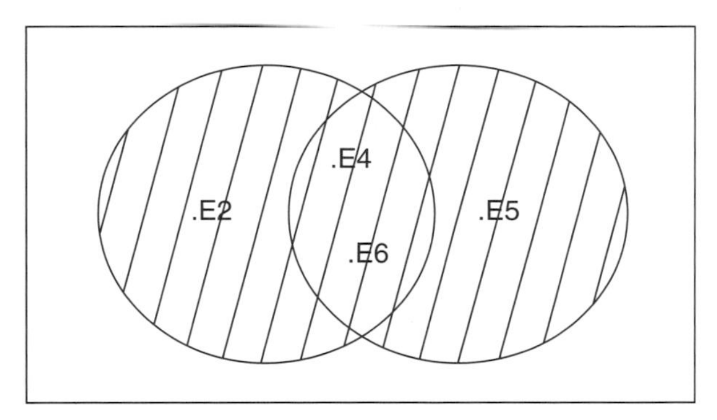

Figure 5.5 Venn diagram related to the union of sets A and B.

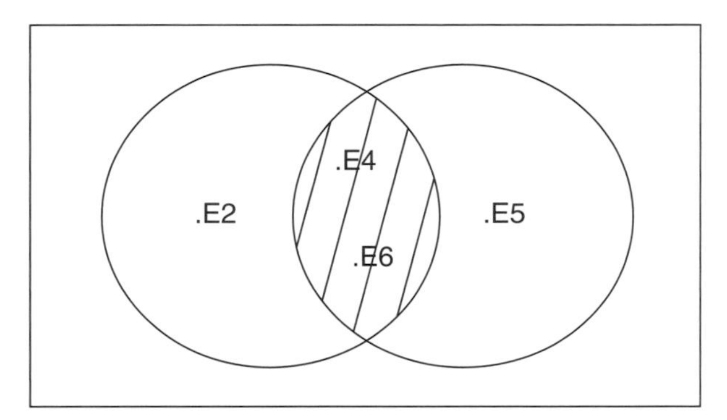

Figure 5.6 Venn diagram related to the intersection of sets A and B.

SOLUTION: Let A stand for students who signed up for managerial statistics and let B stand for students who signed up for the managerial economics course. Then $A \cup B = 60 + 70 - 40 = 90$. Therefore, 90 students signed up for either the managerial statistics or managerial economics courses or both. Why did we subtract 40? To avoid counting the intersection twice! ■

5.4.1 Complement of a Given Set A

The *complement of A* consists of all points in the sample space that are not in set A. For example, in the coin tossing experiment, the event "tail" is the complement of the event "head." Likewise, in the experiment involving the rolling of a die, all the even scores form the complement set of the set of all odd scores. The complement of a set A is usually denoted by $\bar{A}$ (or ~A) (Figure 5.7). Note that $A \cup \bar{A} = S$, $\bar{S} = \emptyset$, $\bar{\emptyset} = S$, and $A\bar{A} = \emptyset$.

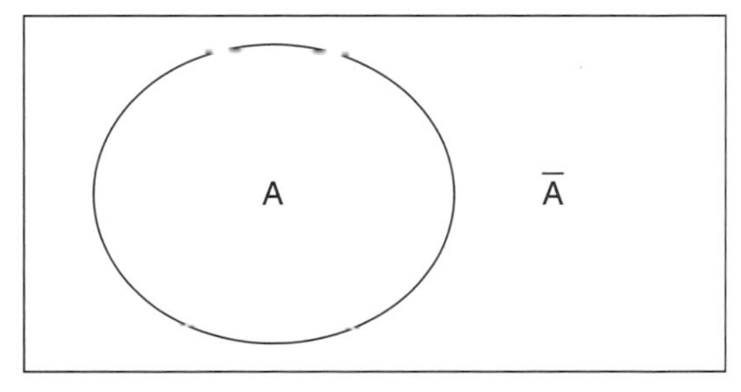

Figure 5.7 Venn diagram describing set A and its complement $\bar{A}$.

5.4.2 Mutually Exclusive Events

The events A and B are called *mutually exclusive* if the occurrence of A precludes the occurrence of B. For example, heads and tails are mutually exclusive events when we toss a coin once. Likewise, odd scores and even scores are mutually exclusive when rolling a die once.

Note that two complementary events are always mutually exclusive, but the converse is not always true. For example, in a deck of 52 cards, the clubs and spades are always mutually exclusive, but they are not complementary, since they do not collectively exhaust the whole deck of 52 cards.

5.5 PROBABILITY LAWS

5.5.1 General Rule of Addition

The *general addition rule* is useful in applications when we are interested in finding the likelihood or the chance of occurrence of at least one of two events. Here the phrases "at least one of the two" or "one of the two" refer to the "union" of both the events.

If A and B are two sets of events, generally,

$$P(A \cup B) = P(A) + P(B) - P(A \cap B),$$

where P(A) and P(B) are known (simple) probabilities. More particularly, these are also known as *marginal probabilities* because each one considers exclusively an individual set. $P(A \cap B)$ or $P(B \cap A)$ are known as *joint probabilities*, which are identical. These are also written as P(AB) or P(BA) without the "$\cap$" symbol. They represent the probability of the common part of sets A and B.

Here, when we add P(A) and P(B), the common part of A and B is counted twice, once as part of A and a second time as part of B. So, the common part is subtracted once to avoid double counting.

EXAMPLE 5.7

In an undergraduate business program at a certain college, there are a total of 200 sophomore students. From this group, a total of 120 take the business statistics course, 90 take the first accounting course, and 60 students take both courses. Find the probability that a randomly selected student takes at least one of the two courses. Interpret the result.

SOLUTION: Let A stand for the group of students taking the business statistics course and B stand for the students taking the first accounting course. Then, using the general addition rule of probability,

$$P(A \cup B) = P(A) + P(B) - P(A \cap B)$$

or

$$P(A \cup B) = \frac{120}{200} + \frac{90}{200} - \frac{60}{200} = \frac{150}{200} = \frac{3}{4} = 0.75 \text{ or } 75\%.$$

Interpretation: Suppose we randomly selected (without replacement) 100 students from the total pool of sophomores. Then we would find that 75% of them take at least one of the two aforementioned courses. ■

EXAMPLE 5.8

Suppose 40% of the customers of a financial company invest in stocks, 30% invest in bonds, and 20% invest in both stocks and bonds. What is the probability of a randomly chosen customer investing in either stocks or bonds or both?

SOLUTION: Let A stand for customers investing in stocks and let B stand for customers investing in bonds. Then, using the addition rule,

$$P(A \cup B) = P(A) + P(B) - P(A \cap B) = 40\% + 30\% - 20\% = 70\% - 20\% = 50\%.$$

Therefore, 50% of the customers of the company invest in either stocks or bonds or both.

We next look to the *special addition rule*.

If A and B are two mutually exclusive sets (i.e., $A \cap B = \phi$), then

$$P(A \cup B) = P(A) + P(B).$$

This becomes a special case of the general addition rule since $P(A \cap B) = 0$ when A and B are mutually exclusive. ■

EXAMPLE 5.9

Find the probability of drawing a card from a deck of 52 cards and finding that it is either an ace of hearts or an ace of diamonds.

SOLUTION: Let A denote the outcome of drawing an ace of hearts and B the outcome of drawing an ace of diamonds. Using the special rule of addition,

$$P(A \cup B) = P(A) + P(B),$$

where

$$P(A) = \frac{1}{52}, \quad P(B) = \frac{1}{52}.$$

Then,

$$P(A \cup B) = \frac{1}{52} + \frac{1}{52} = \frac{1}{26} = 0.03846 \text{ or } 3.846\%.$$

Note that since A and B are mutually exclusive, the intersection set $A \cap B$ is the null set ϕ, and thus $P(\phi) = 0$. ■

EXAMPLE 5.10

In one of the 50 states in the United States, it is estimated that 30% are registered Republicans and 40% are registered Democrats. For this state, find the probability of selecting either a registered Republican or a registered Democrat from the population of both registered Republicans and Democrats.

SOLUTION: Let R stand for registered Republican and D for registered Democrat. Here, R and D are mutually exclusive and, therefore, by applying the special addition rule, we obtain

$$P(R \cup D) = P(R) + P(D) = 0.3 + 0.4 = 0.7 \text{ or } 70\%.$$

■

5.5.2 Rule of Complements

If two sets A and B are complements to each other, then

$$P(A) + P(B) = 1 \text{ or } P(A) = 1 - P(B).$$

This is because A and B constitute the whole sample space, S, and $P(S) = 1$.

This rule is written more commonly as $P(A) = 1 - P(\overline{A})$, where $\overline{A}$ is the complement of A. It is known as the *rule of complements*. This is because A and $\overline{A}$ together constitute the whole sample space, S, and $P(S) = 1$. From the preceding equality, we may write $P(\overline{A}) = 1 - P(A)$, that is, the probability that event A does not occur ($\overline{A}$) is 1 minus the probability that it does occur.

EXAMPLE 5.11

In Example 5.7, which introduced data pertaining to business statistics and accounting courses, what is the probability of a sophomore student taking neither of the two courses?

SOLUTION: Using the same notation as in Example 5.7 above, $P(\text{None}) = 1 - P(A \cup B) = 1 - 0.75 = 0.25$ or 25%. ■

5.5.3 Conditional Probability

The probability of a particular event A occurring given that some other event B has definitely occurred is written as $P(A|B)$, and is known as the *conditional probability* of A given B. Here "|" is read "given." This concept is often useful when there is the need to consider only a segment of the population, for example, products that are designed only for college students, male/female students, and retirees, among others.

In general, $P(A|B) = P(A \cap B)/P(B)$, provided $P(B)$ is different from 0.

Similarly, $P(B|A) = P(A \cap B)/P(A)$, provided $P(A)$ is different from 0.

Some examples are as follows:

The probability of randomly selecting a queen from a deck of cards, given that only diamonds and hearts are considered, is $(2/52)/(26/52) = 2/26 = 1/13$.

The probability of rolling a die once to get a score of 5, given that only odd scores are considered, is $(1/6)/(3/6) = 1/3$.

At a particular university campus, there are a total of 500 undergraduate business students among whom 200 are female and the rest are male. Out of these 200 female students, 30 participate actively in Division 1 sports in some form or another. Given this information, the probability of randomly selecting a Division 1 female sports participant given that a female student is chosen is (30/500)/(200/500) = 30/200 or 3/20.

5.5.4 General Rule of Multiplication (Product Rule)

At times it is necessary to find joint probabilities in business applications and elsewhere. This need arises when we are interested in finding the probability that two particular events, such as A and B, occur together. Remember that if A and B occur jointly, then $A \cap B \neq \phi$, that is, A and B occur together and thus have elements in common. This is like a professional tennis player competing in both singles and doubles in major tournaments.

From the conditional probabilities described above, we can readily obtain the joint probability from the *general multiplication rule* as

$$P(A \cap B) = P(A \mid B) \cdot P(B) = P(B \mid A) \cdot P(A).$$

EXAMPLE 5.12

In a deck of 52 playing cards, find the probability of randomly selecting two kings, one after the other, without replacing the first card drawn.

SOLUTION: Let K_1 denote "obtaining a king in the first draw," and let K_2 represent "getting a king on the second draw." Then,

$$P(K_1 \cap K_2) = P(K_1) \cdot P(K_2 \mid K_1) = \frac{4}{52} \times \frac{3}{51} = \frac{12}{2652} = 0.0045 = 0.45\%.$$

■

EXAMPLE 5.13

In a class consisting of 14 female and 16 male students, find the probability of

a. selecting two female students, one after the other, without replacement;

b. selecting two male students in a row (again without replacement);

c. a female student first followed by a male student (without replacement).

SOLUTION: Let F_1 stand for "female student in the first selection" and let F_2 depict a "female student in the second selection." Similarly, let M_1 and M_2 stand for "male students obtained on the first and second selections," respectively. Then,

a. $P(F_1 \cap F_2) = P(F_1) \cdot P(F_2 \mid F_1) = \frac{14}{30} \times \frac{13}{29} = \frac{182}{870} = 0.209 = 20.9\%;$

b. $P(M_1 \cap M_2) = P(M_1) \cdot P(M_2 \mid M_1) = \frac{16}{30} \times \frac{15}{29} = \frac{240}{870} = 0.276 = 27.6\%;$

c. $P(F_1 \cap M_2) = P(F_1) \cdot P(M_2 \mid F_1) = \frac{14}{30} \times \frac{16}{29} = \frac{224}{870} = 0.257 = 25.7\%.$

■

5.5.5 Independent Events

Two events A and B are said to be *independent* if and only if $P(A|B) = P(A)$ or $P(B|A) = P(B)$. Thus, the probability of occurrence of one of the two events is unaffected by the occurrence of the other event. For example, the probability of tossing a coin twice and observing heads on the second toss is independent of the outcome obtained of the first toss. In this instance, the general multiplication rule assumes the following special form. If A and B are independent events, then their joint probability can be calculated from the *special multiplication rule* as

$$P(A \cap B) = P(A) \cdot P(B).$$

Hence, in this special case, the joint probability becomes the product of the individual or marginal probabilities.

What is the distinction between events (A and B) that are mutually exclusive and events that are independent? Mutually exclusive events cannot occur together, that is, $A \cap B$ cannot occur or $A \cap B = \phi$ and thus, $P(A \cap B) = 0$. Independent events can occur together: it is just that the occurrence of one does not affect the probability of occurrence of the other, that is, $A \cap B$ can occur with $P(A \cap B) = P(A) \cdot P(B)$.

EXAMPLE 5.14

Find the probability of drawing two kings in sequence, assuming that the first card selected is replaced into the deck before the second card is drawn.

SOLUTION: Obviously, these are two independent events since the probability of selecting a king on the second draw does not depend on the selection obtained on the first draw, because we are sampling with replacement. Therefore,

$$P(K_1 \cap K_2) = P(K_1) \cdot P(K_2) = \frac{4}{52} \times \frac{4}{52} = \frac{16}{2704} = 0.006 = 0.6\%. \quad ■$$

EXAMPLE 5.15 A

Obtain the probability of two tails in a row when a fair coin is tossed twice.

SOLUTION: Let T_1 and T_2 represent tails on the first and second tosses, respectively. Obviously, these are two independent events because the probability of obtaining tails on the second toss is independent of the result obtained on the first toss. Therefore,

$$P(T_1 \cap T_2) = \frac{1}{2} \times \frac{1}{2} = \frac{1}{4} = 0.25 = 25\%. \quad ■$$

5.5.6 Probability Tree Approach

A different way of answering the preceding question is shown below. In fact, this approach is much more general as it presents all options for determining any desired

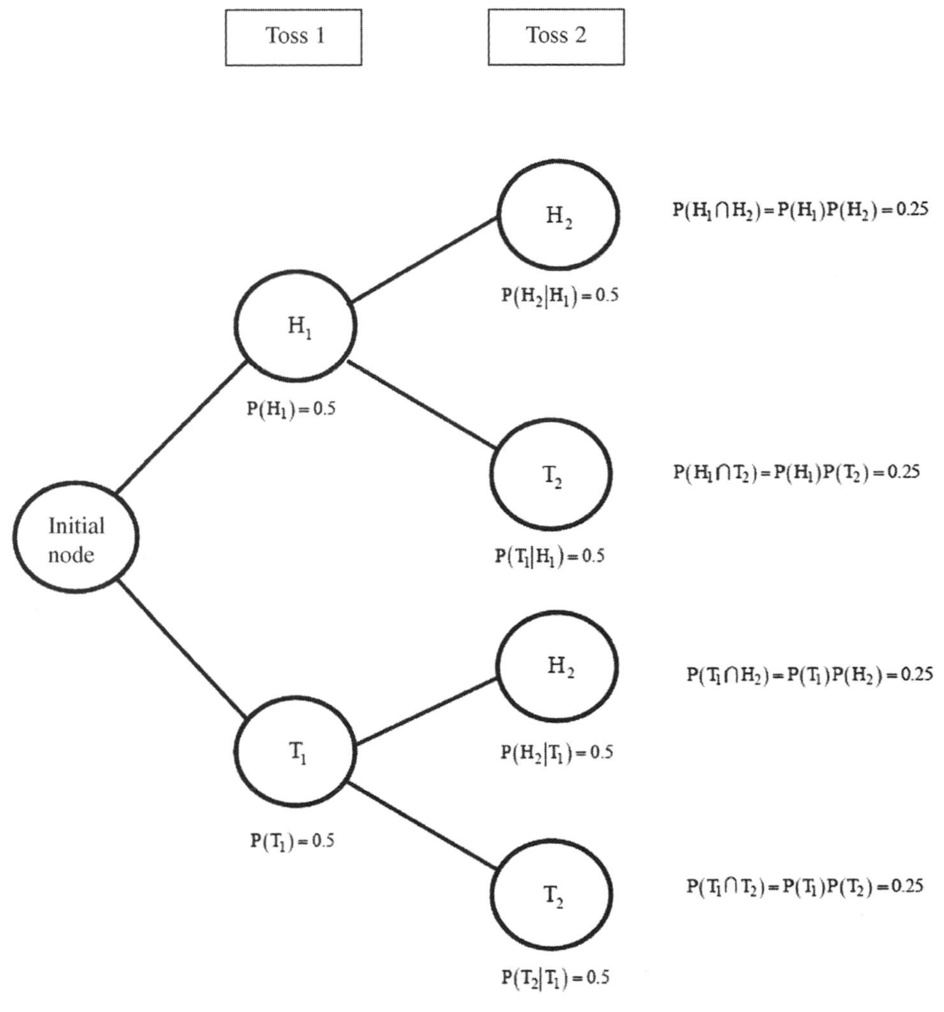

Figure 5.8 Probability tree: tossing a coin twice.

probabilities. Let H_1 and H_2 represent heads on the first and second tosses, respectively. Note that $P(H_i) = P(T_i) = 0.5$ for $i = 1, 2$. From an initial node, the tree starts with two branches representing either heads or tails on the first toss. From each of these two nodes, we branch again into two terminal nodes once the second toss is made. In total, there are four different toss-two nodes, as shown in the tree diagram in Figure 5.8.

From this diagram, one can see that $P(T_1 \cap T_2) = (0.5)(0.5) = 0.25$ or 25%. Other sequential trial probabilities are determined in a similar fashion.

EXAMPLE 5.15 B

Toss a coin three times or toss three coins simultaneously. Develop the probability tree indicating all possible sequential outcomes and find the following:

a. The probability of tossing two tails.

b. Two heads and one tail.

c. Probability of all three tails.

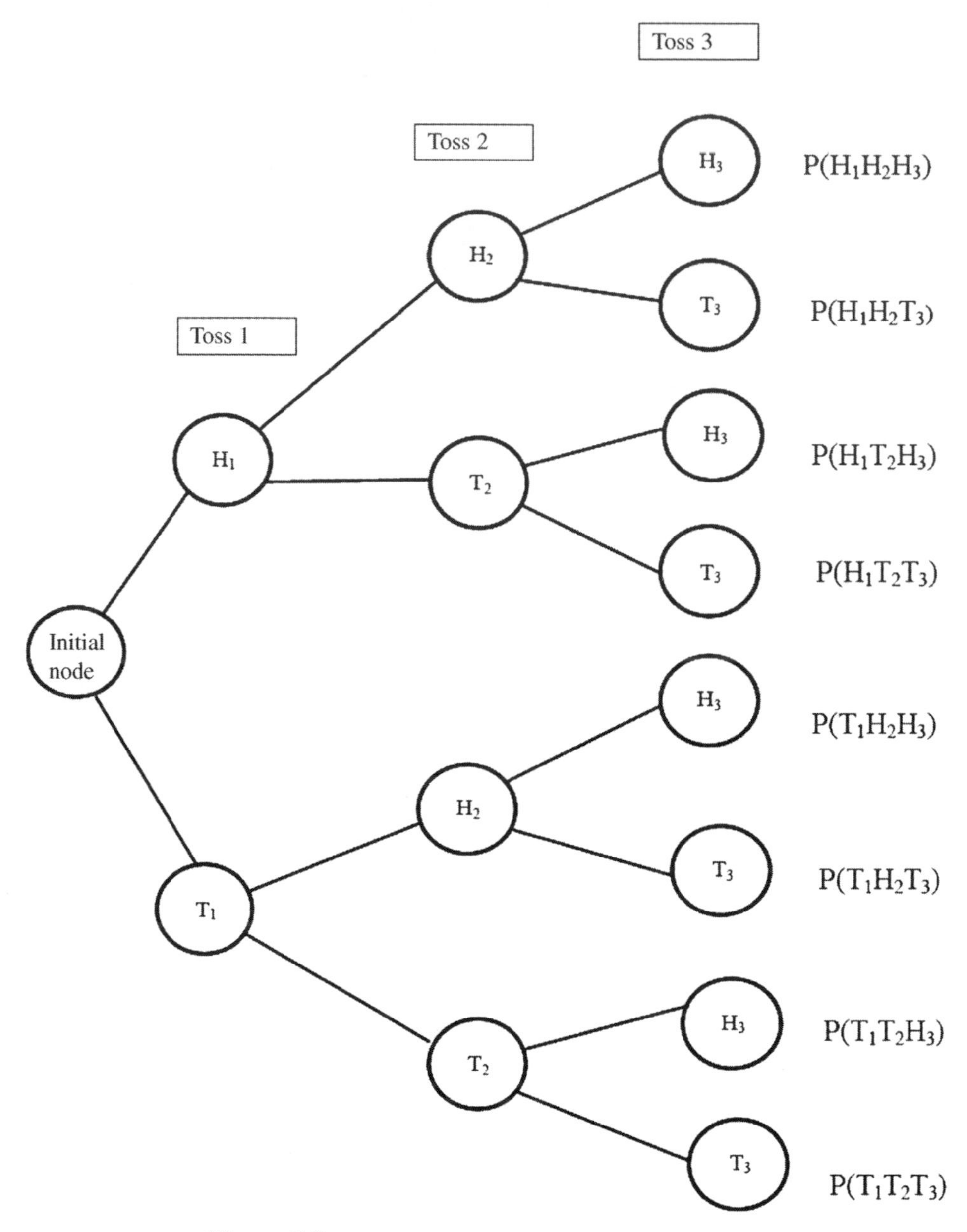

Figure 5.9 Probability tree: tossing a coin three times.

SOLUTION: *Note:* $P(H_i) = P(T_i) = 0.5, i = 1, 2, 3$. Here, i represents the number of the toss. Using Figure 5.9:

a. $P(H_1T_2T_3) + P(T_1H_2T_3) + P(T_1T_2H_3) = (0.5)(0.5)(0.5) + (0.5)(0.5)(0.5) + (0.5)(0.5)(0.5) = 0.125 + 0.125 + 0.125 = 0.375$ (37.5%).

b. $P(H_1H_2T_3) + P(H_1T_2H_3) + P(T_1H_2H_3) = (0.5)(0.5)(0.5) + (0.5)(0.5)(0.5) + (0.5)(0.5)(0.5) = 0.125 + 0.125 + 0.125 = 0.375$ (37.5%).

c. $P(T_1T_2T_3) = (0.5)(0.5)(0.5) = 0.125$ (12.5%).

Note: In this example, the special case of the product rule for independent events is applied. For example, $P(ABC) = P(A)P(B)P(C)$ under independence. ■

Table 5.1 State of the Economy

Change in the Value of the Fund	Boom (B)	Normal (N)	Recession (R)
Increase (I)	0.7	0.5	0.4
Stays same (S)	0.20	0.3	0.3
Decline (D)	0.1	0.2	0.3
Total	1.0	1.0	1.0

EXAMPLE 5.15 C

After the recent recession from 2008 to 2010, it is estimated that in the next 2 years, 2011 and 2012, the economy can assume the states of fast recovery (boom), stay the same (normal), or slip into a downturn (recession) with probabilities of 60, 30, and 10%, respectively. At the same time, an investor determined in Table 5.1 of the probabilities regarding changes in the aggregate value of a stock mutual fund.

Develop a probability tree indicating the different possible outcomes (Figure 5.10). Using the probability tree, find the following terminal probabilities:

a. The stock mutual fund value increases during the recession.

b. The stock mutual fund value declines during the boom period.

c. The stock mutual fund value stays the same in the recession.

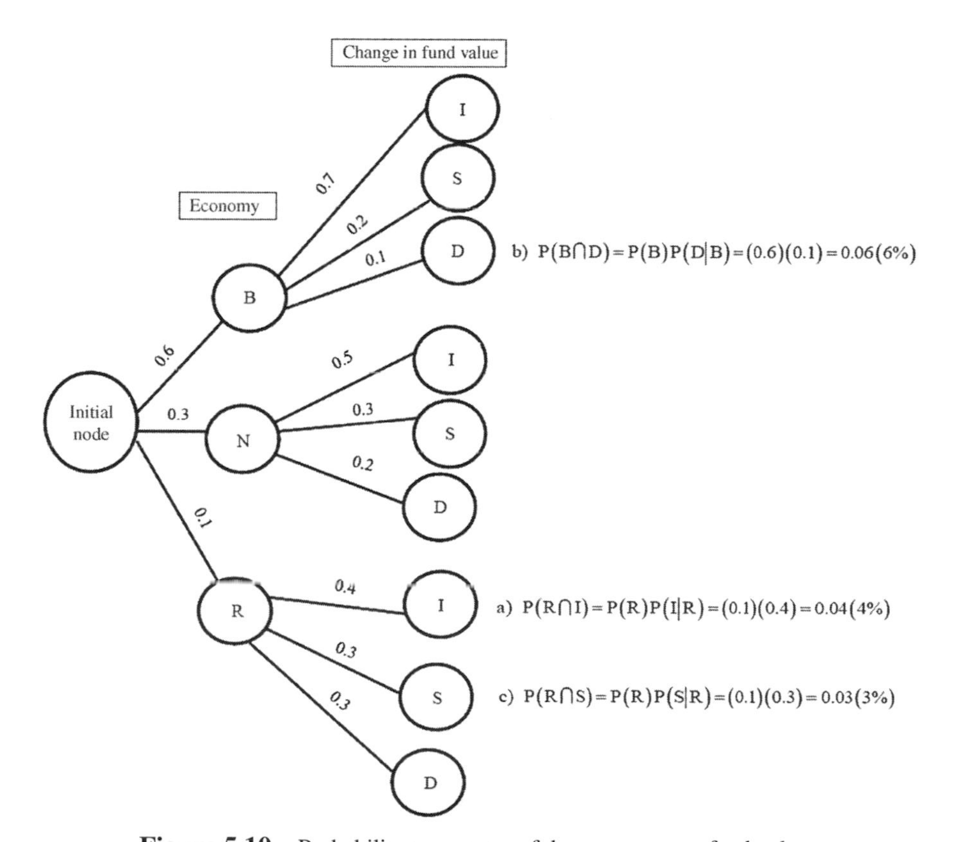

Figure 5.10 Probability tree: states of the economy vs. fund value.

SOLUTION: Using the accompanying table, we list the different options as follows:

B: The state of the economy in boom.

N: The state of the economy is normal.

R: The economy is in recession.

I: The stock mutual fund increases in its value.

S: No change in the value of the stock mutual fund.

D: The stock mutual fund declines in its value.

a. $P(R \cap I) = P(R)P(I|R) = (0.1)(0.4) = 0.04(4\%)$,

b. $P(B \cap D) = P(B)P(D|B) = (0.6)(0.1) = 0.06(6\%)$,

c. $P(R \cap S) = P(R)P(S|R) = (0.1)(0.3) = 0.03(3\%)$. ■

EXAMPLE 5.15 D

Based on historical information, it is reported that within New York state, 76% have passed the Bar exam on the first attempt and 34% passed in subsequent attempts. Develop a probability tree indicating the different "passing" possibilities (Figure 5.11). Based on this tree, find the chance of

a. passing the bar exam on the first attempt;

b. passing the bar exam on the second attempt; and

c. passing the bar exam on the third attempt.

Assume that the probability of passing or failing on subsequent attempts is independent of prior attempts.

SOLUTION: Based on the above tree diagram, the required probabilities are obtained as

a. $P(P_1) = 0.76(= 76\%)$.

b. $P(F_1 \cap P_2) = 0.24 \times 0.34 = 0.0816(= 8.16\%)$.

c. $P(F_1 \cap F_2 \cap P_3) = 0.24 \times 0.66 \times 0.34 = 0.053856(= 5.3856\%)$. ■

5.6 CONTINGENCY TABLE

A rectangular array or table consisting of (survey) data arranged in a specified number of rows and columns is known as a *contingency table*. As we shall now see, the contingency table is useful for determining probabilities of various types of events. Such calculations are illustrated below.

EXAMPLE 5.16

Table 5.2 describes the data on students graduating from a university in the northeast in May, 2013.

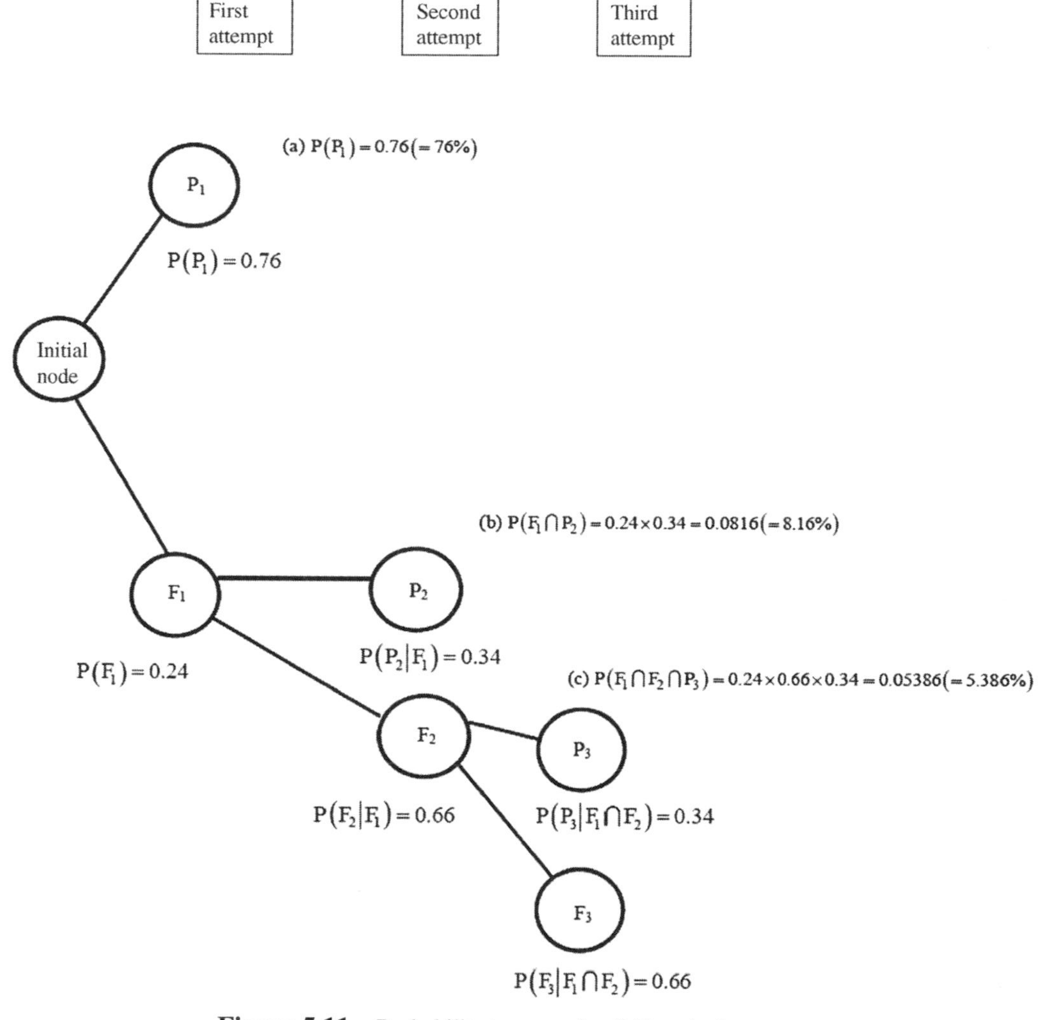

Figure 5.11 Probability tree: passing/failing the bar exam.

Find the probability of the following events:

a. A student graduating is female.

b. A student is female or a graduate from one of the professional schools. Which rule of probability did you use?

c. The student is from the School of Education & Health Sciences given that the student selected is female.

d. Two graduating students are selected without replacement. What is the probability that both of them are female? What is the probability that both are female if sampling is done with replacement? Which rules of probability did you apply?

e. Determine if "gender" and "type of college" are statistically independent.

SOLUTION: *Notation:* Let R_i denote row i, where $i=1$ for female and $i=2$ for male. Likewise, let C_j represent column j, where $j=1$ (Arts & Sciences), $j=2$ (Professional Schools), $j=3$ (Art & Music), and $j=4$ (Education & Health School).

Table 5.2 Students Graduating in May 2013

	Arts & Sciences (C_1)	Professional Schools (C_2)	Art & Music (C_3)	Education & Health (C_4)	Total
Female (R_1)	211	54	92	87	444
Male (R_2)	133	158	84	13	388
Total	344	212	176	100	832

a. Using the objective method of finding probability,

$$P(R_1) = \frac{444}{832} = 0.5337 = 53.37\%.$$

b. $P(R_1 \cup C_2) = P(R_1) + P(C_2) - P(R_1 \cap C_2)$

$$= \frac{444}{832} + \frac{212}{832} - \frac{54}{832} = 0.5337 + 0.2548 - 0.0649 = 0.7236 \quad \text{or} \quad 72.36\%.$$

This result is obtained by applying the general addition rule of probability.

c. $P(C_4 \mid R_1) = \dfrac{87}{444} = 0.1959 = 19.59\%.$

This can also be calculated as

$$P(C_4 \mid R_1) = P(C_4R_1)/P(R_1) = \frac{87/832}{444/832} = \frac{87}{444} = 0.1959 = 19.59\%.$$

d. Define F_1 as getting a female graduate on draw 1 and F_2 as getting a female graduate on the second draw. Then,

$$P(F_1 \cap F_2) = P(F_1) \cdot P(F_2 \mid F_1) = \frac{444}{832} \times \frac{443}{831} = 0.5337 \times 0.5331 = 0.2845 \quad \text{or} \quad 28.45\%.$$

This result is obtained by using the multiplication rule without replacement. If we sample with replacement, then

$$P(F_1 \cap F_2) = P(F_1)P(F_2) = \frac{444}{832} \times \frac{444}{832} = 0.2848.$$

e. There are two ways of testing for independence of "gender" and "type of college."

Method 1

If $P(R_1 \cap C_2)$, which we calculated in part (b) above, equals the product of $P(R_1)$ and $P(C_2)$ or $P(R_1 \cap C_2) = P(R_1) \cdot P(C_2)$, then gender and type of college graduation are independent. Here, $P(R_1 \cap C_2) = 0.0649 = 6.49\%$ and

$$P(R_1) \times P(C_2) = 0.5337 \times 0.2548 = 0.1360 = 13.6\%.$$

Clearly, the preceding equality is violated.

Therefore, gender and type of college graduation are not independent events.

Method 2

If $P(R_1 \mid C_2) = P(R_1)$ or $P(C_2 \mid R_1) = P(C_2)$, then we can say that gender and type of college graduation are independent. Here,

$$P(R_1 \mid C_2) = \frac{54}{212} = 0.2547 = 25.47\%$$

and

$$P(R_1) = \frac{444}{832} = 0.5337 = 53.57\%.$$

Since $0.2547 \neq 0.5337$, gender and college type are not independent. Also,

$$P(C_2 \mid R_1) = \frac{54}{444} = 0.1216 = 12.16\%$$

and

$$P(C_2) = \frac{212}{832} = 0.2548 = 25.48\%.$$

Again, since $0.1216 \neq 0.2548$, gender and type of college are not independent.

Note: Any one row and any one column can be selected to establish independence. Establishing lack of independence with respect to any one row and any one column is sufficient proof that they are dependent events. This process need not be repeated for each cell. ■

EXAMPLE 5.17

Table 5.3 gives data on educational attainment in thousands by gender in 2000 for persons of age 25 or older.

a. What is the probability of selecting a female?

b. What is the probability of selecting a female or a person with an educational level lower than having an undergraduate degree? What probability rule did you apply?

c. Given that a male is selected, what is the probability that this person has at least a master's degree?

d. Two people are selected at random without replacement. What is the probability that both have some college, no degree?

e. Are gender and educational attainment statistically independent?

Table 5.3 Educational Attainment by Gender in Thousands (Persons of Age 25 or Older)

	High School Grad or Less	Some College, No Degree	Undergraduate Degree	Masters' Degree and Higher	Total
	(C_1)	(C_2)	(C_3)	(C_4)	
Female (R_1)	46085	23913	14934	6688	91620
Male (R_2)	39882	20485	14883	8361	83611
Total	85967	44398	29817	15049	175231

Source: US Census Bureau, Statistical Abstract of the United States, 2002.

SOLUTION:

a. $P(R_1) = \dfrac{91620}{175231} = 0.5229$ or 52.29%.

b. $$P(R_1 \cup (C_1 \cup C_2)) = P(R_1) + P(C_1 \cup C_2) - P(R_1 \cap (C_1 \cup C_2))$$
$$= \frac{91620}{175231} + \frac{85967 + 44398}{175231} - \frac{46085 + 23913}{175231} = 0.8674 \quad \text{or} \quad 86.74\%.$$

It is instructive to note that, more formally,

$$\begin{aligned} P(R_1 \cup (C_1 \cup C_2)) &= P(R_1) + P(C_1 \cup C_2) - P(R_1 \cap (C_1 \cup C_2)) \\ &= P(R_1) + P(C_1) + P(C_2) - P((R_1 \cap C_1) \cup (R_1 \cap C_2)) \\ &= P(R_1) + P(C_1) + P(C_2) - [P(R_1 \cap C_1) + P(R_1 \cap C_2)] \end{aligned}$$

since C_1 and C_2, as well as $R_1 \cap C_1$ and $R_1 \cap C_2$, are mutually exclusive events. The reader should verify that this alternative calculation also yields 0.8674.

Here we used both the general and special addition rules of probability.

c. $P(C_4 \mid R_2) = \dfrac{8361}{83611} = 0.1$ or 10%.

This can also be calculated as

$$P(C_4 \mid R_2) = \frac{P(C_4 R_2)}{P(R_2)} = \frac{8361/175231}{83611/175231} = \frac{8361}{83611} = 0.1 \quad \text{or} \quad 10\%.$$

d. Let P_1 and P_2 represent two people selected at random and without replacement. The probability that each has some college but no degree is

$$\begin{aligned} P(P_1 \cap P_2) = P(P_1) \cdot P(P_2 \mid P_1) &= \frac{44398}{175231} \times \frac{44397}{175230} = 0.25337 \times 0.25336 \\ &= 0.06419 \quad \text{or} \quad 6.419\%. \end{aligned}$$

Therefore, the probability of selecting without replacement two persons with some college education but no college degree is 6.419%. Under sampling with replacement,

$$P(P_1 \cap P_2) = P_1 \cdot P_2 = 0.06420.$$

e. To find whether gender and educational attainment are statistically independent, we can pursue either of the following two methods, as in the preceding example.

Method 1

Is $P(C_4 \mid R_2) = P(C_4)$ or $P(R_2 \mid C_4) = P(R_2)$? If so, gender and educational attainment are independent. Otherwise, they are not independent.

Since $P(C_4 \mid R_2) = 0.1$ (calculated above in part (c)) and $P(C_4) = 15049/175231 = 0.08588$ are obviously not equal, we therefore conclude that gender and educational attainment are not independent events.

Similarly,

$$P(R_2 \mid C_4) = 8361/15049 = 0.5556 \neq P(R_2) = 83611/175231 = 0.4771.$$

Here we must also conclude that gender and educational attainment are not independent.

Method 2

Check to see if $P(C_4 \cap R_2) = P(C_4) \cdot P(R_2)$. If equality is obtained, then gender and educational attainment are statistically independent. Otherwise, they are not independent. Here,

$$P(C_4 \cap R_2) = 8361/175231 = 0.0477,$$

$$P(C_4) = 0.08588 \text{ (calculated in method 1 above), and}$$
$$P(R_2) = 0.4771 \text{ (calculated in method 1 above).}$$

Since

$$P(C_4) \cdot P(R_2) = 0.08588 \times 0.4771 = 0.04097 \neq P(C_4 R_2) = 0.0477,$$

we again conclude that gender and educational attainment are not statistically independent. ■

5.7 EXCEL APPLICATIONS

EXAMPLE 5.18

Find solutions to Example 5.16 using Excel.

SOLUTION: Instructions:

Reproduce Table 5.2 on an Excel spread sheet. Then divide all the frequencies (numbers) by the grand total number of 832 and form a similar table, which may be called a *probability table*. The interior probabilities are joint probabilities, and the row end/column end probabilities are called marginal probabilities. Then, the answers to the questions can be found as described in the following Excel table.

7		C1	C2	C3	C4	Total
8	R1	0.253605769	0.064903846	0.110576923	0.104567308	0.533654
9	R2	0.159855769	0.189903846	0.100961538	0.015625	0.466346
10	Total	0.413461538	0.254807692	0.211538462	0.120192308	1
11						
12	Ans: a	0.533653846	→		P(R1), Marginal Probability = CellF8.	
13	Ans: b	0.723557692	→		P(R1) + P(C2) -P(R1∩C2) Addition rule	
14					= CellF8 + CellC10 - Cell C8	
15	Ans: c	0.195945946	→		P(C4/R1) = P(C4∩R1)/P(R1) = Cell E8/Cell F8 :	
16					Using Conditional Rule of probability	
17	Ans: d	0.284786428	→		P(R1)*P(R1) Muliplication rule and selection	
18					with replacement = Cell F8*Cell F8	
19	Ans: e	0.135979105	0.064903846	→	Since P(R1)*P(C2) ≠ P(R1∩C2) gender	
20					and type of college are not independent	
21					i.e. CellF8*CellC10 ≠ CellC8.	

Method 2	P(R1|C2) = 0.25471	P(R1)=0.533654 →	Since P(R1/C2) ≠ P(R1), gender
			and type of college are not independent,
			i.e., CellC8/CellC10 ≠ CellF8
or			
	P(C2|R1) = 0.12162	P(C2) = 0.25480769 →	Since P(C2/R1) ≠ P(C2), gender
			and type of college are not independent,
			i.e., CellC8/CellF8 ≠ CellC10

■

EXAMPLE 5.19

Find solutions to Example 5.17 using Excel.

SOLUTION: Instructions:

Reproduce Table 5.3 on an Excel spread sheet. Then divide all the frequencies (numbers) by the grand total number of 175,231 and form a similar table, which may be called a probability table. The interior probabilities are joint probabilities, and the row end/column end probabilities are called marginal probabilities. Then, the answers to the questions can be found as described in the following Excel table.

	A	B	C	D	E	F	G	H	I
10	**Probability Table**								
11		High	Some	Undergraduate	Masters'	Total			
12		school grad	College, no	degree	degree and				
13		or less	degree		higher				
14		C1	C2	C3	C4				
15	Female(R1)	0.2629957	0.13646558	0.085224646	0.0381668	0.522853			
16	Male (R2)	0.22759672	0.11690283	0.084933602	0.0477142	0.477147			
17	Total	0.49059242	0.25336841	0.170158248	0.0858809	1			
18									
19	Ans (a)	0.52285269		p(R1) Or Cell F15					
20	Ans (b)	0.8674		P(R1) + P(C1) + P(C2) - P(R1∩C1) - P(R1∩C2) = F15 + B17 + C17 -B15 -C15					
21	Ans (c)	0.1		P(C4/R2) = P(C4∩R2)/P(R2) = E16/F16					
22	Ans (d)	0.0642		P(C2)*P(C2) = C17*C17 assuming sampling with replacement.					
23	Ans (e)	Since P(C4/R2) = 0.1 shown in Ans (c) above ≠ P(C4) = 0.08588 (E17 cell) we can conclude							
24		that gender and educational attainment are not independent.							

or:	P(R2|C4) = 0.555586; and P(R2) =0.477147. Since P(R2|C4) = (CellE16/CellE17) ≠ P(R2) (CellF16)
	we can conclude that gender and educational attainment are not independent.
	(The above two ways of answering part (e) relate to Method 2.
	Now, using the special rule of independence (Method 1),
	P(R2∩C4) = 0.0477142. However, P(R2)*P(C4) = 0.040978.
	Since P(R2∩C4) (CellE16) ≠ P(R2) (cellF16)*P(C4)(cellE17).
	we can conclude that gender and educational attainment are not independent.

■

CHAPTER 5 REVIEW

You should be able to:

1. Give at least two examples of a random experiment.
2. Distinguish between simple and compound events.
3. Explain what is meant by a sample space.
4. Explain the difference between objective and subjective probability.
5. Distinguish between classical probability and the relative frequency view of probability.
6. Explain how the words "or" and "and" translate into probability calculations.

7. Describe how the phrase "an event does not occur" translates into set notation. (*Hint:* Think about the rule of complements.)
8. Distinguish between mutually exclusive and independent events.
9. Determine when the general addition rule for calculating probability is used.
10. Distinguish between marginal, joint, and conditional probabilities.
11. Determine when a probability tree is a useful device for calculating probabilities.
12. Determine when a contingency table is a useful device for calculating probabilities.

Key Terms and Concepts:

certain event, 196
classical probability, 196
complement of an event, 199
compound event, 195
conditional probability, 202
contingency table, 208
event, 195
general addition rule, 200
impossible event, 197
independent events, 204
intersection of events, 198
joint probability, 200
marginal probability, 200
mutually exclusive events, 200
probability tree, 204
random experiment, 195
relative frequency, 196
rule of complements, 202
sample space, 195
simple event, 195
special addition rule, 201
union of events, 198

EXERCISES

1. Find the probability of observing three heads when you toss a coin three times or three coins simultaneously. Also, find the probability of getting four heads if you toss a coin four times or four coins simultaneously.
2. Find the probability of rolling a face value of 4 with a six-sided die.
3. Find the probability of picking any card at random with a numerical value less than 10 in a deck of 52 playing cards. Consider an ace as having a numeric value of 1.
4. **a.** Find the probability of randomly picking two cards sequentially that have a total value of 21 (also called Black Jack) from a deck of 52 playing cards. Consider an ace having a value of 11 and all face cards having a score of 10 each.

   b. Recalculate this probability using a probability tree.
5. Find the probability of selecting three even-numbered balls out of a collection of 59 balls numbered from 1 to 59.
6. In a deck of 52 playing cards, find the probability of drawing a card that is either a king or a diamond.
7. Suppose 60% of MBA graduates read the *Wall Street Journal*, 50% read the *New York Times,* and 30% read both. Find the probability that a randomly selected MBA graduate reads at least one of them. Also, find the probability that this graduate reads neither of them.

8. In rolling a six-sided die once, find the probability of obtaining a face value of 3 or a face value of 4.

9. In 2014, the U.S. Senate was made up of 45 Republicans, 53 Democrats, and 2 Independents. Find the probability of selecting two Republican senators sequentially (one after another) with and without replacement.

10. In a college basketball team of five players, three are graduating seniors. Find the probability of selecting three graduating seniors sequentially without replacement.

11. In a college basketball team of five players, having three graduating seniors, find the probability of selecting three graduating seniors sequentially assuming replacement.

12. **a.** Find the probability of obtaining the sum of the face scores equaling 9, when a pair of fair dice is rolled once.

b. Recalculate this probability using a probability tree.

13. The following contingency table relates data on educational attainment by people of age 25 or older to ethnicity in the United States in the year 2000.

Educational Attainment versus Ethnicity (2000)

	High School Graduate, or Less	Some College, No Degree	Undergraduate Degree	Master's Degree and Higher	Total
White	71,327	37,355	25,443	12,942	147,067
Black	11,360	5,370	2,284	1,022	20,036
Others	3,275	1,731	2,048	1,073	8,127
Total	85,963	44,456	29,775	15,036	175,230

Source: U.S. Census Bureau, Statistical Abstract of the United States, 2002.

Answer the following questions based on the data provided in the table above:

a. What is the probability of selecting a black person?

b. What is the probability of selecting a white person or a person with an educational level lower than an undergraduate degree?

c. What is the probability that a randomly selected black person has a master's degree and higher?

d. Two people are selected at random and without replacement. What is the probability that both are "high school graduates or less"? What is your answer if sampling is done with replacement?

e. Are ethnicity and educational attainment statistically independent?

14. The following contingency table gives data (in thousands) on resident population by region and ethnicity in the United States in the year 2000.

U.S. Resident Population by Region versus Ethnicity (2000)

	Northeast	Midwest	South	West	Total
White	39,327	52,386	65,928	36,912	194,553
Black	6,100	6,500	18,982	3,077	34,659
Hispanic	5,254	3,125	11,587	15,341	35,307
Others	2,913	2,382	3,740	7,868	16,903
Total	53,594	64,393	100,237	63,198	281,422

Source: U.S. Census Bureau, Statistical Abstract of the United States, 2002, Table 24.

Using these data, answer the following questions:

a. What is the probability of selecting a white person living in the Northeast?

b. What is the probability of selecting a person living in the Midwest or a Hispanic person nationwide? Which rule of probability do you apply?

c. Given that a person selected is Hispanic, what is the probability that this person is living in the West? Which rule of probability do you apply?

d. Two people are sequentially selected at random and with replacement. What is the probability that both are in the Midwest?

e. Are ethnicity and residency statistically independent?

15. The following contingency table gives locations of different Wal-Mart stores in the United States in the year 2002.

Wal-Mart Stores in the United States (2002)

	Discount Stores	Supercenters	SAM's Club	Neighborhood Markets	Total
Northeast	243	83	60	0	386
Midwest	460	266	146	0	872
South	591	790	238	48	1667
West	274	119	81	1	475
Total	1568	1258	525	49	3400

Source: Wal-Mart 2002 Annual Report.

Based on these data, answer the following questions:

a. What is the probability of selecting a store in the Northeast?

b. What is the probability of selecting a store in the Midwest or a discount store nationwide?

c. If a store in the South is selected at random, what is the probability that this store is Sam's Club?

d. Two stores are selected at random and without replacement. What is the probability that both are in the West?

Excel Applications

16. Solve Exercise 14 using Excel.

17. Find solutions to Exercise 15 using Excel.

Chapter 6

Random Variables and Probability Distributions

6.1 INTRODUCTION

In the previous chapters, we considered the analysis of an observed data set using graphical and descriptive numerical techniques. The grouping of data, along with constructing frequency and relative frequency distributions, was also presented. Most of this analysis focused on the past or the given data, but the actual process of generating such data was not considered. It is equally important to consider the process of actually generating data, including the role of chance factors in the occurrence of data values. This leads us to the concept of a random variable and its associated probability distribution. A *random variable* is formally defined as a numerical-valued function defined over the chance outcomes in a sample space S. It must be noted that the random experiment generating the sample outcomes or events is repeated a large number of times and under uniform conditions. In simple terms, a random variable takes a particular event in S and associates with it a numerical value. A random variable can be classified into two specific categories, namely, discrete or continuous. A random variable that assumes a finite or countably infinite number of values is defined as a *discrete random variable*. For example, the number of cars traveling on a given stretch of a highway over a given time period, the number of babies born in a certain city in 2011, the number of stars visible in the night sky, and the number of daily transactions of the Bank of America branches in the month of May are some discrete cases.

A *continuous random variable* is one that assumes any value in a specified range of outcomes. Obviously, the number of values assumed by a continuous random variable is infinite. For example, the amount of gasoline used by a motorist in a week, the final value of goods and services (GDP) produced in the United States in the year 2011, the grade point average earned by a student in the Fall of 2011, the height of a randomly selected student, and the amount of snowfall recorded in 2011 in Connecticut are all examples of continuous random variables.

Introduction to Quantitative Methods in Business: With Applications Using Microsoft® Office Excel®, First Edition. Bharat Kolluri, Michael J. Panik, and Rao Singamsetti.

Companion website: www.wiley.com/go/Kolluri/QuantitativeMethods

6.2 PROBABILITY DISTRIBUTION OF A DISCRETE RANDOM VARIABLE X

The display of all possible values of a discrete random variable X along with their associated probabilities is called a *discrete probability distribution*. As far as its properties are concerned, $0 \leq P(X_i) \leq 1$ and $\sum P(X_i) = 1$ for all values X_i of X.

EXAMPLE 6.1

Find the probability distribution of a discrete random variable X, where X represents the number of heads obtained in the random experiment of tossing a coin twice.

SOLUTION: Table 6.1a illustrates the possible outcomes, corresponding X values, and the associated probabilities. T = Tail, H = Head.

From this worktable, the probability distribution of the discrete random variable X is depicted in Table 6.1b.

The graphical representation of this distribution, called a *probability mass function*, is provided in Figure 6.1.

Notice that there are two possible outcomes where one head is observed, and they are mutually exclusive. So we added the two corresponding probabilities, 0.25 and 0.25, to get the probability of the outcome of X = 1 as 0.50. Also, note that each of the probabilities is nonnegative, less than 1, and all the probabilities add up to unity, which satisfy the properties of a discrete probability distribution described above.

Table 6.1a All Possible Outcomes of Tossing Two Coins

Possible Outcome No.	First Toss	Second Toss	X (No. of heads)	P(X)
1	T	T	0	0.25
2	T	H	1	0.25
3	H	T	1	0.25
4	H	H	2	0.25

Table 6.1b Number of Heads and Their Associated Probabilities

X	P(X)
0	0.25
1	0.50
2	0.25
Total	1.00

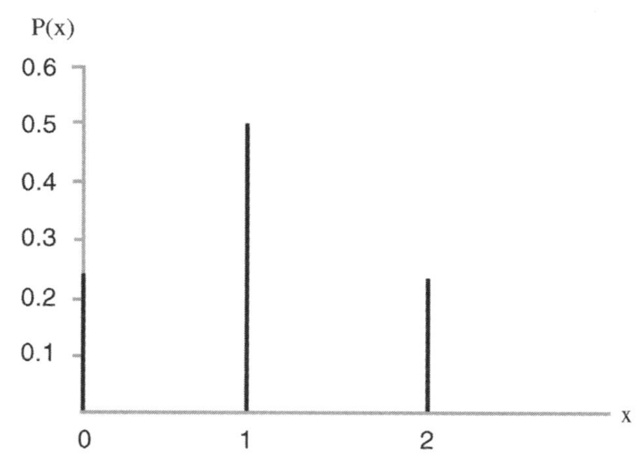

Figure 6.1 Probability distribution of number of heads, X. ■

EXAMPLE 6.2

Given a new microchip device, assume that a company is now able to produce powerful laptops and desktop computers with a hard drive capacity of 100 GB. The company has estimated that the chance of producing a defective laptop is 20% and the chance of producing a defective desktop is 10%. The manufacturer conducted a quality control inspection process by randomly selecting a laptop computer and a desktop computer.

a. Determine the discrete probability distribution of the random variable, "number of defectives," associated with this experiment.

b. What is the probability that only the desktop computer is nondefective?

c. What is the probability that both the desktop and laptop computers are nondefective?

d. What is the probability that only the laptop computer is nondefective?

e. What is the probability that at least one of them is nondefective?

SOLUTION: Let C_N and L_N stand for nondefective desktop and laptop computers, respectively. Also, let C_D and L_D stand for defective desktop and laptop computers, respectively. Then, from the given information, we know that

$$P(C_D) = 10\% \text{ and so } P(C_N) = 90\%. \text{ Similarly, } P(L_D) = 20\% \text{ and so } P(L_N) = 80\%.$$

Remember that our random experiment amounts to selecting a desktop and a laptop computer and observing if each is or is not defective. (*Note:* These selections are independent events.) The random variable X is then the *number* of defective computers detected, that is, neither is defective ($X = 0$); or only one is defective ($X = 1$); or both are defective ($X = 2$).

Looking to the sample space S, we have four possible outcome combinations: (C_D, L_D), (C_N, L_N), (C_D, L_N), and (C_N, L_D). Their associated probabilities (under independence) are as follows:

$$\begin{aligned}
P(C_D \cap L_D) &= P(C_D) \cdot P(L_D) = 0.02.\\
P(C_N \cap L_N) &= P(C_N) \cdot P(L_N) = 0.72.\\
P(C_D \cap L_N) &= P(C_D) \cdot P(L_N) = 0.08.\\
P(C_N \cap L_D) &= P(C_N) \cdot P(L_D) = 0.18.
\end{aligned}$$

Table 6.2 Number of Defectives and Their Associated Probabilities

X	P(X)
0	0.72
1	0.26 (= 0.08 + 0.18)
2	0.02
Total	1.00

a. On the basis of these calculations, the associated discrete probability distribution is given in Table 6.2.

b. $P\,(C_N \cap L_D) = P\,(C_N) \cdot P(L_D) = 0.9 \times 0.2 = 0.18$.

c. $P\,(C_N \cap L_N) = P\,(C_N) \cdot P(L_N) = 0.9 \times 0.8 = 0.72$.

d. $P\,(C_D \cap L_N) = P\,(C_D) \cdot P(L_N) = 0.1 \times 0.8 = 0.08$.

e. $P(C_N \cap L_D) + P(C_D L_N) + P(C_N L_N) = 0.18 + 0.8 + 0.72 = 0.98$.

Alternatively,
$P\,(C_N \cup L_N) = P(C_N) + P(L_N) - P(C_N \cap L_N) = 0.9 + 0.8 - 0.72 = 0.98$ (using the general addition rule); or, in general, P (at least one) = 1 − P (none). Hence, the probability that at least one of them is nondefective is $1 - P\,(C_D \cap L_D) = 1 - 0.02 = 0.98$ (by using the probability rule of complements). ■

6.3 EXPECTED VALUE, VARIANCE, AND STANDARD DEVIATION OF A DISCRETE RANDOM VARIABLE X

If X is a discrete random variable, then the *expected value* of X is denoted by

$$\mu = E(X) = \sum X_i \cdot P(X_i)$$

over all values X_i of X. This is also denoted by μ_x. It is to be noted that the simple or arithmetic mean developed previously is actually a special case of this formula and results when all of the probabilities are equal, that is, $P(X_i) = 1/N$ for all X_i values in the case of ungrouped data while the probabilities are approximated by $f_j/\sum f_j$ in the case of grouped data. It is obvious that the expected value is a weighted mean, where probabilities or relative frequencies serve as weights. As explained previously, this is a measure of "center" of the discrete probability distribution. To provide a comprehensive picture of a discrete probability distribution, we also present a measure of variability around the mean or expectation.

First, we start with the *variance*, defined by the formula

$$\sigma^2 = \sum (X_i - \mu)^2 P(X_i) \quad \text{(long formula)}$$

or

$$\sigma^2 = \sum X_i^2 P(X_i) - \mu^2 \quad \text{(short formula).}$$

Note that this is the mean or the expected value of the squared deviations of the actual measurements from their mean value μ. Mathematically, it is written as $E[(X-\mu)^2]$. Second, as defined earlier, the *standard deviation* is the positive square root of the variance. Thus,

$$\sigma = \sqrt{\sigma^2} = \sqrt{\sum (X_i - \mu)^2 P(X_i)} = \sqrt{\sum X_i^2 P(X_i) - \mu^2}.$$

EXAMPLE 6.3

Let the probability distribution of X correspond to the number of heads obtained in tossing a fair coin twice, as given in Example 6.1. Find the following:

a. Expected value of *X*

b. Variance of X

c. Standard deviation of X

SOLUTION: Table 6.3 houses the detailed calculations: The expected value is

$$E\,(X) = \sum X_i P\,(X_i) = \mu = 1;$$

the variance is

$$\sigma^2 = \sum (X_i - 1)^2 P(X_i) = 0.5 \quad \text{(using the long formula)}$$

or

$$\sigma^2 = \sum X_i^2 P(X_i) - \mu^2 = 1.5 - 1^2 = 0.5 \quad \text{(using the short formula);}$$

and the standard deviation is

$\sigma = \sqrt{0.5} = 0.7071$. Here, σ is a measure of average variability about the mean. ■

Table 6.3 Computations Related to Mean and Variance

X	P(X)	X P(X)	(X − 1)	$(X-1)^2$	$(X-1)^2P(X)$	X^2	$X^2P(X)$
0	0.25	0	−1	1	0.25	0	0
1	0.50	0.5	0	0	0	1	0.5
2	0.25	0.5	1	1	0.25	4	1.0
Total	1.00	1.0 = μ			$0.5 = \sigma^2$		1.5

EXAMPLE 6.4

Table 6.4a displays the probability distribution of a discrete random variable, X.

Find the mean and the standard deviation by both the long and short formulas for this distribution.

SOLUTION: Looking to Table 6.4b, the mean or the expected value is $E(X) = \mu = 6.0$ and the standard deviation is

$$\sigma = \sqrt{\sigma^2} = \sqrt{16.8} = 4.0988 \quad \text{(via the long formula)}$$

or

$$\sigma = \sqrt{52.8 - 36} = \sqrt{16.8} = 4.0988 \quad \text{(via the short formula).}$$

■

6.3.1 Some Basic Rules of Expectation

1. E (A) = A, where A is a constant.
2. E(A + BX) = E(A) + E(BX) = A + BE (X), where A and B are constants and X is a random variable. In general, for random variables $X_1, \ldots, X_k$,

$$E\left(\sum_{i=1}^{k} X_i\right) = \sum_{i=1}^{k} E(X_i).$$

(It is assumed that $E(X_i)$ exists for all i = 1, . . . , k.)

Table 6.4a Probability Distribution

X	P(X)
−2	0.15
3	0.2
8	0.3
9	0.2
10	0.15

Table 6.4b Computations on Mean and Standard Deviation

X	P(X)	X P(X)	(X − 6)	$(X-6)^2$	$(X-6)^2P(X)$	X^2	X^2 P(X)
−2	0.15	−0.3	−8	64	9.6	4	0.6
3	0.2	0.6	−3	9	1.8	9	1.8
8	0.3	2.4	2	4	1.2	64	19.2
9	0.2	1.8	3	9	1.8	81	16.2
10	0.15	1.5	4	16	2.4	100	15.0
Total	1.00	6.0 = Mean			16.8 = Variance		52.8

6.3.2 Some Useful Properties of Variance of X

By definition, variance of the random variable X is the mean or expected value of the squared deviations from the expectation of X. Therefore, V(X) can be written as

$$\sigma^2 = V(X) = E[X - E(X)]^2 = \sum [X_i - E(X_i)]^2 P(X_i),$$

where we sum over all values of X. The shorter version of this expression is given by

$$\sigma^2 = V(X) = E(X^2) - [E(X)]^2 = \sum X_i^2 P(X_i) - \left[\sum X_i P(X_i)\right]^2.$$

Moreover, for A and B constants,

$$V(A) = 0;$$
$$V(A + BX) = 0 + B^2 V(X) = B^2\sigma^2.$$

6.3.3 Applications of Expected Values

EXAMPLE 6.5

Find the expected value and standard deviation of

$$Z = \frac{X - \mu}{\sigma}, \quad \text{where } \mu = E(X) \text{ and } \sigma = \sqrt{V(X)} \neq 0.$$

SOLUTION: Note that Z can be written as

$$\frac{1}{\sigma}(X - \mu).$$

Therefore,

$$E(Z) = E\left[\frac{1}{\sigma}(X - \mu)\right] = \frac{1}{\sigma}[E(X) - E(\mu)] = \frac{1}{\sigma}(\mu - \mu) = \frac{1}{\sigma}.(0) = 0 \text{ since } 1/\sigma \text{ is constant.}$$

Similarly,

$$V(Z) = V\left[\frac{1}{\sigma}(X - \mu)\right] = \left(\frac{1}{\sigma}\right)^2 [V(X) - V(\mu)] = \frac{1}{\sigma^2}(\sigma^2 - 0) = \frac{\sigma^2}{\sigma^2} = 1 \text{ since } \mu \text{ is}$$

constant. ■

EXAMPLE 6.6

Table 6.5a describes the probability distribution of the number of children (X) per married couple in Asia.

Find the expected number of children per married couple. Interpret your result.

Table 6.5a Probability Distribution

X	0	1	2	3	4	5
P(X)	0.1	0.17	0.3	0.35	0.05	0.03

Table 6.5b Computations on Expected Value

X	P(X)	X P(X)
0	0.1	0
1	0.17	0.17
2	0.3	0.6
3	0.35	1.05
4	0.05	0.2
5	0.03	0.15
Total	1.00	2.17

SOLUTION: First prepare the following worksheet (Table 6.5b).
Then, the expected number of children per married couple is $E(X) = 2.17$.

It is to be noted that no married couple at any time will have a fraction of child. At the same time, we cannot round it off to a whole number (2.0) due to the fact that this leads to an underestimate of the average. However, $100\ (2.17) = 217$. Thus, the average number of children per 100 married couples is 217. ■

EXAMPLE 6.7

Suppose the consumption expenditure of low- and middle-income families in a community is estimated by the relationship $C = 14{,}000 + 0.75X$, where C is consumption and X is disposable income of a typical family. In addition, Table 6.6a describes the disposable income distribution in the community. Find the expected annual consumption and calculate its variability (by way of the standard deviation) for these families.

SOLUTION: First we find the expected value and variance of disposable income of this group of families using the Table 6.6b worksheet.

Table 6.6a Disposable Income Distribution

X (in thousand dollars)	30	40	50	60	70
P(X)	0.20	0.25	0.35	0.15	0.05

Table 6.6b Computations on Expected Value and Variance

X	P(X)	X P(X)	(X − 46)	$(X-46)^2$	$(X-46)^2$ P(X)
30	0.2	6	−16	256	51.2
40	0.25	10	−6	36	9.0
50	0.35	17.5	4	16	5.6
60	0.15	9	14	196	29.4
70	0.05	3.5	24	576	28.8
Total	1.00	46			124

Now, from this worksheet, $E(X) = \$46{,}000$ and $V(X) = 124\ (\$1000)^2$.
So, the expected annual consumption is given by

$$\begin{aligned} E(C) &= E[14,000 + 0.75X] = E(14,000) + E(0.75X) = 14,000 + 0.75E(X) \\ &= 14,000 + 0.75(46,000) = \$48,500. \end{aligned}$$

To compute the standard deviation of consumption, we first determine the variance of consumption as

$$\begin{aligned} V(C) &= V[14,000 + 0.75X] = V(14,000) + V(0.75X) = 0 + (0.75)^2 V(X) = (0.75)^2(124) \\ &= (0.5625)(124) = 69.75. \end{aligned}$$

Then, the standard deviation of consumption is

$$\begin{aligned} \sqrt{V(C)} &= \sqrt{69.75} \\ &= \$8351.60 \text{ (thousands of dollars)}. \end{aligned}$$

■

EXAMPLE 6.8

An auto insurance company estimated the probability of an individual dying in an auto accident to be 0.1%. Find the premium (p) for a life insurance policy of \$1 million if the insurance company wants to break-even on average.

SOLUTION: The probability distribution of X (gain or loss) for the insurance company is provided in Table 6.7.

Hence, the expected gain or loss is $E(X) = p - 1000$, and to break-even we set $E(X) = p - 1000 = 0$ and solve for p.

Therefore, a premium of $p = \$1000$ should be charged to cover the policy. ■

EXAMPLE 6.9

A grocer has to decide each morning how much of a perishable commodity (such as fish) to buy from a wholesale dealer in order to run a profitable operation. Assume that each unit is bought at \$14.00 and can be sold for \$20.00. At the end of the day, any leftover units have no value. Since the demand for the product is random, it is estimated that the demand (X) has the following probability distribution (Table 6.8a):

Table 6.7 Probability Distribution

Event	X	P(X)	X P(X)
No death	P	0.999	0.999p
Death	−(1,000,000 − p)	0.001	−(1,000,000 − p) × 0.001
Total		1.000	0.999p − 1,000 + 0.001p = p − 1,000

Table 6.8a Probability Distribution

X	1	2	3
P(X)	0.5	0.3	0.2

The grocer has three choices (possible actions) to follow: Purchase 1 unit (A_1), purchase 2 units (A_2), or purchase 3 units (A_3). Given this information, answer the following questions:

a. Find the *payoff table* consisting of net profit values corresponding to each demand/action combination.

b. Find the optimal decision using the concept of expected payoff.

c. Given the above information, construct the opportunity loss table.

d. Find the optimal decision using the concept of expected opportunity loss.

SOLUTION:

a. Table 6.8b shows net profit values, known as *payoff values*, corresponding to a given demand and action taken. For example, given a demand of $X = 1$ in column 2, if the action taken is A(2) in row 5, then the net profit is given by revenue − cost = 20 − 28 = −8.

b. Now, using probabilities corresponding to each level of demand, one can obtain the *expected payoff* values of each action as follows:

$$\begin{aligned} E[A(1)] &= 6(0.5) + 6(0.3) + 6(0.2) = 6, \\ E[A(2)] &= -8(0.5) + 12(0.3) + 12(0.2) = 2, \text{ and} \\ E[A(3)] &= -22(0.5) + -2(0.3) + 18(0.2) = -8. \end{aligned}$$

Clearly, the optimal decision in this case is to buy only one unit because this action, A(1), gives maximum expected payoff of 6.

c. *Opportunity loss* in a particular cell (i, j) is defined as the difference between maximum profit that could be realized if demand j ($j = 1, 2, 3$) occurs and the net profit level appears in the payoff table. As a first step in calculating the *opportunity loss table*, we present below the *maximum potential profit table* (Table 6.8c).

Next, the opportunity loss for each cell is obtained by subtracting the cell value of the payoff table from the corresponding cell value in the maximum potential profit table. To this end, the opportunity loss table appears as Table 6.8d.

Table 6.8b Payoff Table

P(X)	0.5	0.3	0.2
X	1	2	3
Action (purchase)			
A(1)	6	6	6
A(2)	−8	12	12
A(3)	−22	−2	18

Table 6.8c Maximum Potential Profit Values

X	1	2	3
A(1)	6	12	18
A(2)	6	12	18
A(3)	6	12	18

Table 6.8d Opportunity Loss Values

X	1	2	3
A(1)	$6-6=0$	$12-6=6$	$18-6=12$
A(2)	$6-(-8)=14$	$12-12=0$	$18-12=6$
A(3)	$6-(-22)=28$	$12-(-2)=14$	$18-18=0$

Note that the opportunity loss is never negative, but it could be zero if the demand outcome coincides with the action taken.

d. To find the optimal decision based on the opportunity loss table, we compute the expected opportunity loss corresponding to each possible action, A(1), A(2), and A(3), as follows:

$$E[A(1)] = 0(0.5) + 6(0.3) + 12(0.2) = 0 + 1.8 + 2.4 = 4.2,$$
$$E[A(2)] = 14(0.5) + 0(0.3) + 6(0.2) = 7 + 0 + 1.2 = 8.2,$$
$$E[A(3)] = 28(0.5) + 14(0.3) + 0(0.2) = 14 + 4.2 + 0 = 18.2.$$

The optimal decision is simply to buy one unit because this action gives a minimum expected loss of $4.2. It is important to note that the optimal action should be the same by either method (expected payoff versus expected loss). ■

EXAMPLE 6.10

A car rental agency next to an airport has to decide on three rental options: 100 cars, 200 cars, or 400 cars in a working day. The demand for rental cars is random and it depends on the specific needs of passengers at the airport. Three levels of demand for rental cars are identified and the probability distribution is given in Table 6.9a.

The total cost, including maintenance and office and other expenses, is given as $30.00 per car per day. The average rental fee per car per day is $50.00. Find the optimal decision in terms of maintaining 100 cars or 200 or 400 cars, based on the payoff table.

Table 6.9a Demand Probability Distribution

Demand (X)	Low (100 Cars)	Medium (200 Cars)	High (400 Cars)
P(X)	0.2	0.3	0.5

Table 6.9b Payoff Values

Demand (X)	Low (100 Cars)	Medium (200 Cars)	High (400 Cars)
P(X)	0.2	0.3	0.5
A(100)	2000	2000	2000
A(200)	−1000	4000	4000
A(400)	−7000	−2000	8000

SOLUTION: Table 6.9b presents the payoff values. For example, for a medium demand of 200 cars and an action of renting 400 cars per day, the payoff value is given by revenue − cost $= 200 \times 50 - 400 \times 30 = 10{,}000 - 12{,}000 = -2000$. Thus, the net profit is −$2000 or loss of $2000.

So, let us find expected payoff for each possible action:

$$\begin{aligned} E[A\ (100)] &= 2000(0.2) + 2000(0.3) + 2000(0.5) = 2000, \\ E[A\ (200)] &= -1000(0.2) + 4000(0.3) + 4000(0.5) = 3000, \\ E[A\ (400)] &= -7000(0.2) - 2000(0.3) + 8000(0.5) = 2000. \end{aligned}$$

The optimal decision is to take the action corresponding to maximum expected payoff or $3000, which is to maintain 200 cars per day. ■

6.4 CONTINUOUS RANDOM VARIABLES AND THEIR PROBABILITY DISTRIBUTIONS

A probability distribution is *continuous* if its underlying random variable (X) is *continuous*, that is, X can assume any value over some interval from, say, a to b. Moreover, the interval can be open (denoted $(a,b) = \{a < X < b\}$) or closed (written as $[a, b] = \{a \leq X \leq b\}$) or half-open.

The probability distribution of a continuous random variable is generally described by a *probability density function*, say f(x). As in the discrete case, f(x) is nonnegative. In addition, the area under the curve of f(x) over any particular interval is interpreted as probability. In this regard, the total area (probability) over the entire range of the random variable must be unity. So, for the event $a \leq X \leq b$,

$$P(a \leq X \leq b) = \int_a^b f(x)dx.$$

(Note that for a continuous distribution, $P(a \leq X \leq b) = P(a < X < b)$ since $P(A) = P(B) = 0$).

In this regard, probability can be described as the shaded area under f(x) from a to b (Figure 6.2) and can be calculated by using integral calculus methods. For instance, suppose X follows a *uniform probability distribution* of the form

$$f(x) = \begin{cases} \dfrac{1}{\beta - \alpha}, & \alpha < x < \beta, \\ 0, & \text{elsewhere.} \end{cases}$$

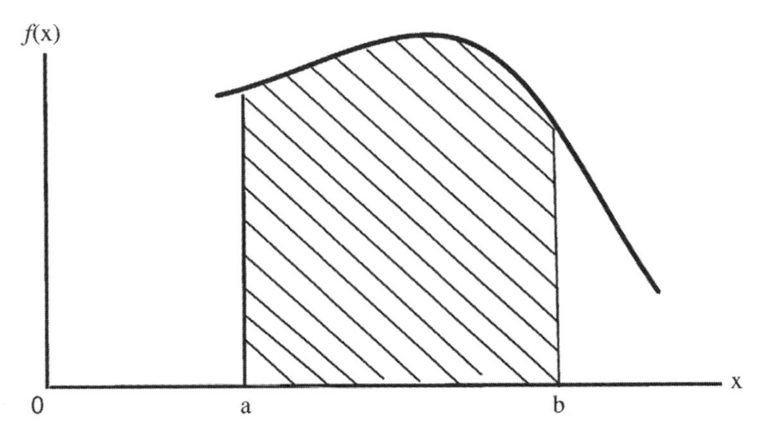

Figure 6.2 Probability shown as the shared area.

(See Figure 6.3a.) For $\alpha = 0$ and $\beta = 3$,

$$f(x) = \begin{cases} 0, & x \leq 0, \\ \frac{1}{3}, & 0 < x < 3, \\ 0, & x \geq 3. \end{cases}$$

(See Figure 6.3b.) What is the probability that $0.75 \leq X \leq 2$? To answer this, we directly calculate

$$P(0.75 \leq x \leq 2) = \int_{0.75}^{2} \frac{1}{3} dx = \frac{1}{3} x \bigg]_{0.75}^{2} = \frac{2}{3} - \frac{1}{4} = 0.4167.$$

Hence, P $(0.75 \leq X \leq 2) = 0.4167$ is the area under f(x) from 0.75 to 2. What is P $(X = 1)$?

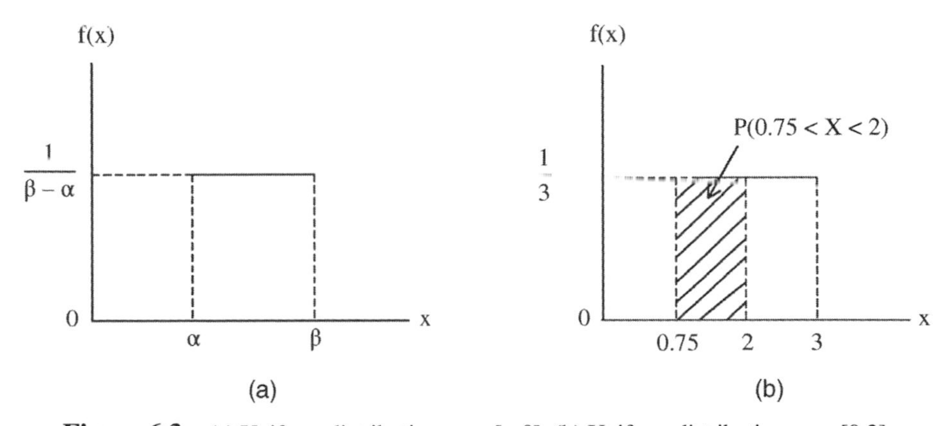

Figure 6.3 (a) Uniform distribution over [α,β]. (b) Uniform distribution over [0,3].

6.5 A SPECIFIC DISCRETE PROBABILITY DISTRIBUTION: THE BINOMIAL CASE

There are several useful discrete probability distributions: the binomial, Poisson, and hypergeometric distributions, among others. However, in this text, we deal with only the most important one, namely, the binomial distribution.

6.5.1 Binomial Probability Distribution

Let a random experiment be classified as a *simple alternative*—there are only two possible (mutually exclusive) outcomes: "success" or "failure." (Note that these terms are used without any specific connotation regarding "good" or "bad.") Alternatively, a simple alternative experiment is also described as a *binomial random experiment.* Some examples of this type of experiment are: a manufactured product is labeled as defective or nondefective, a randomly selected student either passes or fails an exam, a financial firm makes either a profit or incurs a loss, a candidate either wins or loses an election, and so on. The characteristics of a binomial experiment are as follows:

- **i.** There are only two possible (mutually exclusive) outcomes of a random experiment. These are described as "success" or "failure."
- **ii.** The binomial random variable (X) is the (discrete) number of successes observed in a given number (n) of independent and identical trials of a binomial experiment.
- **iii.** The probability of success (failure) is constant from trial to trial (under sampling with replacement).

An experimental process describing all of the preceding characteristics is known as a *Bernoulli process* (after John Bernoulli (1654–1705), a Swiss mathematician).

EXAMPLE 6.11

We articulate the importance of the binomial distribution by introducing an example from a large manufacturing company. Suppose an aerospace company is contemplating the production of an updated version of an aircraft that uses less fuel and has a larger capacity relative to an older model. Assume that the company estimates the probability of a defective aircraft to be 30% after the completion of a production run. Let us find the probability or chance of detecting two defective aircraft out of a randomly selected sample of five aircraft.

Let D stand for defective and N for nondefective aircraft. Out of a sample of five aircraft produced, two defective ones can be manufactured in any one of the following 10 possible sequences:

$$
\begin{array}{c}
DDNNN \\
DNDNN \\
DNNDN \\
DNNND \\
NDDNN \\
NNDDN \\
NNNDD \\
NNDND \\
NDNND \\
NDNDN
\end{array}
$$

In these mutually exclusive sequences, it is important to note that the order of defective or non-defective aircraft is not considered, that is, the two defective aircraft appearing in any order are treated as indistinguishable. The same is true with nondefective aircraft. For example, the arrangement D_1D_2 is the same as D_2D_1, where the subscripts refer to the specific defective aircraft 1 or 2. These arrangements are described as *combinations* or groupings of items without regard to order.

In general, the formula for computing the number of combinations is

$$_{n}C_{r} = \frac{n!}{r!(n-r)!}, \quad r \leq n,$$

where n stands for the total number of items, r stands for the number of items selected in the arrangement, and C stands for combinations. Sometimes another notation $\binom{n}{r}$ is used instead of $_{n}C_{r}$. In either case, it should be read as "the number of combinations of n distinct items taken r at a time." Here n! is pronounced "n factorial" and is calculated as

$$n! = n \cdot (n-1) \cdot (n-2) \cdots 3 \times 2 \times 1,$$

with $0! \equiv 1$. (The symbol " $\equiv$ " reads "is defined as.")

Note that $_{n}C_{n} = 1$. Here, $_{n}C_{n}$ is the number of combinations of n distinct items taken altogether.

We just noted that a combination is an arrangement of objects without regard to order. But if order is important, then, although D_1D_2 and $D_2\ D_1$ depict the same combination of defective aircraft, they represent different "ordered arrangements." In fact, order is considered important in situations such as "who gets ranked in first place and who gets ranked in second place?" We shall refer to such ordered arrangements as *permutations*. In this regard, the number of permutations of n distinct items taken r at a time can be calculated as

$$_{n}P_{r} = \frac{n!}{(n-r)!}, r \leq n.$$

Additionally, $_{n}P_{n} = n!$ (the number of permutations of n distinct items taken altogether).

To illustrate, let us consider our example of defective and nondefective aircraft, where n = 5 and r = 2. Then, the number of combinations is given by

$$_{5}C_{2} = \frac{5!}{2!(5-2)!} = \frac{5!}{2!(3!)} = \frac{5 \times 4 \times 3 \times 2 \times 1}{2 \times 1 \times 3 \times 2 \times 1} = 10,$$

and the number of permutations is

$$_{5}P_{2} = \frac{5!}{(5-2)!} = \frac{5!}{(3!)} = \frac{5 \times 4 \times 3 \times 2 \times 1}{3 \times 2 \times 1} = 20.$$

Note that there are fewer combinations than permutations. Will this always be the case for fixed n and r? The answer is yes since $_{n}C_{r} = \frac{n!}{r!(n-r)!} = \frac{_{n}P_{r}}{r!}$.

Going back to our example, we want to find the probability of obtaining exactly two defective aircraft in any one of the 10 mutually exclusive possibilities. This is given by the following expression:

$$\begin{aligned} P(2,5) &= (DDNNN \cup DNDNN \cup \cdots \cup NDNDN) \\ &= P(DDNNN) + P(DNDNN) + \cdots + P(NDNDN) \end{aligned}$$

(*since these sequences are all mutually exclusive*),

$$P(D)P(D)P(N)P(N)P(N) + P(D)P(N)P(D)P(N)P(N) + \cdots + P(N)P(D)P(N)P(D)P(N)$$

(*using the multiplicative probability law for independent events*),

$$(0.3)(0.3)(0.7)(0.7)(0.7) + (0.3)(0.7)(0.3)(0.7)(0.7) + \cdots + (0.7)(0.3)(0.7)(0.3)(0.7),$$

where 0.3 is the probability of manufacturing a defective aircraft and 0.7 is the probability of manufacturing a nondefective aircraft (found by using the probability rule of complements or $1-0.3=0.7$). Then, since there are 10 equal terms of the form $(0.3)^2(0.7)^3$ on the right-hand side of this expression, it follows that

$$\begin{aligned} P(2,5) &= (0.3)^2(0.7)^3 + (0.3)^2(0.7)^3 + \cdots + (0.3)^2(0.7)^3 \\ &= 10(0.3)^2(0.7)^3 \\ &= {}_5C_2(0.3)^2(0.7)^{5-2}. \end{aligned}$$

Thus, if *n* stands for the number of units produced, r stands for number of defective items, p stands for the probability of obtaining a defective item, and $q = 1 - p$ stands for probability of getting a nondefective item, then the probability of obtaining r defective items out of the n items is written as

$$P(r; n, p) = {}_nC_r p^r (1-p)^{n-r} = {}_nC_r p^r q^{n-r}$$

and termed the *binomial probability function*. In general, think of P(r; n, p) as the probability of obtaining r successes in n independent trials of a simple alternative experiment.

To obtain the binomial probability distribution, let us define the random variable X as the number of successes obtained in n independent trials of a simple alternative experiment. Then, as X varies from 0 to n, we obtain the sequence of binomial probabilities generated by the binomial probability function:

$$P(X; n, p) \quad = \quad {}_nC_X p^X (1-p)^{n-X} = \frac{n!}{X!(n-X)!} p^X (1-p)^{n-X}, X = 0, 1, \ldots, n,$$

and thus the *binomial probability distribution*, where n and p are called the *binomial parameters*. So, once n and p are known, the entire binomial probability distribution is completely specified. ■

EXAMPLE 6.12

Returning to Example 6.11, where $n = 5$ and $p = 0.3$, determine the complete binomial probability distribution for $X = 0$, 1,2,3,4, and 5 defectives. Also, show the probability distribution graphically.

SOLUTION:

$$P(X; 5, 0.3) = {}_5C_x(0.3)^x(0.7)^{5-x}, \quad X = 0, 1, 2, 3, 4, 5.$$

Table 6.10 Binomial Probability Distribution of Defectives

X	0	1	2	3	4	5
P(X)	0.1681	0.3602	0.3087	0.1323	0.0284	0.0024

Then,

$$P(0;5,0.3) = {}_5C_0(0.3)^0(0.7)^5 = \frac{5!}{0!(5-0)!}1(0.7)^5 = \frac{5!}{5!}(0.7)^5 = 1(0.7)^5 = 0.1681,$$

$$P(1;5,0.3) = {}_5C_1(0.3)^1(0.7)^4 = \frac{5!}{1!(5-1)!}(0.3)(0.7)^4 = 5(0.3)(0.7)^4 = 0.3602.$$

Similarly, we can show that

$P(2;5,0.3) = 0.3087$, $P(3;5,0.3) = 0.1323$, $P(4;5,0.3) = 0.0284$, $P(5,0.3) = 0.0024$, and so on. These probabilities can be arranged in a tabular fashion so as to depict the complete binomial probability distribution (Table 6.10).

All of the probabilities add up to 1 (approximately).

The graph of the above probability distribution of defectives (called the *binomial probability mass function)* is as shown in Figure 6.4. ■

EXAMPLE 6.13

Suppose we toss a fair coin five times in succession and define a success (denoted X) as getting heads. Clearly, the probability of a success is $p = \frac{1}{2}$.

a. Do we have a binomial experiment?

b. What is the implied binomial probability distribution? (Hint: find P(X;n, p), where X = 0,1, . . . ,5.).

c. Find the mean and variance of the distribution.

d. Find $P(X \leq 1)$; $P(X \geq 2)$; $P(2 < X \leq 4)$; $P(2 < X < 4)$.

e. Does $P(2 \leq X \leq 4) = P(X \leq 4) - P(X \leq 1)$?

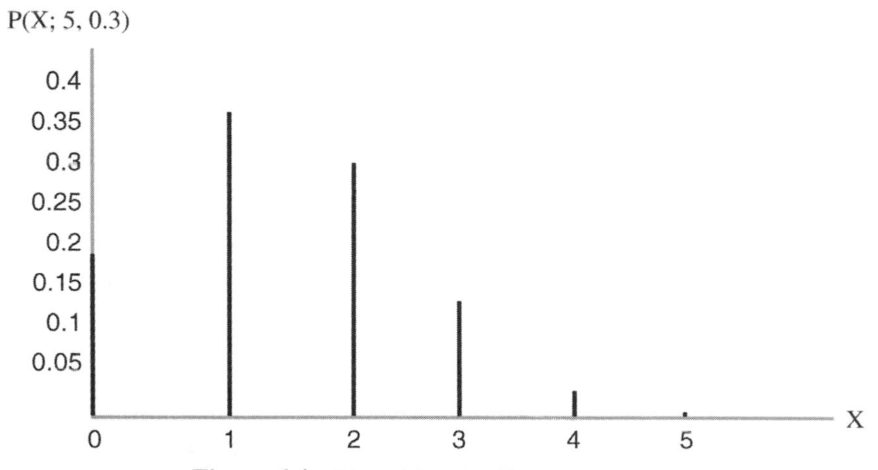

Figure 6.4 Binomial probability mass function.

SOLUTION:

a. Yes. We have a simple alternative random experiment, the trials are independent, X is discrete, and p is constant from trial to trial.

b. $P(X; n,p) = {}_nC_X\, p^X (1-p)^{n-X}$, X = 0, 1, . . . ,5. Then,

$$P\left(0;5,\frac{1}{2}\right) = \frac{5!}{0!5!}\left(\frac{1}{2}\right)0\left(\frac{1}{2}\right)5 = \frac{1}{32},$$

$$P\left(1;5,\frac{1}{2}\right) = \frac{5!}{1!4!}\left(\frac{1}{2}\right)1\left(\frac{1}{2}\right)4 = \frac{5}{32},$$

$$P\left(2;5,\frac{1}{2}\right) = \frac{5!}{2!3!}\left(\frac{1}{2}\right)2\left(\frac{1}{2}\right)3 = \frac{10}{32}.$$

Since $p = \frac{1}{2}$, the binomial distribution is symmetrical, as the accompanying table illustrates.

X	0	1	2	3	4	5
P(X; n, p)	$\frac{1}{32}$	$\frac{5}{32}$	$\frac{10}{32}$	$\frac{10}{32}$	$\frac{5}{32}$	$\frac{1}{32}$

c. E(X) = np = 5/2, V(X) = np(1 − p) = 5/4.

d. $P(X \le 1) = \frac{6}{32}$; $P(X \ge 2) = \frac{26}{32}$ or $P(X \ge 2) = 1 - P(X \le 1) = 1 - \frac{6}{32} = \frac{26}{32}$;
$P(2 < X \le 4) = \frac{10}{32} + \frac{5}{32} = \frac{15}{32}$; $P(2 < X < 4) = P(X = 3) = \frac{10}{32}$.

e. Yes. $P(2 \le X \le 4) = P(X \le 4) - P(X \le 1) = \frac{31}{32} - \frac{6}{32} = \frac{25}{32}$. ■

EXAMPLE 6.14.A

It is estimated that about 40% of drivers use a cell phone while driving, in spite of the fact that it is illegal to do so in some states. In a randomly selected group of 10 drivers behind the wheel, find the following probabilities, both by the binomial formula and by using the binomial probability tables appended at the end of this chapter: P(0; 10, 0.4), P(1; 10,0.4), P(6; 10, 0.4), and P(10; 10, 0.4).

SOLUTION: Let X stand for a binomial random variable. In this case, X represents the number of drivers using a cell phone while driving, n = 10, and the probability of using a cell phone while driving is p = 0.4. In general,

$$P(X; 10, 0.4) = {}_nC_X(0.4)^X(1-0.4)^{10-X}, X = 0, 1, \cdots, 10.$$

Using this formula,

$$P(0; 10, 0.4) = {}_{10}C_0(0.4)^0(1-0.4)^{10-0} = 1 \times 1 \times 0.6^{10} = 0.0060,$$

$$P(1; 10, 0.4) = {}_{10}C_1(0.4)^1(1-0.4)^{10-1} = 10 \times (0.4)^1 \times 0.6^9 = 0.0403,$$

$$P(6; 10, 0.4) = {}_{10}C_6(0.4)^6(1-0.4)^{10-6} = 210 \times (0.4)^6 \times 0.6^4 = 0.1115,$$

$$P(10; 10, 0.4) = {}_{10}C_{10}(0.4)^{10}(1-0.4)^{10-10} = 1 \times (0.4)^{10} \times 0.6^0 = 0.0001.$$

Table 6.11 Binomial Probabilities, n = 10 and x Is the Number of Successes

p								
n	x	0.20	0.25	0.30	0.35	0.40	0.45	0.50
10	0	0.1074	0.0563	0.0282	0.0135	**0.0060**	0.0025	0.0010
	1	0.2684	0.1877	0.1211	0.0725	**0.0403**	0.0207	0.0098
	2	0.3020	0.2816	0.2335	0.1757	0.1209	0.0763	0.0439
	3	0.2013	0.2503	0.2668	0.2522	0.2150	0.1665	0.1172
	4	0.0881	0.1460	0.2001	0.2377	0.2508	0.2384	0.2051
	5	0.0284	0.0584	0.1029	0.1536	0.2007	0.2340	0.2461
	6	0.0055	0.0162	0.0368	0.0689	**0.1115**	0.1596	0.2051
	7	0.0608	0.0031	0.0090	0.0212	0.0425	0.0746	0.1172
	8	0.0001	0.0004	0.0014	0.0043	0.0106	0.0229	0.0439
	9	0.0000	0.0000	0.0001	0.0005	0.0016	0.0042	0.0098
	10	0.0000	0.0000	0.0000	0.0000	**0.0001**	0.0003	0.0016

Alternatively, the binomial probability table (see Table A.1) can be used to find binomial probabilities rather than performing lengthy and time-consuming computations. These tables are provided for some specific popular values of n and p. These probabilities can also be determined from the computer using the Excel function, BINOM.DIST, under the function category "Statistical." Here, we deal with only the binomial probability tables appended at the end of this chapter. A portion of the table, for n = 10, is shown here for your convenience (Table 6.11).

From this table, we can obtain the same highlighted probabilities for n = 10 and p = 0.4 that were calculated above. In fact, you can see the whole binomial probability distribution for n = 10 and p = 0.4 under the 0.4 column. ■

6.5.2 Mean and Standard Deviation of the Binomial Random Variable

It can be shown that the mean, variance, and standard deviation of a binomial random variable, X, are, respectively,

$$E(X) = \mu = np,$$

$V(X) = \sigma^2 = np(1 - p) = npq$, where $q = 1 - p$, and the standard deviation of X is

$$\sigma = \sqrt{V(X)} = \sqrt{npq}.$$

It is important to present the mean and standard deviation of any binomial distribution along with all of its probabilities. These taken together present a comprehensive picture of the distribution.

EXAMPLE 6.14.B

In Example 6.14.A regarding cell phone usage by drivers, find the mean and standard deviation of X, where X is the number of drivers that use a cell phone while driving.

SOLUTION: E(X) = μ = np = 10(0.4) = 4 and the standard deviation of X is

$$\sigma = \sqrt{V(X)} = \sqrt{npq} = \sqrt{10 \times 0.4 \times 0.6} = \sqrt{2.4} = 1.5492.$$

■

6.5.3 Cumulative Binomial Probability Distribution

Oftentimes it is important to obtain *cumulative binomial probabilities*, such as the probability of obtaining "at most" a given number of defective items in a given random sample. Likewise, it is equally important to find the probability of a basketball team winning "at least" a specific number of games during a particular season. In general, the cumulative probability of a binomial random variable X can be written as

$$P(X \leq x) = \sum_{0}^{x} P(X; n, p),$$

where P(X; n, p) is the binomial probability function of X. Sometimes we may need to find a binomial probability of the form $P(X > x)$; and it can be found using the probability rule of complements $1 - P(X \leq x)$, where $P(X \leq x)$ is the cumulative binomial probability. A table of cumulative binomial probabilities for specific n and p values is also appended at the end of this chapter (Table 6.A.2). The use of cumulative binomial probabilities is illustrated by the following examples.

EXAMPLE 6.15

In Example 6.14.A, involving cell phone usage by drivers, use the information provided to find the probability of at most four drivers using a cell phone while driving. Also, find the probability of at least five drivers using the cell phone while driving.

SOLUTION: We desire to find

$$P(X \leq 4) = P(0) + P(1) + P(2) + P(3) + P(4)$$

and, from Table 6.A.1,

$$P(X \leq 4) = 0.0060 + 0.0403 + 0.1209 + 0.2150 + 0.2508 = 0.6330 \text{ (or 63.3\%)}.$$

This cumulative probability can also be determined from the cumulative probability table (Table 6.A.2) corresponding to $n = 10$ and $p = 0.4$. Any small difference is due to rounding off. Next,

$P(X \geq 5) = 1 - P(X \leq 4) = 1 - 0.6330 = 0.3670$ (or 36.7%). Here the probability rule of complements was used.

Note that for a discrete distribution, it matters whether one calculates $P(X \leq r)$ or $P(X < r)$. For instance, looking at Table 6.11,

$$P(X \leq 3) = 0.0060 + 0.0403 + 0.1209 + 0.2150 = 0.3822,$$

while

$$P(X < 3) = P(X \leq 2) = 0.0060 + 0.0403 + 0.1209 = 0.1672.$$ ■

EXAMPLE 6.16

In a college that has a division 1 basketball team, the probability of being selected to play on the team with full tuition waiver is only 5%. A group of four students applied to play on the team. For $n = 4$ and $p = 0.05$, find the probability of the following events using the binomial probability formula:

a. Probability that none got selected?

b. Probability that exactly one was selected?

c. Probability that two or more were selected?

SOLUTION: $P(X; 4, 0.05) = {}_4C_X\ (0.05)^x (0.95)^{4-0}$ for $X = 0,1,2,3,4$, where X stands for the number of the applicants selected to play on the division 1 basketball team.

a. $P(X = 0) = {}_4C_0\ (0.05)^0 (0.95)^{4-0} = (0.95)^4 = 0.8145 = 81.45\%$.

b. $P(X = 1) = {}_4C_1 (0.05)^1 (0.95)^{4-1} = 4(0.05)\ (0.95)^3 = (0.20)(0.8574) = 0.1715 = 17.15\%$.

c. $P(X \geq 2) = 1 - P(X \leq 1) = 1 - [P(0) + P(1)] = 1 - [0.8145 + 0.1715]$
$= 1 - (0.986) = 0.014 = 1.4\%$. ■

EXAMPLE 6.17

Most of the universities and colleges in the United States have study-abroad programs and very few offer fully paid scholarships to complete a 1-year study-abroad program. As it is highly competitive, there is only a 5% selection rate for these fully paid scholarships. If a talented group of six undergraduate students from a particular university apply for this special study-abroad scholarship, find the probabilities of the following events using the binomial probability formula:

a. Exactly three students win the study-abroad scholarship

b. At most two receive the scholarship

c. Three or more students receive the scholarship

SOLUTION: Let X be the number of students selected for a fully paid study-abroad scholarship program.

Given $n = 6$ and $p = 0.05$,

$$P[X; 6, 0.05] = {}_6C_x (0.05)^x (0.95)^{6-x} \quad \text{for } X = 0, 1, 2, \ldots, 6.$$

a. $P(X = 3) = {}_6C_3 (0.05)^3 (0.95)^3 = 20(0.000125)(0.857375) = 0.0021 = 0.21\%$.

b. $P(X \leq 2) = P(0) + P(1) + (P2)$, where

$$P(0) = {}_6C_0(0.05)^0(0.95)^6 = (0.95)^6 = 0.7351 = 73.51\%,$$

$$P(1) = {}_6C_1(0.05)(0.95)^5 = (0.30)(0.95)^5 = (0.30)(0.7738) = 0.2312 = 23.21\%, \text{ and}$$

$$P(2) = {}_6C_2(0.05)^2(0.95)^4 = 15(0.0025)(0.95)^4 = (0.0375)(0.8145) = 0.0305.$$

Thus, $P(X \leq 2) = 0.7351 + 0.2321 + 0.0305 = 0.9977$.

c. $P(X \geq 3) = 1 - P(X \leq 2) = 1 - 0.9977 = 0.0023 = 0.23\%$. ■

EXAMPLE 6.18

A salesman with a cell phone company makes 10 telephone calls per hour to prospective customers. Based on his experience, it is believed that there is a 20% chance that any such call results in a successful deal. Assume that the salesman spends 2hr in calling prospective customers to sell cell phone contracts. Find answers for the following questions:

a. Determine if this example satisfies the three fundamental characteristics of the binomial distribution (or Bernoulli process).

b. What are the values of the binomial parameters?

c. What is the probability of exactly three customers signing up for a contract?

d. Find the probability that at most seven customers sign up for a contract.

e. Find the probability that at least eight customers will sign up for a contract.

f. Find the mean and standard deviation of this distribution and interpret these values.

SOLUTION:

a. First, notice that there are only two possible outcomes in this example: the customer either accepts the contract (success) or does not (failure), after each telephone call from the salesman. Second, the binomial random variable X is the discrete number of successes or customers accepting the contract. The customers are independent of each other and they are given identical information by the sales agent. Third, the probability of a customer accepting the contract is constant and remains so throughout the 2 h calling period.

b. Here, $p = 20\%$ or 0.2 and $n = 20$ (the number of calls made during the 2hr period) are the two parameters of the binomial probability function.

c. In general, the binomial probability function of X given the above parameters appears as

$$P(X; 20, 0.2) = {}_{20}C_x(0.2)^x(1 - 0.2)^{20-x}, \quad \text{for } X = 0, 1, \ldots, 20.$$

In particular, for $X = 3$,

$$P(3; 20, 0.2) = {}_{20}C_3(0.2)^3(1 - 0.20)^{20-3} = \frac{20!}{3!17!}(0.2)^3(0.8)^{17} = 0.2054.$$

Note: This answer is easily obtained by using the binomial probability table (Table 6.A.1) at the end of the chapter. For example, for $n = 20$, $p = 0.2$, and $X = x = 3$, P(3) can be read as 0.2054. It is advisable in cases like this, where we have to find the probability values for numbers with high exponents such as $(0.8)^{17}$, to use the binomial probability table in order to avoid tedious calculations.

d. $P(X \leq 7) = P(0) + P(1) + P(2) + P(3) + P(4) + P(5) + P(6) + P(7)$

$= 0.0115 + 0.0576 + 0.1369 + 0.2054 + 0.2182 + 0.1746 + 0.1091 + 0.0545 = 0.9678$

(from Table 6.A.1).

This answer can also be obtained from the cumulative binomial probability table (Table 6.A.2) at the end of this chapter. For n = 20, X = 7, and p = 0.2, we obtain 0.9679 (the minor difference in the fourth decimal place is due to rounding off).

e. $P(X \geq 8) = 1 - P(1 \leq 7) = 1 - 0.9679$ (calculated above) $= 0.0321$ or 3.21%. As expected, the probability of getting eight or more customers to accept the contract is very low.

f. Mean = E(X) = np = 20(0.2) = 4, while the standard deviation of X is

$$\sqrt{npq} = \sqrt{20(0.2)(0.8)} = 1.7889.$$

Based on this value for the mean, we can say that, on average, 4 customers will accept the contract in a randomly selected group of 20 individuals, given the historical probability of 20%.

The standard deviation reflects the average variability of the binomial random variable X about its mean. ■

6.6 EXCEL APPLICATIONS

EXAMPLE 6.19

Using Excel, find the solutions for Example 6.12.

SOLUTION: Open Excel spread sheet and type X in A1, 0 in A2, 1 in A3, and so on, and finally 5 in A7.

Type P(X) in B1 and highlight B2 cell and click on f_x in the formula bar to get a dialog box. Select a category "Statistical" and select a function "BINOM.DIST" and click on OK. In the Function Arguments dialog box, highlight the cell A2 against "Number_s" white box, type "5"in Trials white box, type "0.3" against "Probability_s" white box, and type "0" for false in the white box for "Cumulative." Then click the "OK" button. You will see "0.16807" in the B2 cell, which represents $P(X = 0|n = 5, p = 0.3)$. Highlight the B2 cell, go to right bottom corner of the B2 cell, where the cursor becomes black cross and click and drag down to cell B7 to copy the formula. Then you will see the probability distribution, as given below. Notice that the values of P(X) add up to 1, approximately.

	A	B
1	X	P(X)
2	0	0.16807
3	1	0.36015
4	2	0.3087
5	3	0.1323
6	4	0.02835
7	5	0.00243

■

EXAMPLE 6.20

Using Excel, find the solutions for Example 6.16.

SOLUTION: Opcn Excel spread sheet and type X in A1 cell and type "≤4" in A2 cell.

Type P(X) in B1 and highlight the B2 cell and click on f_x in the formula bar to get a dialog box. Select a category "Statistical" and select a function "BINOM.DIST" and click on OK. In the Function Arguments dialog box, type "4" against "Numbers" white box, type "10"in Trials white box, type "0.4" against "Probabilities" white box, and type "1" for true in the white box for "Cumulative." Then click the "OK" button. You will see "0.633103" in the B2 cell, which represents $P(X \leq 4|n=10,\ p=0.4)$ or the probability of at most four drivers using a cell phone while driving.

To answer the other question, type "≥5" in cell A3. Highlight B3 and enter a formula (Rule of complements) by typing "=1 – highlight B2 cell" and hit the Enter button. Then you see 0.366897 in the B3 cell. The complete solution on the Excel spreadsheet is given below:

	A	B
1	X	P(X)
2	≤4	0.633103
3	≥5	0.366897

■

EXAMPLE 6.21

Answer the following questions for the "Gross Rent" data given in Table 6.12.

For the discrete probability distribution shown, compute the mean, variance, and standard deviation using the Function Key as well as the long and the short formulas in Excel.

Note: The starred values of X, corresponding to the class marks are used since the first and last classes are open-ended.

Profile of Selected Housing Characteristics

Table 6.12 Gross Rent

Gross Rent ($)	X ($)	Renter-Occupied Units (Frequency)	Probability – P(X) (Relative Frequency)
Less than 200	100*	1,447,098	0.04
200–299	250	1,537,171	0.05
300–499	400	6,733,722	0.20
500–749	625	11,905,100	0.35
750–999	875	7,058,164	0.21
1,000–1,499	1,250	3,835,217	0.11
1,500 or more	1,750*	1,369,610	0.04
Total		33,886,082	1.00

Source: 2001 Supplementary Survey—U.S. Census Bureau, Profile of Selected Housing Characteristics.

SOLUTION:

= B6 * C6 = B6 − 706.5 =power(E6,2) = C6 * F6 = power(B6,2) = C6 * H6

X	P(X)	XP(X)	X − μ	(X − μ)²	(X − μ)²P(X)	X²	X²P(X)
100	0.04	4	-606.5	367842.25	14713.69	10000	400
250	0.05	12.5	-456.5	208392.25	10419.6125	62500	3125
400	0.2	80	-306.5	93942.25	18788.45	160000	32000
625	0.35	218.75	-81.5	6642.25	2324.7875	390625	136718.75
875	0.21	183.75	168.5	28392.25	5962.3725	765625	160781.25
1250	0.11	137.5	543.5	295392.25	32493.1475	1562500	171875
1750	0.04	70	1043.5	1088892.25	43555.69	3062500	122500
		μ = 706.5			128257.75		627400

= sum(D6:D12) = sum(G6:G12) = sum(I6:I12)

For copying the formula to other cells in column I you have to click and drag from the right bottom corner of cell I6, where you see the black plus sign (called fill handle) to the last value, which is I12.

VARIANCE

Long formula:

$$\sigma^2 = \sum (X_i - \mu)^2 P(X_i) = 128,257.75.$$

Short formula:

$$\sigma^2 = \sum X_i^2 P(X_i) - \mu^2 = 627,400 - (706.5)^2 = 627,400 - 499,142.25 = 128,257.75.$$

STANDARD DEVIATION

Long formula:

$$\sigma = \sqrt{\sum (X_i - \mu)^2 P(X)} = \sqrt{128,257.75} = 358.13.$$

Short formula:

$$\sigma = \sqrt{\sum X_i^2 P(X_i) - \mu^2} = \sqrt{128,257.75} = 358.13.$$

■

EXAMPLE 6.22

Answer the following questions for the "Monthly Owner Costs" data given in Table 6.13.

For the following discrete probability distribution, compute the mean, variance, and standard deviation using the Function Key as well as the long and the short formulas in Excel.

Note: The starred X values, corresponding to the class marks, are used since the first and last classes are open-ended.

Profile of Selected Housing Characteristics

Table 6.13 Mortgage Level and Related Monthly Owner Costs

Housing Units with a Mortgage	X ($) (Class Mark)	Housing Units (Frequency)	P(X) (Relative Frequency)
Less than 300	200*	200,695	0.01
300–499	400	1,718,878	0.04
500–699	600	4,272,986	0.11
700–999	850	9,163,792	0.23
1,000–1,499	1,250	12,140,627	0.31
1,500–1,999	1,750	6,216,065	0.16
2,000 or more	2,250*	5,334,777	0.14
Total		39,047,820	1.00

Source: 2001 Supplementary Survey—U.S. Census Bureau, Profile of Selected Housing Characteristics.

SOLUTION:

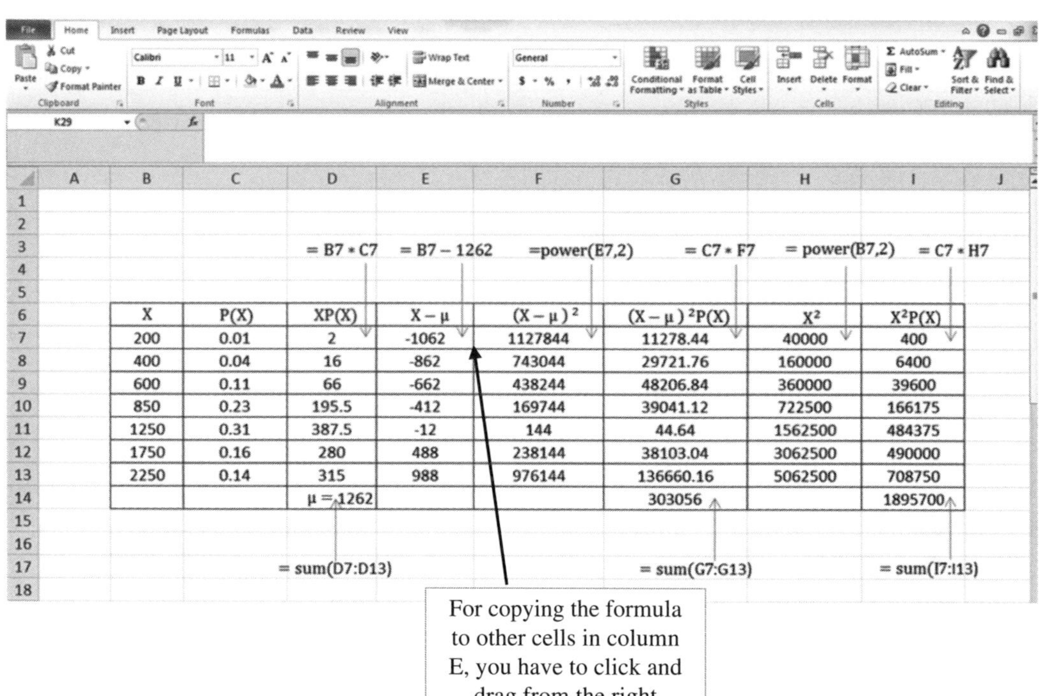

X	P(X)	XP(X)	X − μ	(X − μ)²	(X − μ)²P(X)	X²	X²P(X)
200	0.01	2	-1062	1127844	11278.44	40000	400
400	0.04	16	-862	743044	29721.76	160000	6400
600	0.11	66	-662	438244	48206.84	360000	39600
850	0.23	195.5	-412	169744	39041.12	722500	166175
1250	0.31	387.5	-12	144	44.64	1562500	484375
1750	0.16	280	488	238144	38103.04	3062500	490000
2250	0.14	315	988	976144	136660.16	5062500	708750
		μ = 1262			303056		1895700

VARIANCE

Long formula:

$$\sigma^2 = \sum (X_i - \mu)^2 P(X_i) = 303,056.$$

Short formula:

$$\sigma^2 = \sum X_i^2 P(X_i) - \mu^2 = 1,895,700 - (1262)^2 = 1,895,700 - 1,592,644 = 303,056.$$

STANDARD DEVIATION

Long formula:

$$\sigma = \sqrt{\sum (X_i - \mu)^2 P(X_i)} = \sqrt{303,056} = 550.505.$$

Short formula:

$$\sigma = \sqrt{\sum X_i^2 P(X_i) - \mu^2} = \sqrt{303,056} = 550.505.$$

■

CHAPTER 6 REVIEW

You should be able to:

1. Distinguish between discrete and continuous random variables.
2. State the properties of a discrete probability distribution.
3. State the meaning of the term "expected value" of a discrete random variable.
4. Explain the role of payoff tables in decision making.
5. State the properties of a probability density function.
6. State the characteristics of a simple alternative experiment.
7. Distinguish between a permutation and a combination.
8. Define the concept of a binomial random variable.
9. Identify the (two) parameters of a binomial probability distribution.

Key Terms and Concepts:

Bernoulli process, 232
binomial probability, 232
binomial random experiment, 232
combination, 233
continuous random variable, 219
cumulative binomial probability, 238
discrete probability distribution, 220
discrete random variable, 219
expected profit, 247
expected value of a discrete random variable, 222
standard deviation of a discrete random variable, 222
maximum potential profit, 228
opportunity loss, 228
payoff table, 228
payoff values, 228
permutation, 233
probability density function, 230
simple alternative experiment, 232
uniform probability distribution, 230

EXERCISES

1. Determine which of the following are legitimate discrete probability distributions. Justify your answers.

a.

X	P(X)
−3	0.8
2	0.2
4	0.1
10	−0.1

b.

X	P(X)
1	0.4
3	0.3
5	0.3
7	0.2

c.

X	P(X)
0	0.2
2	0.3
3	0.3
4	0.2

2. According to the U.S. Census Bureau, health insurance coverage for 1999 had the following characteristics: 16.5% of males and 14.5% of females were without health insurance in United States. Answer the following questions, assuming that a male and a female were selected at random.

a. Find the probability that both a male and a female were covered by health insurance.

b. Find the probability that only a male had health insurance.

c. What is the probability that one of them had health insurance?

(*Hint:* Use the notation: M_c stands for a male covered by health insurance, M_n for a male that was not covered by health insurance. Similarly, let F_c stand for a female covered and F_n for a female not covered by health insurance.)

3. Find the expected value, variance, and standard deviation of the number of heads obtained in tossing a fair coin three times (or three coins once).

4. The probability distribution of the daily closing stock price of Google, based on data from August and September of 2015, is estimated in the table below.

(This table was formed from 2 months of daily data.)

Daily Closing Stock Price of Google (August and September 2015)

Closing price (X)	591	610	629	648	667	686
P(X)	0.10	0.21	0.40	0.17	0.10	0.02

Source: NASDAQ, August and September 2015.

a. Show that this table is actually a probability distribution.

b. Find the expected closing price per share of the Google stock.

c. Find the expected gain (or loss) per share, assuming that the investor paid $410 per share. Also find the expected percentage return per share.

d. Another similar technology stock produced an expected price of $640 for the same period with a standard deviation of $15. Which stock appears to be a better investment if the goal is to minimize the risk?

5. Some insurance agents are paid a basic salary plus a commission based on the number of customers. Assume that the annual salary is given by the relation $Y(\$) = 80{,}000 + 0.4(100X)$, where X is the number of customers serviced by the agent and each customer pays $100 to the company from which the agent gets 40%. Based on prior experience, the underwriting department of the insurance company estimated the probability distribution of X as follows.

Number of Customers

X (in thousands)	1	2	3	4
P(X)	0.2	0.3	0.3	0.2

Find the expected annual salary of an insurance agent and its standard deviation.

6. Generally, contractors participate in any bidding process only if the expected return exceeds a predetermined cost. Suppose a contractor has estimated that the cost of preparing a proposal for a highway bridge project with an expected net return of at least $10 million is $200,000. Based on his experience, the contractor determines that his chance of winning the contract is 30%. Find the expected net income after all expenses in preparing the proposal.

The concept of expected value can also be used in decision making under uncertainty. The decision criterion could be selecting the strategy that maximizes the expected profit (payoff) or minimizes the expected loss (or opportunity loss).

7. Answer the following questions for the Gross Rent data given below.

a. We have generated the following discrete probability distribution. Compute the mean, variance, and standard deviation for this distribution by using the Function Key of Excel.

b. Using Excel, draw the probability mass function for this distribution. Use the X values on the X-axis and probability P(X) on the Y-axis.

Gross Rent

Gross Rent ($)	X($) (Class Mark)	Renter-Occupied Units (Frequency)	Probability P(X) (Relative Frequency)
Less than 200	100*	1,447,098	0.04
200–299	250	1,537,171	0.05
300–499	400	6,733,722	0.20
500–749	625	11,905,100	0.35
750–999	875	7,058,164	0.21
1,000–1,499	1,250	3,835,217	0.11
1,500 or more	1,750*	1,369,610	0.04
Total		33,886,082	1.00

Note: The starred values of X, the class marks, are used due to open-ended classes.

Source: 2001 Supplementary Survey—U.S. Census Bureau, Table 4, Profile of Selected Housing Characteristics. Available at www.census.gov/acs/www/products/profiles/single/2001/ss01/tabular/010/01000us4.htm.

8. Answer the following questions for the income and benefits data given in the accompanying table.

a. We have generated the following discrete probability distribution. Compute the mean, variance, and standard deviation for this distribution using the Function Key and the long and the short formulas using Excel.

b. Using Excel, draw the probability mass function for this distribution. Use the X values on the X-axis and probability P(X) on the Y-axis.

Income and Benefits (in 2001 Inflation-Adjusted Dollars)

Income Benefits (in $)	X($) (Class Mark)	Number of Households (Frequency)	Probability P(X) (Relative Frequency)
Less than 10,000	5,000*	9,778,717	0.09188
10,000–14,999	12,500	6,873,366	0.064582
15,000–24,999	20,000	13,724,781	0.128957
25,000–34,999	30,000	13,471,179	0.126574
35,000–49,999	42,500	17,302,272	0.162571
50,000–74,999	62,500	20,649,208	0.194019
75,000–99,999	87,500	11,063,380	0.103951
100,000–149,999	125,000	8,745,884	0.082176
150,000–199,999	175,000	2,482,210	0.023323
200,000 or more	225,000*	2,337,860	0.021966
Total households		106,428,857	1.00

Note: X values with * are used due to open-ended classes.

Source: 2001 Supplementary Survey—U.S. Census Bureau, Table 3, Profile of Selected Economic Characteristics. Available at www.census.gov/acs/www/products/profiles/single/2001/ss01/tabular/010/01000us3.htm.

9. Answer the following questions for the housing characteristics data given below.

a. We have generated the following discrete probability distribution. Compute the mean, variance, and standard deviation for this distribution by using the Function Key and the long and the short formulas using Excel.

b. Using Excel, draw the probability mass function for this distribution. Use the X values on the X-axis and probability P(X) on the Y-axis.

Housing Characteristics: Rooms

Rooms (X)	Number of Houses (Frequency)	Probability P(X) (Relative Frequency)
1	1,856,987	0.01576
2	4,400,070	0.03734
3	11,630,310	0.09871
4	20,958,754	0.17788
5	25,986,847	0.22056
6	21,886,935	0.18576
7	13,898,567	0.11796
8	8,876,069	0.07533
9 or more (let X = 10)	8,329,595	0.07070
Total	117,824,134	1.00

Source: 2001 Supplementary Survey—U.S. Census Bureau, Table 4, Profile of Selected Housing Characteristics. Available at www.census.gov/acs/www/products/profiles/single/2001/ss01/tabular/010/01000us4.htm.

10. Answer the following questions for the monthly owner costs data given below:

a. We have generated the following discrete probability distribution. Compute the mean, variance, and standard deviation for this distribution by using the Function Key of Excel.

b. Using Excel, draw the probability mass function for this distribution. Use the X values on the X-axis and probability, P(X), on the Y-axis.

Mortgage Level and Related Monthly Owner Costs

Housing Units with a Mortgage ($)	X ($) (Class Mark)	Frequency (Housing Units)	Probability P(X) (Relative Frequency)
Less than 300	200*	200,695	0.01
300–499	400	1,718,878	0.04
500–699	600	4,272,986	0.11
700–999	850	9,163,792	0.23
1,000–1,499	1,250	12,140,627	0.31
1,500–1,999	1,750	6,216,065	0.16
2,000 or more	2,250*	5,334,777	0.14
Total		39,047,820	1.00

Note: The X values with * are used due to open-ended classes.

Source: 2001 Supplementary Survey—U.S. Census Bureau, Table 4, Profile of Selected Housing Characteristics. Available at www.census.gov/acs/www/products/profiles/single/2001/ss01/tabular/010/01000us4.htm.

11. A contractor builds expensive homes, each costing $1.1 million, and lists them at a sale price of $1.5 million. His options are to build subdivisions of 10, 20, or 30 homes at a time, depending on the demand for real estate housing. The demand for real estate housing depends on the state of the economy. The probabilities of different states of the economy are estimated as shown below.

State of the Economy

State of the Economy (X)	P(X)
Down	0.2
Same as now	0.3
Boom	0.5

Here X represents the demand for luxury homes.

a. Construct a payoff table.

b. Find the optimal decision based on expected payoff.

c. Construct an opportunity loss table.

d. What is your optimal decision based on expected opportunity loss?

12. An investor is considering investing in one of three avenues of investment. The table below presents the estimated rates of return (%), depending on the nature of the market (X). It is estimated that the market is down 20% of the time, stays the same about 40% of the time, and is up 40% of the time.

Investment Options and Market Conditions

X	Low	Stays the Same	Up
P(X)	0.2	0.4	0.4
Investment avenues			
Stocks	4	10	16
Bonds	3	4	6
Money market	2	5	7

Find the investor's optimal decision.

13. The stock price of a company goes up or down by the close of the day every day. It is known from past experience that, on the average, the price goes up 20% of the time. Find the probability that the stock price goes up (a) exactly 2 days, (b) at least 2 days, (c) at most 2 days, and (d) none in the last 4 days, using the binomial formula.

14. It is reported that 70% of flights operated by JetBlue Airways from Fort Lauderdale to New York City arrived on time during August 15, 2015–October 15, 2015. In a sample of three flights, find the probability that a flight will be on time: (a) exactly once, (b) three times, (c) not more than two times, and (d) never. What is the average number of flights that are on time in the sample of three flights?

15. Based on historical information, it was found that approximately 10% of female employees with an MBA degree and with job experience of at least 10 years have become the CEOs of major corporations. Consider a group of 15 female employees in U.S. major corporations with an MBA degree and having10 years of job experience. Answer the following questions.

a. Explain how this example problem fits the binomial distribution via the three basic characteristics of the binomial distribution.

b. Find the parameters of the binomial distribution.

c. Find the probability that exactly three out of the 15 will become CEOs of major corporations.

d. Find the probability that at most three will become CEOs.

e. Find the probability that at least four will become CEOs.

f. Find the mean and standard deviation of the implied distribution and interpret your values.

Hint: First show the binomial probability function for $n = 15$ and $p = 0.1$ and then find the desired probabilities using the tables for binomial probabilities.

16. As per recent surveys, approximately 15% of the U.S. population does not have health insurance coverage. For a randomly selected group of 10 individuals, answer the following:

a. Find the probability that exactly four do not have health insurance coverage.

b. Find the probability that no one is covered by health insurance.

c. Find the probability that all 10 are covered by health insurance.

d. Find the probability that at least two individuals are covered by health insurance.

e. Find the mean and standard deviation and interpret these values.

APPENDIX 6.A

Table 6.A.1

Binomial Probabilities (X is $b(X; n, p)$)

Given n and p, the table gives the binomial probability of $X = r$ successes or $P(X = r; n, p) = \binom{n}{r} p^r (1 - p)^{n-r}, r = 0, 1, \ldots, n$. For $p > 0.50$, $P(X = r; n, p) = P(X = n - r; n, 1 - p)$ (e.g., $P(3; 10, 0.40) = 0.2150$; $P(4; 10, 0.70) = P(6; 10, 0.30) = 0.0368$).

		p									
n	r	**0.05**	**0.10**	**0.15**	**0.20**	**0.25**	**0.30**	**0.35**	**0.40**	**0.45**	**0.50**
1	**0**	0.9500	0.9000	0.8500	0.8000	0.7500	0.7000	0.6500	0.6000	0.5500	0.5000
	1	0.0500	0.1000	0.1500	0.2000	0.2500	0.3000	0.3500	0.4000	0.4500	0.5000
2	**0**	0.9025	0.8100	0.7225	0.6400	0.5625	0.4900	0.4225	0.3600	0.3025	0.2500
	1	0.0950	0.1800	0.2550	0.3200	0.3750	0.4200	0.4550	0.4800	0.4950	0.5000
	2	0.0025	0.0100	0.2550	0.0400	0.0625	0.0900	0.1225	0.1600	0.2025	0.2500
3	**0**	0.8574	0.7290	0.6141	0.5120	0.4219	0.3430	0.2746	0.2160	0.1664	0.1250
	1	0.1354	0.2430	0.3251	0.3840	0.4219	0.4410	0.4436	0.4320	0.4084	0.3750
	2	0.0071	0.0270	0.0574	0.0960	0.1406	0.1890	0.2389	0.2880	0.3341	0.3750
	3	0.0001	0.0010	0.0034	0.0080	0.0156	0.0270	0.0429	0.0640	0.0911	0.1250
4	**0**	0.8145	0.6561	0.5220	0.4098	0.3164	0.2401	0.1785	0.1296	0.0915	0.0625
	1	0.1715	0.2916	0.3685	0.4096	0.4219	0.4116	0.3845	0.3456	0.2995	0.2500
	2	0.0135	0.0486	0.0975	0.1536	0.2109	0.2646	0.3105	0.3456	0.3675	0.3750
	3	0.0005	0.0036	0.0115	0.0256	0.0469	0.0756	0.1115	0.1536	0.2005	0.2500
	4	0.0000	0.0001	0.0005	0.0016	0.0039	0.0081	0.0150	0.0256	0.0410	0.0625
5	**0**	0.7738	0.5905	0.4437	0.3277	0.2373	0.1681	0.1160	0.0778	0.0503	0.0312
	1	0.2036	0.3280	0.3916	0.4096	0.3955	0.3602	0.3124	0.2592	0.2059	0.1562
	2	0.0214	0.0729	0.1382	0.2048	0.2637	0.3087	0.3364	0.3456	0.3369	0.3125
	3	0.0011	0.0081	0.0244	0.0512	0.0879	0.1323	0.1811	0.2304	0.2757	0.3125
	4	0.0000	0.0004	0.0022	0.0064	0.0146	0.0284	0.0488	0.0768	0.1128	0.1562
	5	0.0000	0.0000	0.0001	0.0003	0.0010	0.0024	0.0053	0.0102	0.0185	0.0312
6	**0**	0.7351	0.5314	0.3771	0.2621	0.1780	0.1176	0.0754	0.0467	0.0277	0.0156
	1	0.2321	0.3543	0.3993	0.3932	0.3560	0.3025	0.2437	0.1866	0.1359	0.0938
	2	0.0305	0.0984	0.1762	0.2458	0.2966	0.3241	0.3280	0.3110	0.2780	0.2344
	3	0.0021	0.0146	0.0415	0.0819	0.1318	0.1852	0.2355	0.2785	0.3032	0.3125
	4	0.0001	0.0012	0.0055	0.0154	0.0330	0.0595	0.0951	0.1382	0.1861	0.2344
	5	0.0000	0.0001	0.0004	0.0015	0.0044	0.0102	0.0205	0.0369	0.0609	0.0938
	6	0.0000	0.0000	0.0000	0.0001	0.0002	0.0007	0.0018	0.0041	0.0083	0.0156
7	**0**	0.6983	0.4783	0.3206	0.2097	0.1335	0.0824	0.0490	1.0280	0.0152	0.0078
	1	0.2573	0.3720	0.3960	0.3670	0.3115	0.2471	0.1848	0.1306	0.0872	0.0547
	2	0.0406	0.1240	0.2097	0.2753	0.3115	0.3177	0.2985	0.2613	0.2140	0.1641
	3	0.0036	0.0230	0.0617	0.1147	0.1730	0.2269	0.2679	0.2903	0.2918	0.2734
	4	0.0002	0.0026	0.0109	0.0287	0.0577	0.0972	0.1442	0.1935	0.2388	0.2734
	5	0.0000	0.0002	0.0012	0.0043	0.0115	0.0250	0.0466	0.0774	0.1172	0.1641
	6	0.0000	0.0000	0.0001	0.0004	0.0013	0.0036	0.0084	0.0172	0.0320	0.0547
	7	0.0000	0.0000	0.0000	0.0000	0.0001	0.0002	0.0006	0.0016	0.0037	0.0078

Source: Adapted from M. J. Panik, *Statistical Inference: A Short Course*. New Jersey: John Wiley & Sons, Inc., 2012, Tables A5 and A.6, pp. 334–341.

n	r	p									
		0.05	**0.10**	**0.15**	**0.20**	**0.25**	**0.30**	**0.35**	**0.40**	**0.45**	**0.50**
8	**0**	0.6634	0.4305	0.2725	0.1678	0.1001	0.0576	0.0319	0.0168	0.0084	0.0039
	1	0.2793	0.3826	0.3847	0.3355	0.2760	0.1977	0.1373	0.0896	0.0548	0.0312
	2	0.0515	0.1488	0.2376	0.2936	0.3115	0.2965	0.2587	0.2090	0.1569	0.1094
	3	0.0054	0.0331	0.0839	0.1468	0.2076	0.2541	0.2786	0.2787	0.2568	0.2188
	4	0.0004	0.0046	0.0185	0.0459	0.0865	0.1361	0.1875	0.2322	0.2627	0.2734
	5	0.0000	0.0004	0.0026	0.0092	0.0231	0.0467	0.0808	0.1239	0.1719	0.2188
	6	0.0000	0.0000	0.0002	0.0011	0.0038	0.0100	0.0217	0.0413	0.0703	0.1094
	7	0.0000	0.0000	0.0000	0.0001	0.0004	0.0012	0.0033	0.0079	0.0164	0.0312
	8	0.0000	0.0000	0.0000	0.0000	0.0000	0.0001	0.0002	0.0007	0.0017	0.0039
9	**0**	0.6302	0.3874	0.2316	0.1342	0.0751	0.0404	0.0277	0.0101	0.0046	0.0020
	1	0.2985	0.3874	0.3679	0.3020	0.2253	0.1556	0.1004	0.0605	0.0339	0.0176
	2	0.0629	0.1722	0.7260	0.3020	0.3003	0.2668	0.2162	0.1612	0.1110	0.0703
	3	0.0077	0.0446	0.1069	0.1762	0.2336	0.2668	0.2716	0.2508	0.2119	0.1641
	4	0.0006	0.0074	0.0283	0.0661	0.1168	0.1715	0.2194	0.2508	0.2600	0.2461
	5	0.0000	0.0008	0.0050	0.0165	0.0389	0.0735	0.1181	0.1672	0.2128	0.2461
	6	0.0000	0.0001	0.0006	0.0028	0.0087	0.0210	0.0424	0.0743	0.1160	0.1641
	7	0.0000	0.0000	0.0000	0.0003	0.0012	0.0039	0.0098	0.0212	0.0407	0.0703
	8	0.0000	0.0000	0.0000	0.0000	0.0001	0.0004	0.0013	0.0035	0.0083	0.0176
	9	0.0000	0.0000	0.0000	0.0000	0.0000	0.0000	0.0001	0.0003	0.0008	0.0020
10	**0**	0.5987	0.3487	0.1969	0.1074	0.0563	0.0282	0.0135	0.0060	0.0025	0.0010
	1	0.3151	0.3874	0.3474	0.2684	0.1877	0.1211	0.0725	0.0403	0.0207	0.0098
	2	0.0746	0.1937	0.2759	0.3020	0.2816	0.2335	0.1757	0.1209	0.0763	0.0439
	3	0.0105	0.0574	0.1298	0.2013	0.2503	0.2668	0.2522	0.2150	0.1665	0.1172
	4	0.0010	0.0112	0.0401	0.0881	0.1460	0.2001	0.2377	0.2508	0.2384	0.2051
	5	0.0001	0.0015	0.0085	0.0284	0.0584	0.1029	0.1536	0.2007	0.2340	0.2461
	6	0.0000	0.0001	0.0012	0.0055	0.0162	0.0368	0.0689	0.1115	0.1596	0.2051
	7	0.0000	0.0000	0.0001	0.0608	0.0031	0.0090	0.0212	0.0425	0.0746	0.1172
	8	0.0000	0.0000	0.0000	0.0001	0.0004	0.0014	0.0043	0.0106	0.0229	0.0439
	9	0.0000	0.0000	0.0000	0.0000	0.0000	0.0001	0.0005	0.0016	0.0042	0.0098
	10	0.0000	0.0000	0.0000	0.0000	0.0000	0.0000	0.0000	0.0001	0.0003	0.0016
11	**0**	0.5688	0.3138	0.1673	0.0859	0.0422	0.0198	0.0088	0.0036	0.0014	0.0005
	1	0.3293	0.3835	0.3248	0.2362	0.1549	0.0932	0.0518	0.0266	0.0125	0.0054
	2	0.0867	0.2131	0.2866	0.2953	0.2581	0.1998	0.1395	0.0887	0.0513	0.0269
	3	0.0137	0.0710	0.1517	0.2215	0.2581	0.2568	0.2254	0.1774	0.1259	0.0806
	4	0.0014	0.0158	0.0536	0.1107	0.1721	0.2201	0.2428	0.2365	0.2060	0.1611
	5	0.0001	0.0025	0.0132	0.0388	0.0803	0.1231	0.1830	0.2207	0.2360	0.2256
	6	0.0000	0.0003	0.0023	0.0097	0.0268	0.0566	0.0985	0.1471	0.1931	0.2256
	7	0.0000	0.0000	0.0003	0.0017	0.0064	0.0173	0.0379	0.0701	0.1128	0.1611
	8	0.0000	0.0000	0.0000	0.0002	0.0011	0.0037	0.0102	0.0234	0.0462	0.0806
	9	0.0000	0.0000	0.0000	0.0000	0.0001	0.0005	0.0018	0.0052	0.0126	0.0269
	10	0.0000	0.0000	0.0000	0.0000	0.0000	0.0000	0.0002	0.0007	0.0021	0.0054
	11	0.0000	0.0000	0.0000	0.0000	0.0000	0.0000	0.0000	0.0000	0.0002	0.0005

n	r	p									
		0.05	0.10	0.15	0.20	0.25	0.30	0.35	0.40	0.45	0.50
12	0	0.5404	0.2824	0.1422	0.0687	0.0317	0.0138	0.0057	0.0022	0.0008	0.0002
	1	0.3413	0.3766	0.3012	0.2062	0.1267	0.0712	0.0368	0.0174	0.0075	0.0029
	2	0.0988	0.2301	0.2924	0.2835	0.2323	0.1678	0.1088	0.0639	0.0339	0.0161
	3	0.0173	0.0852	0.1720	0.2362	0.2581	0.2397	0.1954	0.1419	0.0923	0.0537
	4	0.0021	0.0213	0.0683	0.1329	0.1936	0.2311	0.2367	0.2128	0.1700	0.1208
	5	0.0002	0.0038	0.0193	0.0532	0.1032	0.1585	0.2039	0.2270	0.2225	0.1934
	6	0.0000	0.0005	0.0040	0.0155	0.0401	0.0792	0.1281	0.1766	0.2124	0.2256
	7	0.0000	0.0000	0.0006	0.0033	0.0115	0.0291	0.0591	0.1009	0.1489	0.1934
	8	0.0000	0.0000	0.0001	0.0005	0.0024	0.0078	0.0199	0.0420	0.0762	0.1208
	9	0.0000	0.0000	0.0000	0.0001	0.0004	0.0015	0.0048	0.0125	0.0277	0.0537
	10	0.0000	0.0000	0.0000	0.0000	0.0000	0.0002	0.0008	0.0025	0.0068	0.0161
	11	0.0000	0.0000	0.0000	0.0000	0.0000	0.0000	0.0001	0.0003	0.0010	0.0029
	12	0.0000	0.0000	0.0000	0.0000	0.0000	0.0000	0.0000	0.0000	0.0001	0.0002
13	0	0.5133	0.2542	0.1209	0.0550	0.0238	0.0097	0.0037	0.0013	0.0004	0.0001
	1	0.3512	0.3672	0.2774	0.1787	0.1029	0.0540	0.0259	0.0113	0.0045	0.0016
	2	0.1109	0.2448	0.2937	0.2680	0.2059	0.1388	0.0836	0.0453	0.0220	0.0095
	3	0.0214	0.0997	0.1900	0.2457	0.2517	0.2181	0.1651	0.1107	0.0660	0.0349
	4	0.0028	0.0277	0.0838	0.1535	0.2097	0.2337	0.2222	0.1845	0.1350	0.0873
	5	0.0003	0.0055	0.0266	0.0691	0.1258	0.1803	0.2154	0.2214	0.1989	0.1571
	6	0.0000	0.0008	0.0063	0.0230	0.0559	0.1030	0.1546	0.1968	0.2169	0.2095
	7	0.0000	0.0001	0.0011	0.0058	0.0186	0.0442	0.0833	0.1312	0.1775	0.2095
	8	0.0000	0.0000	0.0001	0.0011	0.0047	0.0142	0.0336	0.0656	0.1089	0.1571
	9	0.0000	0.0000	0.0000	0.0001	0.0009	0.0034	0.0101	0.0243	0.0495	0.0873
	10	0.0000	0.0000	0.0000	0.0000	0.0001	0.0006	0.0022	0.0065	0.0162	0.0349
	11	0.0000	0.0000	0.0000	0.0000	0.0000	0.0001	0.0003	0.0012	0.0036	0.0095
	12	0.0000	0.0000	0.0000	0.0000	0.0000	0.0000	0.0000	0.0001	0.0005	0.0016
	13	0.0000	0.0000	0.0000	0.0000	0.0000	0.0000	0.0000	0.0000	0.0000	0.0001
14	0	0.4877	0.2288	0.1028	0.0440	0.0178	0.0068	0.0024	0.0008	0.0002	0.0001
	1	0.3593	0.3559	0.2539	0.1539	0.0832	0.0407	0.0181	0.0073	0.0027	0.0009
	2	0.1229	0.2570	0.2912	0.2501	0.1802	0.1134	0.0634	0.0317	0.0141	0.0056
	3	0.0259	0.1142	0.2056	0.2501	0.2402	0.1943	0.1366	0.0845	0.0462	0.0222
	4	0.0037	0.0349	0.0998	0.1720	0.2202	0.2290	0.2022	0.1549	0.1040	0.0611
	5	0.0004	0.0078	0.0352	0.0860	0.1468	0.1963	0.2178	0.2066	0.1701	0.1222
	6	0.0000	0.0013	0.0093	0.0322	0.0734	0.1262	0.1759	0.2066	0.2088	0.1833
	7	0.0000	0.0002	0.0019	0.0092	0.0280	0.0618	0.1082	0.1574	0.1952	0.2095
	8	0.0000	0.0000	0.0003	0.0020	0.0082	0.0232	0.0510	0.0918	0.1398	0.1833
	9	0.0000	0.0000	0.0000	0.0003	0.0018	0.0066	0.0183	0.0408	0.0762	0.1222
	10	0.0000	0.0000	0.0000	0.0000	0.0003	0.0014	0.0049	0.0136	0.0312	0.0611
	11	0.0000	0.0000	0.0000	0.0000	0.0000	0.0002	0.0010	0.0033	0.0093	0.0222
	12	0.0000	0.0000	0.0000	0.0000	0.0000	0.0000	0.0001	0.0005	0.0019	0.0056
	13	0.0000	0.0000	0.0000	0.0000	0.0000	0.0000	0.0000	0.0001	0.0002	0.0009
	14	0.0000	0.0000	0.0000	0.0000	0.0000	0.0000	0.0000	0.0000	0.0000	0.0001

n	r	p 0.05	0.10	0.15	0.20	0.25	0.30	0.35	0.40	0.45	0.50
15	**0**	0.4633	0.2059	0.0874	0.0352	0.0134	0.0047	0.0016	0.0005	0.0001	0.0000
	1	0.3658	0.3432	0.2312	0.1319	0.0668	0.0305	0.0126	0.0047	0.0016	0.0005
	2	0.1348	0.2669	0.2856	0.2309	0.1559	0.0916	0.0476	0.0219	0.0090	0.0032
	3	0.0307	0.1285	0.2184	0.2501	0.2252	0.1700	0.1110	0.0634	0.0318	0.0139
	4	0.0049	0.0428	0.1156	0.1876	0.2252	0.2186	0.1792	0.1268	0.0780	0.0417
	5	0.0006	0.0105	0.0449	0.1032	0.1651	0.2061	0.2123	0.1859	0.1404	0.0916
	6	0.0000	0.0019	0.0132	0.0430	0.0917	0.1472	0.1906	0.2066	0.1914	0.1527
	7	0.0000	0.0003	0.0030	0.0138	0.0393	0.0811	0.1319	0.1771	0.2013	0.1964
	8	0.0000	0.0000	0.0005	0.0035	0.0131	0.0348	0.0710	0.1181	0.1647	0.1964
	9	0.0000	0.0000	0.0001	0.0007	0.0034	0.0116	0.0298	0.0612	0.1048	0.1527
	10	0.0000	0.0000	0.0000	0.0001	0.0007	0.0030	0.0096	0.0245	0.0515	0.0916
	11	0.0000	0.0000	0.0000	0.0000	0.0001	0.0006	0.0024	0.0074	0.0191	0.0417
	12	0.0000	0.0000	0.0000	0.0000	0.0000	0.0001	0.0004	0.0016	0.0052	0.0139
	13	0.0000	0.0000	0.0000	0.0000	0.0000	0.0000	0.0001	0.0003	0.0010	0.0032
	14	0.0000	0.0000	0.0000	0.0000	0.0000	0.0000	0.0000	0.0000	0.0001	0.0005
	15	0.0000	0.0000	0.0000	0.0000	0.0000	0.0000	0.0000	0.0000	0.0000	0.0000
16	**0**	0.4401	0.1853	0.0743	0.0281	0.0100	0.0033	0.0010	0.0003	0.0001	0.0000
	1	0.3706	0.3294	0.2097	0.1126	0.0535	0.0228	0.0087	0.0030	0.0009	0.0002
	2	0.1463	0.2745	0.2775	0.2111	0.1336	0.0732	0.0353	0.0150	0.0056	0.0018
	3	0.0359	0.1423	0.2285	0.2463	0.2079	0.1465	0.0888	0.0468	0.0215	0.0085
	4	0.0061	0.0514	0.1311	0.2001	0.2252	0.2040	0.1553	0.1014	0.0572	0.0278
	5	0.0008	0.0137	0.0555	0.1201	0.1802	0.2099	0.2008	0.1623	0.1123	0.0667
	6	0.0001	0.0028	0.0180	0.0550	0.1101	0.1649	0.1982	0.1983	0.1684	0.1222
	7	0.0000	0.0004	0.0045	0.0197	0.0524	0.1010	0.1524	0.1889	0.1969	0.1746
	8	0.0000	0.0001	0.0009	0.0055	0.0197	0.0487	0.0923	0.1417	0.1812	0.1964
	9	0.0000	0.0000	0.0001	0.0012	0.0058	0.0185	0.0442	0.0840	0.1318	0.1746
	10	0.0000	0.0000	0.0000	0.0002	0.0014	0.0056	0.0167	0.0392	0.0755	0.1222
	11	0.0000	0.0000	0.0000	0.0000	0.0002	0.0013	0.0049	0.0142	0.0337	0.0667
	12	0.0000	0.0000	0.0000	0.0000	0.0000	0.0002	0.0011	0.0040	0.0115	0.0278
	13	0.0000	0.0000	0.0000	0.0000	0.0000	0.0000	0.0002	0.0008	0.0029	0.0085
	14	0.0000	0.0000	0.0000	0.0000	0.0000	0.0000	0.0000	0.0001	0.0005	0.0018
	15	0.0000	0.0000	0.0000	0.0000	0.0000	0.0000	0.0000	0.0000	0.0001	0.0002
	16	0.0000	0.0000	0.0000	0.0000	0.0000	0.0000	0.0000	0.0000	0.0000	0.0000
17	**0**	0.4181	0.1668	0.0631	0.0225	0.0075	0.0023	0.0007	0.0002	0.0000	0.0000
	1	0.3741	0.3150	0.1893	0.0957	0.0426	0.0169	0.0060	0.0019	0.0005	0.0001
	2	0.1575	0.2800	0.2673	0.1914	0.1136	0.0581	0.0260	0.0102	0.0035	0.0010
	3	0.0415	0.1556	0.2359	0.2393	0.1893	0.1245	0.0701	0.0341	0.0144	0.0052
	4	0.0076	0.0605	0.1457	0.2093	0.2209	0.1868	0.1320	0.0796	0.0411	0.0182
	5	0.0010	0.0175	0.0668	0.1361	0.1914	0.2081	0.1849	0.1379	0.0875	0.0472
	6	0.0001	0.0039	0.0236	0.0680	0.1276	0.1784	0.1991	0.1839	0.1432	0.0944
	7	0.0000	0.0007	0.0065	0.0267	0.0668	0.1201	0.1685	0.1927	0.1841	0.1484
	8	0.0000	0.0001	0.0014	0.0084	0.0279	0.0644	0.1143	0.1606	0.1883	0.1855
	9	0.0000	0.0000	0.0003	0.0021	0.0093	0.0276	0.0611	0.1070	0.1540	0.1855
	10	0.0000	0.0000	0.0000	0.0004	0.0025	0.0095	0.0263	0.0571	0.1008	0.1484

		p									
n	*r*	**0.05**	**0.10**	**0.15**	**0.20**	**0.25**	**0.30**	**0.35**	**0.40**	**0.45**	**0.50**
17	**11**	0.0000	0.0000	0.0000	0.0001	0.0005	0.0026	0.0090	0.0242	0.0525	0.0944
	12	0.0000	0.0000	0.0000	0.0000	0.0001	0.0006	0.0024	0.0081	0.0215	0.0472
	13	0.0000	0.0000	0.0000	0.0000	0.0000	0.0001	0.0005	0.0021	0.0068	0.0182
	14	0.0000	0.0000	0.0000	0.0000	0.0000	0.0000	0.0001	0.0004	0.0016	0.0052
	15	0.0000	0.0000	0.0000	0.0000	0.0000	0.0000	0.0000	0.0001	0.0003	0.0010
	16	0.0000	0.0000	0.0000	0.0000	0.0000	0.0000	0.0000	0.0000	0.0000	0.0001
	17	0.0000	0.0000	0.0000	0.0000	0.0000	0.0000	0.0000	0.0000	0.0000	0.0000
18	**0**	0.3972	0.1501	0.0536	0.0180	0.0058	0.0016	0.0004	0.0001	0.0000	0.0000
	1	0.3763	0.3002	0.1704	0.0811	0.0338	0.0126	0.0042	0.0012	0.0003	0.0001
	2	0.1683	0.2835	0.2558	0.1723	0.0958	0.0458	0.0190	0.0069	0.0022	0.0006
	3	0.0473	0.1680	0.2406	0.2297	0.1704	0.1046	0.0547	0.0246	0.0095	0.0031
	4	0.0093	0.0700	0.1592	0.2153	0.2130	0.1681	0.1104	0.0614	0.0291	0.0117
	5	0.0014	0.0218	0.0787	0.1507	0.1988	0.2017	0.1664	0.1146	0.0666	0.0327
	6	0.0002	0.0052	0.0310	0.0818	0.1436	0.1873	0.1941	0.1655	0.1181	0.0708
	7	0.0000	0.0010	0.0091	0.0350	0.0820	0.1376	0.1792	0.1892	0.1657	0.1214
	8	0.0000	0.0002	0.0022	0.0120	0.0376	0.0811	0.1327	0.1734	0.1864	0.1669
	9	0.0000	0.0000	0.0004	0.0033	0.0139	0.0386	0.0794	0.1284	0.1694	0.1855
	10	0.0000	0.0000	0.0001	0.0008	0.0042	0.0149	0.0385	0.0771	0.1248	0.1669
	11	0.0000	0.0000	0.0000	0.0001	0.0010	0.0046	0.0151	0.0374	0.0742	0.1214
	12	0.0000	0.0000	0.0000	0.0000	0.0002	0.0012	0.0047	0.0145	0.0354	0.0708
	13	0.0000	0.0000	0.0000	0.0000	0.0000	0.0002	0.0012	0.0045	0.0134	0.0327
	14	0.0000	0.0000	0.0000	0.0000	0.0000	0.0000	0.0002	0.0011	0.0039	0.0117
	15	0.0000	0.0000	0.0000	0.0000	0.0000	0.0000	0.0000	0.0002	0.0009	0.0031
	16	0.0000	0.0000	0.0000	0.0000	0.0000	0.0000	0.0000	0.0000	0.0001	0.0006
	17	0.0000	0.0000	0.0000	0.0000	0.0000	0.0000	0.0000	0.0000	0.0000	0.0001
	18	0.0000	0.0000	0.0000	0.0000	0.0000	0.0000	0.0000	0.0000	0.0000	0.0000
19	**0**	0.3774	0.1351	0.0456	0.0144	0.0042	0.0011	0.0003	0.0001	0.0000	0.0000
	1	0.3774	0.2852	0.1529	0.0685	0.0268	0.0093	0.0029	0.0008	0.0002	0.0000
	2	0.1787	0.2852	0.2428	0.1540	0.0803	0.0358	0.0138	0.0046	0.0013	0.0003
	3	0.0533	0.1796	0.2428	0.2182	0.1517	0.0869	0.0422	0.0175	0.0062	0.0018
	4	0.0112	0.0798	0.1714	0.2182	0.2023	0.1491	0.0909	0.0467	0.0203	0.0074
	5	0.0018	0.0266	0.0907	0.1636	0.2023	0.1916	0.1468	0.0933	0.0497	0.0222
	6	0.0002	0.0069	0.0374	0.0955	0.1574	0.1916	0.1844	0.1451	0.0949	0.0518
	7	0.0000	0.0014	0.0122	0.0443	0.0974	0.1525	0.1844	0.1797	0.1443	0.0961
	8	0.0000	0.0002	0.0032	0.0166	0.0487	0.0981	0.1489	0.1797	0.1771	0.1442
	9	0.0000	0.0000	0.0007	0.0051	0.0198	0.0514	0.0980	0.1464	0.1771	0.1762
	10	0.0000	0.0000	0.0001	0.0013	0.0066	0.0220	0.0528	0.0976	0.1449	0.1762
	11	0.0000	0.0000	0.0000	0.0003	0.0018	0.0077	0.0233	0.0532	0.0970	0.1442
	12	0.0000	0.0000	0.0000	0.0000	0.0004	0.0022	0.0083	0.0237	0.0529	0.0961

n	r	p									
		0.05	**0.10**	**0.15**	**0.20**	**0.25**	**0.30**	**0.35**	**0.40**	**0.45**	**0.50**
19	**13**	0.0000	0.0000	0.0000	0.0000	0.0001	0.0005	0.0024	0.0085	0.0233	0.0518
	14	0.0000	0.0000	0.0000	0.0000	0.0000	0.0001	0.0006	0.0024	0.0082	0.0222
	15	0.0000	0.0000	0.0000	0.0000	0.0000	0.0000	0.0001	0.0005	0.0022	0.0074
	16	0.0000	0.0000	0.0000	0.0000	0.0000	0.0000	0.0000	0.0001	0.0005	0.0018
	17	0.0000	0.0000	0.0000	0.0000	0.0000	0.0000	0.0000	0.0000	0.0001	0.0003
	18	0.0000	0.0000	0.0000	0.0000	0.0000	0.0000	0.0000	0.0000	0.0000	0.0000
	19	0.0000	0.0000	0.0000	0.0000	0.0600	0.0000	0.0000	0.0000	0.0000	0.0000
20	**0**	0.3585	0.1216	0.0388	0.0115	0.0032	0.0008	0.0002	0.0000	0.0000	0.0000
	1	0.3774	0.2702	0.1368	0.0576	0.0211	0.0068	0.0020	0.0005	0.0001	0.0000
	2	0.1887	0.2852	0.2293	0.1369	0.0669	0.0278	0.0100	0.0031	0.0008	0.0002
	3	0.0596	0.1901	0.2428	0.2054	0.1339	0.0716	0.0323	0.0123	0.0040	0.0011
	4	0.0133	0.0898	0.1821	0.2182	0.1897	0.1304	0.0738	0.0350	0.0139	0.0046
	5	0.0022	0.0319	0.1028	0.1746	0.2023	0.1789	0.1272	0.0746	0.0365	0.0148
	6	0.0003	0.0089	0.0454	0.1091	0.1686	0.1916	0.1712	0.1244	0.0746	0.0370
	7	0.0000	0.0020	0.0160	0.0545	0.1124	0.1643	0.1844	0.1659	0.1221	0.0739
	8	0.0000	0.0004	0.0046	0.0222	0.0609	0.1144	0.1614	0.1797	0.1623	0.1201
	9	0.0000	0.0001	0.0011	0.0074	0.0271	0.0654	0.1158	0.1597	0.1771	0.1602
	10	0.0000	0.0000	0.0002	0.0020	0.0099	0.0308	0.0686	0.1171	0.1593	0.1762
	11	0.0000	0.0000	0.0000	0.0005	0.0030	0.0120	0.0336	0.0710	0.1185	0.1602
	12	0.0000	0.0000	0.0000	0.0001	0.0008	0.0039	0.0136	0.0355	0.0727	0.1201
	13	0.0000	0.0000	0.0000	0.0000	0.0002	0.0010	0.0045	0.0146	0.0366	0.0739
	14	0.0000	0.0000	0.0000	0.0000	0.0000	0.0002	0.0012	0.0049	0.0150	0.0370
	15	0.0000	0.0000	0.0000	0.0000	0.0000	0.0000	0.0003	0.0013	0.0049	0.0148
	16	0.0000	0.0000	0.0000	0.0000	0.0000	0.0000	0.0000	0.0003	0.0013	0.0046
	17	0.0000	0.0000	0.0000	0.0000	0.0000	0.0000	0.0000	0.0000	0.0002	0.0011
	18	0.0000	0.0000	0.0000	0.0000	0.0000	0.0000	0.0000	0.0000	0.0000	0.0002
	19	0.0000	0.0000	0.0000	0.0000	0.0000	0.0000	0.0000	0.0000	0.0000	0.0000
	20	0.0000	0.0000	0.0000	0.0000	0.0000	0.0000	0.0000	0.0000	0.0000	0.0000

Table 6.A.2

Cumulative Distribution Function Values for the Binomial Distribution (X is $b(X;n,p)$)

$$B(x;n,p) = \sum_{i \le x} \binom{n}{i} p^i (1-p)^{n-i} = \sum_{i \le x} b(i;n,p)$$

$B(x;n,p)$ gives the probability that the random variable X assumes a value $\le x$.

		p				
n	x	0.1	0.2	0.3	0.4	0.5
2	0	0.8100	0.6400	0.4900	0.3600	0.2500
	1	0.9900	0.9600	0.9100	0.8400	0.7500
3	0	0.7290	0.5120	0.3430	0.2160	0.1250
	1	0.9720	0.8960	0.7840	0.6480	0.5000
	2	0.9990	0.9920	0.9730	0.9360	0.8750
4	0	0.6561	0.4096	0.2401	0.1296	0.0625
	1	0.9477	0.8192	0.6517	0.4752	0.3125
	2	0.9963	0.9728	0.9163	0.8208	0.6875
	3	0.9999	0.9984	0.9919	0.9744	0.9375
5	0	0.5905	0.3277	0.1681	0.0778	0.0312
	1	0.9185	0.7373	0.5282	0.3370	0.1875
	2	0.9914	0.9421	0.8369	0.6826	0.5000
	3	0.9995	0.9933	0.9692	0.9130	0.8125
	4	1.0000	0.9997	0.9976	0.9898	0.9688
6	0	0.5314	0.2621	0.1176	0.0467	0.0156
	1	0.8857	0.6554	0.4202	0.2333	0.1094
	2	0.9842	0.9011	0.7443	0.5443	0.3438
	3	0.9987	0.9830	0.9295	0.8208	0.6562
	4	0.9999	0.9984	0.9891	0.9590	0.8906
	5	1.0000	0.9999	0.9993	0.9959	0.9844
7	0	0.4783	0.2097	0.0824	0.0280	0.0078
	1	0.8503	0.5767	0.3294	0.1586	0.0625
	2	0.9743	0.8520	0.6471	0.4199	0.2266
	3	0.9973	0.9667	0.8740	0.7102	0.5000
	4	0.9998	0.9953	0.9712	0.9037	0.7734
	5	1.0000	0.9996	0.9962	0.9812	0.9375
	6	1.0000	1.0000	0.9998	0.9984	0.9922
8	0	0.4305	0.1678	0.0576	0.0168	0.0039
	1	0.8131	0.5033	0.2553	0.1064	0.0352
	2	0.9619	0.7969	0.5518	0.3154	0.1445
	3	0.9950	0.9437	0.8059	0.5941	0.3633
	4	0.9996	0.9896	0.9420	0.8263	0.6367
	5	1.0000	0.9988	0.9887	0.9502	0.8555
	6	1.0000	0.9999	0.9987	0.9915	0.9648
	7	1.0000	1.0000	0.9999	0.9993	0.9961

n	x	p 0.1	0.2	0.3	0.4	0.5
9	0	0.3874	0.1342	0.0404	0.0101	0.0020
	1	0.7748	0.4362	0.1960	0.0705	0.0195
	2	0.9470	0.7382	0.4628	0.2318	0.0898
	3	0.9917	0.9144	0.7197	0.4826	0.2539
	4	0.9991	0.9804	0.9012	0.7334	0.5000
	5	0.9999	0.9969	0.9747	0.9006	0.7461
	6	1.0000	0.9997	0.9957	0.9750	0.9102
	7	1.0000	1.0000	0.9996	0.9962	0.9805
	8	1.0000	1.0000	1.0000	0.9997	0.9980
10	0	0.3487	0.1074	0.0282	0.0060	0.0010
	1	0.7361	0.3758	0.1493	0.0464	0.0107
	2	0.9298	0.6778	0.3828	0.1673	0.0547
	3	0.9872	0.8791	0.6496	0.3823	0.1719
	4	0.9984	0.9672	0.8497	0.6331	0.3770
	5	0.9999	0.9936	0.9527	0.8338	0.6230
	6	1.0000	0.9991	0.9894	0.9452	0.8281
	7	1.0000	0.9999	0.9984	0.9877	0.9453
	8	1.0000	1.0000	0.9999	0.9983	0.9893
	9	1.0000	1.0000	1.0000	0.9999	0.9990
11	0	0.3138	0.0859	0.0198	0.0036	0.0005
	1	0.6974	0.3221	0.1130	0.0302	0.0059
	2	0.9104	0.6174	0.3127	0.1189	0.0327
	3	0.9815	0.8389	0.5696	0.2963	0.1133
	4	0.9972	0.9496	0.7897	0.5328	0.2744
	5	0.9997	0.9883	0.9218	0.7535	0.5000
	6	1.0000	0.9980	0.9784	0.9006	0.7256
	7	1.0000	0.9998	0.9957	0.9707	0.8867
	8	1.0000	1.0000	0.9994	0.9941	0.9673
	9	1.0000	1.0000	1.0000	0.9993	0.9941
	10	1.0000	1.0000	1.0000	1.0000	0.9995
12	0	0.2824	0.0687	0.0138	0.0022	0.0002
	1	0.6590	0.2749	0.0850	0.0196	0.0032
	2	0.8891	0.5583	0.2528	0.0834	0.0193
	3	0.9744	0.7946	0.4925	0.2253	0.0730
	4	0.9957	0.9274	0.7237	0.4382	0.1938
	5	0.9995	0.9806	0.8822	0.6652	0.3872
	6	0.9999	0.9961	0.9614	0.8418	0.6128
	7	1.0000	0.9994	0.9905	0.9427	0.8062
	8	1.0000	0.9999	0.9983	0.9847	0.9270
	9	1.0000	1.0000	0.9998	0.9972	0.9807
	10	1.0000	1.0000	1.0000	0.9997	0.9968
	11	1.0000	1.0000	1.0000	1.0000	0.9998

n	x	p = 0.1	0.2	0.3	0.4	0.5
13	**0**	0.2542	0.0550	0.0097	0.0013	0.0001
	1	0.6213	0.2336	0.0637	0.0126	0.0017
	2	0.8661	0.5017	0.2025	0.0579	0.0112
	3	0.9658	0.7473	0.4206	0.1686	0.0461
	4	0.9935	0.9009	0.6543	0.3530	0.1334
	5	0.9991	0.9700	0.8346	0.5744	0.2905
	6	0.9999	0.9930	0.9376	0.7712	0.5000
	7	1.0000	0.9988	0.9818	0.9023	0.7095
	8	1.0000	0.9998	0.9960	0.9679	0.8666
	9	1.0000	1.0000	0.9993	0.9922	0.9539
	10	1.0000	1.0000	0.9999	0.9987	0.9888
	11	1.0000	1.0000	1.0000	0.9999	0.9983
	12	1.0000	1.0000	1.0000	1.0000	0.9999
14	**0**	0.2288	0.0440	0.0068	0.0008	0.0001
	1	0.5846	0.1979	0.0475	0.0081	0.0009
	2	0.8416	0.4481	0.1608	0.0398	0.0065
	3	0.9559	0.6982	0.3552	0.1243	0.0287
	4	0.9908	0.8702	0.5842	0.2793	0.0898
	5	0.9985	0.9561	0.7805	0.4859	0.2120
	6	0.9998	0.9884	0.9067	0.6925	0.3953
	7	1.0000	0.9976	0.9685	0.8499	0.6047
	8	1.0000	0.9996	0.9917	0.9417	0.7880
	9	1.0000	1.0000	0.9983	0.9825	0.9102
	10	1.0000	1.0000	0.9998	0.9961	0.9713
	11	1.0000	1.0000	1.0000	0.9994	0.9935
	12	1.0000	1.0000	1.0000	0.9999	0.9991
	13	1.0000	1.0000	1.0000	1.0000	0.9999
15	**0**	0.2059	0.0352	0.0047	0.0005	0.0000
	1	0.5490	0.1671	0.0353	0.0052	0.0005
	2	0.8159	0.3980	0.1268	0.0271	0.0037
	3	0.9444	0.6482	0.2969	0.0905	0.0176
	4	0.9873	0.8358	0.5155	0.2173	0.5920
	5	0.9978	0.9389	0.7216	0.4032	0.1509
	6	0.9997	0.9819	0.8689	0.6098	0.3036
	7	1.0000	0.9958	0.9500	0.7869	0.5000
	8	1.0000	0.9992	0.9848	0.9050	0.6964
	9	1.0000	0.9999	0.9963	0.9662	0.8491
	10	1.0000	1.0000	0.9993	0.9907	0.9408
	11	1.0000	1.0000	0.9999	0.9981	0.9824
	12	1.0000	1.0000	1.0000	0.9997	0.9963
	13	1.0000	1.0000	1.0000	1.0000	0.9995
	14	1.0000	1.0000	1.0000	1.0000	1.0000

n	x	p 0.1	0.2	0.3	0.4	0.5
16	**0**	0.1853	0.0281	0.0033	0.0003	0.0000
	1	0.5147	0.1407	0.0261	0.0033	0.0003
	2	0.7892	0.3518	0.0994	0.0183	0.0021
	3	0.9316	0.5981	0.2459	0.0651	0.0106
	4	0.9830	0.7982	0.4499	0.1666	0.0384
	5	0.9967	0.9183	0.6598	0.3288	0.1051
	6	0.9995	0.9733	0.8247	0.5272	0.2272
	7	0.9999	0.9930	0.9256	0.7161	0.4018
	8	1.0000	0.9985	0.9743	0.8577	0.5982
	9	1.0000	0.9998	0.9929	0.9417	0.7728
	10	1.0000	1.0000	0.9984	0.9809	0.8949
	11	1.0000	1.0000	0.9991	0.9951	0.9616
	12	1.0000	1.0000	1.0000	0.9991	0.9894
	13	1.0000	1.0000	1.0000	0.9999	0.9979
	14	1.0000	1.0000	1.0000	1.0000	0.9997
	15	1.0000	1.0000	1.0000	1.0000	1.0000
17	**0**	0.1668	0.0225	0.0023	0.0002	0.0000
	1	0.4818	0.1182	0.0193	0.0021	0.0001
	2	0.7618	0.3096	0.0774	0.0123	0.0012
	3	0.9174	0.5489	0.2019	0.0464	0.0064
	4	0.9779	0.7582	0.3887	0.1260	0.0245
	5	0.9953	0.8943	0.5968	0.2639	0.0717
	6	0.9992	0.9623	0.7752	0.4478	0.1662
	7	0.9999	0.9891	0.8954	0.6405	0.3145
	8	1.0000	0.9974	0.9597	0.8011	0.5000
	9	1.0000	0.9995	0.9873	0.9081	0.6855
	10	1.0000	0.9999	0.9968	0.9652	0.8338
	11	1.0000	1.0000	0.9993	0.9894	0.9283
	12	1.0000	1.0000	0.9999	0.9975	0.9755
	13	1.0000	1.0000	1.0000	0.9995	0.9936
	14	1.0000	1.0000	1.0000	0.9999	0.9988
	15	1.0000	1.0000	1.0000	1.0000	0.9999
	16	1.0000	1.0000	1.0000	1.0000	1.0000
18	**0**	0.1501	0.0180	0.0016	0.0001	0.0000
	1	0.4503	0.0991	0.0142	0.0013	0.0001
	2	0.7338	0.2713	0.0600	0.0082	0.0007
	3	0.9018	0.5010	0.1646	0.0328	0.0038
	4	0.9718	0.7164	0.3327	0.0942	0.0154
	5	0.9936	0.8671	0.5344	0.2088	0.0481
	6	0.9988	0.9487	0.7217	0.3743	0.1189
	7	0.9998	0.9837	0.8593	0.5634	0.2403
	8	1.0000	0.9957	0.9404	0.7368	0.4073
	9	1.0000	0.9991	0.9790	0.8653	0.5927
	10	1.0000	0.9998	0.9939	0.9424	0.7597

n	x	0.1	0.2	p 0.3	0.4	0.5
18	**11**	1.0000	1.0000	0.9986	0.9797	0.8811
	12	1.0000	1.0000	0.9997	0.9942	0.9519
	13	1.0000	1.0000	1.0000	0.9987	0.9846
	14	1.0000	1.0000	1.0000	0.9998	0.9962
	15	1.0000	1.0000	1.0000	1.0000	0.9993
	16	1.0000	1.0000	1.0000	1.0000	0.9999
	17	1.0000	1.0000	1.0000	1.0000	1.0000
19	**0**	0.1351	0.0144	0.0011	0.0001	0.0000
	1	0.4203	0.0829	0.0104	0.0008	0.0000
	2	0.7054	0.2369	0.0462	0.0055	0.0004
	3	0.8850	0.4551	0.1332	0.0230	0.0022
	4	0.9648	0.6733	0.2822	0.0696	0.0096
	5	0.9914	0.8369	0.4739	0.1629	0.0318
	6	0.9983	0.9324	0.6655	0.3081	0.0835
	7	0.9997	0.9767	0.8180	0.4878	0.1796
	8	1.0000	0.9933	0.9161	0.6675	0.3238
	9	1.0000	0.9984	0.9674	0.8139	0.5000
	10	1.0000	0.9997	0.9895	0.9115	0.6762
	11	1.0000	1.0000	0.9720	0.9648	0.8204
	12	1.0000	1.0000	0.9994	0.9884	0.9165
	13	1.0000	1.0000	0.9999	0.9969	0.9682
	14	1.0000	1.0000	1.0000	0.9994	0.9904
	15	1.0000	1.0000	1.0000	0.9999	0.9978
	16	1.0000	1.0000	1.0000	1.0000	0.9996
	17	1.0000	1.0000	1.0000	1.0000	1.0000
20	**0**	0.1216	0.0115	0.0008	0.0000	0.0000
	1	0.3917	0.0692	0.0076	0.0005	0.0000
	2	0.6769	0.2061	0.0355	0.0036	0.0002
	3	0.8670	0.4114	0.1071	0.0160	0.0013
	4	0.9568	0.6296	0.2375	0.0510	0.0059
	5	0.9887	0.8042	0.4164	0.1256	0.0207
	6	0.9976	0.9133	0.6080	0.2500	0.0577
	7	0.9996	0.9679	0.7723	0.4159	0.1316
	8	0.9999	0.9900	0.8867	0.5956	0.2517
	9	1.0000	0.9974	0.9520	0.7553	0.4119
	10	1.0000	0.9994	0.9829	0.8725	0.5881
	11	1.0000	0.9999	0.9949	0.9435	0.7483
	12	1.0000	1.0000	0.9987	0.9790	0.8684
	13	1.0000	1.0000	0.9997	0.9935	0.9423
	14	1.0000	1.0000	1.0000	0.9984	0.9793
	15	1.0000	1.0000	1.0000	0.9997	0.9941
	16	1.0000	1.0000	1.0000	1.0000	0.9987
	17	1.0000	1.0000	1.0000	1.0000	0.9998
	18	1.0000	1.0000	1.0000	1.0000	1.0000

Solutions to Odd-Numbered Exercises

Chapter 1

1. $X = 132$
3. $X = 2$
5. $X = 3, Y = 4$
7. $X_1 = 3, X_2 = 3$
9. 0
11. 44
13. 8
15. −9
17. 0
19. 107
21. $\{1, 3, 5, 7, 9, \ldots\}$
23. True
25. False
27. $\{1, 3, 5, 6, 7, 9\}$
29. $A \cup \overline{A} = \{8 \text{ white}, 3 \text{ red}, 7 \text{ blue}\}$
    $A \cap \overline{A} = \varnothing$
    $\overline{U} = \phi$
    $\overline{\phi} = U$
31. We can write this equation in standard form as $y = 10 - 2x$. Prepare a table of some arbitrary x values and the corresponding y values and plot the points on a graph and connect them to get a straight line as shown below.

Introduction to Quantitative Methods in Business: With Applications Using Microsoft® Office Excel®, First Edition. Bharat Kolluri, Michael J. Panik, and Rao Singamsetti.

Companion website: www.wiley.com/go/Kolluri/QuantitativeMethods

x	y
0	10
1	8
2	6
5	0

33.

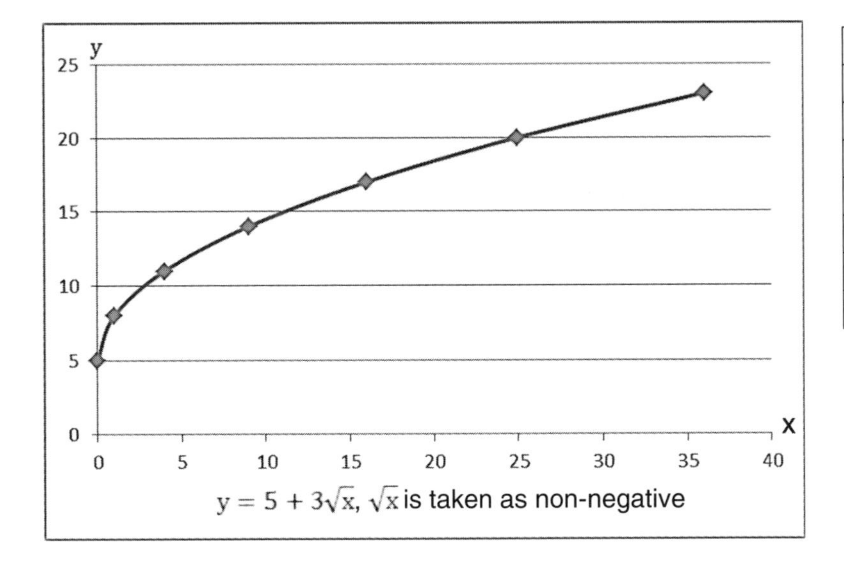

$y = 5 + 3\sqrt{x}$, $\sqrt{x}$ is taken as non-negative

x	y
0	5
1	8
4	11
9	14
16	17
25	20
36	23

35. $\frac{dy}{dx} = -3\left(\frac{1}{x^4}\right)$. At $x = 1, \frac{dy}{dx} = -3$.

37. 0

39. $\frac{dy}{dx} = x^4$

41. $\frac{dy}{dx} = -0.5$

43. $\frac{dy}{dx} = 15x^2 + 8x + 3$

45. **a.** $\frac{dy}{dx} = \frac{12x^2 + 24x + 6}{16x^4 + 8x^3 + x^2}$

b. $\frac{dy}{dx} = \frac{-5x^4 + 10x^3 - 20x + 10}{-x^4 - 2x^3 + x^2}$

47. $-x^{-1} + c$

49. $x^3 - 2x^2 + 8x + c$

51. 1

53. 0.093567

Interpretation: In general, if $r > 0$, X and Y move together; and if $r < 0$, X and Y movein opposite directions. More specifically, if:

(i) $r = -1$, we have perfect negative linear association between X and Y;

(ii) $r = 0$, no linear association between X and Y;

(iii) $r = 1$, we have perfect positive linear association between X and Y.

55. **a.** $b = 0.029183$,

$a = 7 - 0.029183\ (1.8) = 6.947471$

$Y = 6.947471 + 0.029183\ (11) = 7.268484$

b. 0.029183

57. **a.** $\frac{dy}{dx} = 72x^3 + 42x^2 + 9x$

b. $\frac{dy}{dx} = 36x^8 - 24x^7 - 56x + 21$

59. **a.** $\frac{dy}{dx} = \frac{2x^2 + 6x - x^2 + 9}{x^2 + 6x + 9}$

b. $\frac{dy}{dx} = \frac{-2x + 2x^3 - 3x^{-2} + 3 - 6x - 6x^3}{4x^6 + 12x + 9}$

$= \frac{-4x^3 - 3x^{-2} - 8x + 3}{4x^6 + 12x^3 + 9}$

61. a. $r = 0.0936$; b. $r = 0.0936$

63. Exercise 4: $X = 5$, $Y = -7$

Exercise 6: $X = 7$, $Y = 1$

Exercise 8: $P = 8.25$, $Q = 1.25$

A.1 8

A.3 1034

A.5 x^5y^6

A.7 $x/(-4y^2)$

A.9 1/9

A.11 0.375
A.13 1
A.15 1012.5
A.17 0.7071
A.19 x^2y^2
A.21 x^{16}
A.23 undefined
A.25 $x^{15}y^{-20}$
A.27 x^5
A.29 1728
A.31 3/4
A.33 81
A.35 $x^8/8y^2$
A.37 1/27
A.39 9
A.41 $(x - 3)(x - 4)$
A.43 $(x + 4)(x - 3)$
A.45 $(5x + 2)(x + 1)$
A.47 $(x + 3)(x + 2)$
A.49 3/4
A.51 1/4
A.53 LCM of numerator = 15, LCM of denominator = 6. Ans. 4/15
A.55 −6/35
A.57 −10/9
A.59 0
A.61 LCM = $3 \times 5 \times 7$ Ans. 278/105
A.63 99/8
A.65 33/52
A.67 $0.8 + 0.07 + 0.003$
A.69 4.473
A.71 1.97
A.73 0.000005
A.75 −0.2575
A.77 −290
A.79 0.71
A.81 3.1429
A.83 Let x denote number of female students to start with.

Then,

$$\frac{x + 200}{4x} = \frac{2}{3}$$

$$3x + 600 = 8x$$

$$5x = 600$$
$$x = \frac{600}{5}$$
$$x = 120$$

So, male students to start with are 480. After 200 female students joined, the total number of female students is $120 + 200 = 320$

Therefore, the total number of students after the 200 female students are admitted is $320 + 480 = 800$.

A.85 Let's call his salary before taxes "S"

100%-30% = 70%

70% of Mr. John Smith's salary before taxes (S) is equal to 1400.00

$$\frac{70}{100} \times S = 1400, \qquad S = \frac{1400 \times 100}{70} = 2000 \text{ (which is in dollars)}$$

S = $2000.00

A.87

Row	Fraction	Decimal	%, Percentage
1	$\frac{1}{50}$	0.02	2
2	$\frac{1}{8}$	0.125	12.5
3	$\frac{15}{4}$	3.750	375
4	$\frac{1}{16}$	0.0625	6.25
5	$\frac{1}{8}$	0.125	12.5
6	$\frac{3}{2}$	1.5	150
7	$\frac{3}{10}$	0.3	30
8	$\frac{1}{2}$	0.5	50
9	$\frac{1}{10}$	0.1	10
10	2	2.0	200
11	$\left(\frac{1}{4}\right)^2$	0.0625	6.25
12	$\left(\frac{1}{10}\right)^5$	0.00001	0.001

A.89 31.06%

Chapter 2

1.

P in \$'s	0	10	20	30	40
Q in 1000's	8	6	4	2	0

From the data the slope $= \dfrac{\Delta Q}{\Delta P} = \dfrac{6-8}{10-0} = \dfrac{-2}{10} = \dfrac{-1}{5}$

Therefore, we can write the demand equation as $Q = \frac{-1}{5}P + C$, where C is the intercept. Substituting one of the given points from the table above, say (0, 8), we obtain $8 = \frac{-1}{5}(0) + C$. Thus $C = 8$ and the demand equation is given by $Q = \dfrac{-1}{5}P + 8$.

The graph of this equation is given below.

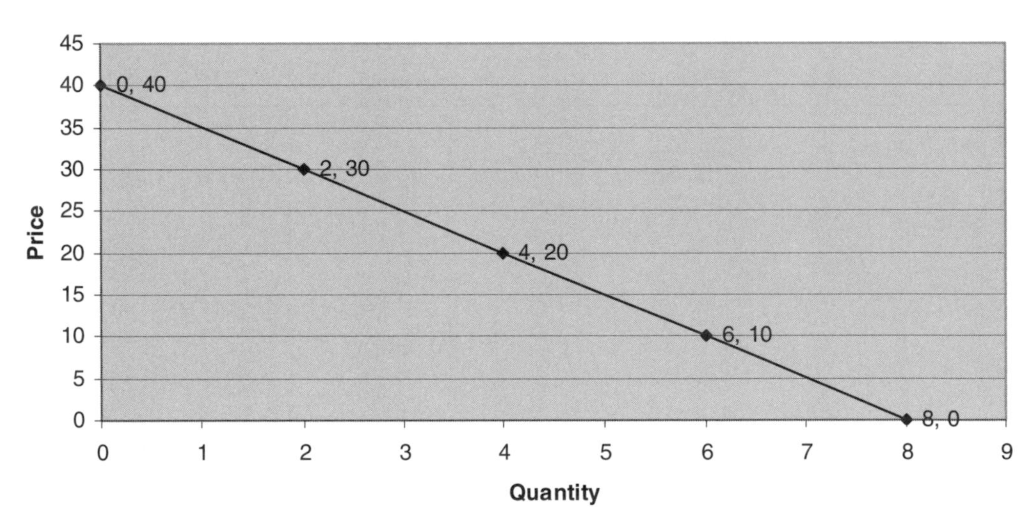

3. a. Here, the breakeven income-level can be found by solving the two equations, namely

$$C = 20 + \frac{2}{3}Y, \text{ and } C = Y$$

Subtracting the second equation from the first, we obtain $0 = 20 + \dfrac{2}{3}Y - Y$ or $0 = 20 + \left(\dfrac{2}{3} - 1\right)Y$ or $0 = 20 + \left(\dfrac{-1}{3}\right)Y$

$$\frac{-1}{3}Y = -20$$

Multiplying by -3 on both sides we obtain $Y = 60$.
Thus, the breakeven level of income is 60.

b. Consumption at $Y = 40$ is

$$C = 20 + \frac{2}{3} \times 40 = 20 + \frac{80}{3} \text{ or}$$

$C = 20 + 26.67 = 46.67 = \$46,666.67.$
Consumption at $Y = 80$ is
$C = 20 + \frac{2}{3} \times 80$
$C = 20 + 53.33 = 73.33 = \$73,333.33$

c. Graphs for a) and b):

$C = 20 + \frac{2}{3}Y$ and $C = Y$

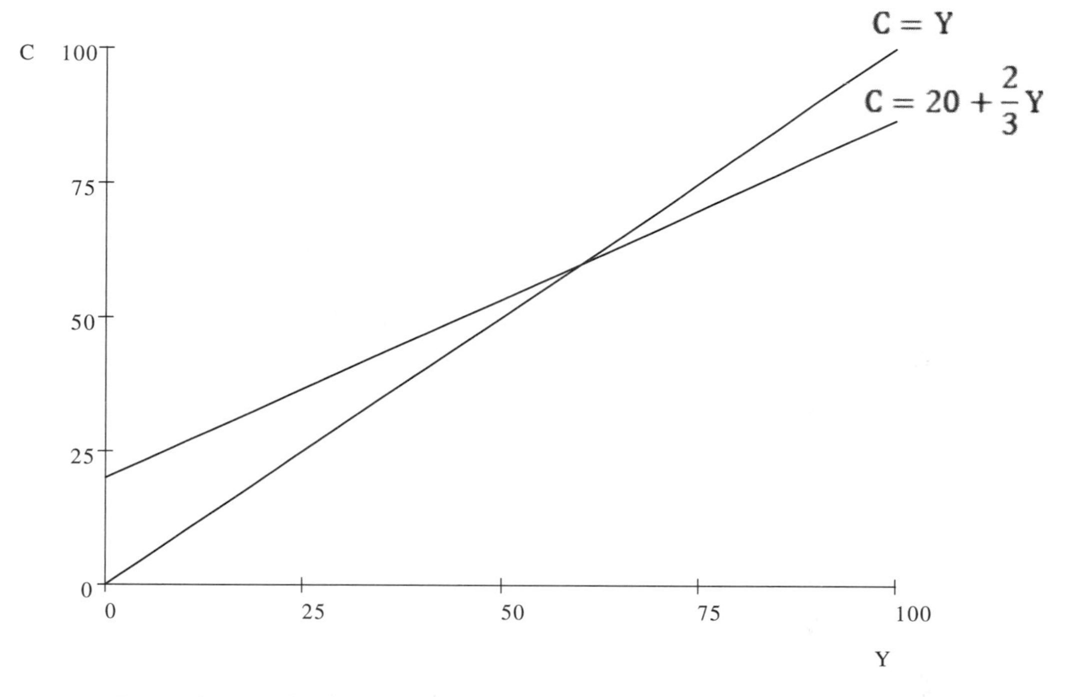

From the graph above, we can see that up to the income level of $Y = 60$, consumption exceeds income. Therefore, the income range of 0 to 60 corresponds to dis-saving, and the income level above 60 indicates saving.

d. Savings function, $S = Y - C = Y - (20 + \frac{2}{3}Y) = -20 + \frac{1}{3}Y.$

To plot the savings function, prepare a table as follows:

Y	S
30	−10
60	0
90	10

Marginal propensity to save (MPS) is the slope of Y in the savings function, S, and it is $\frac{1}{3}$.

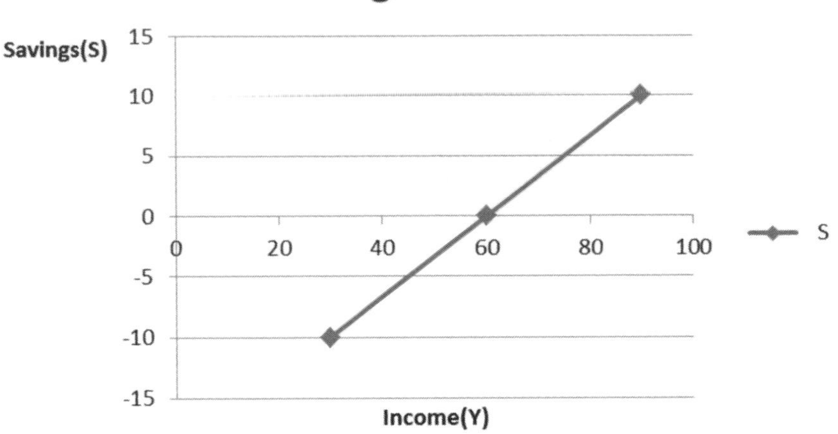

Now, $MPC + MPS = \frac{2}{3} + \frac{1}{3} = 1$. At $Y = 20$, $C = 20 + \frac{2}{3}(20) = 20 + 13.33 = 33.33$ and $S = -20 + \frac{1}{3}(20) = -20 + 6.67 = -13.33$. Hence $C + S = 33.33 - 13.33 = 20 = Y$. Yes, at $Y = 20$, C and S add up to 20. In fact, $Y = C + S$ holds at any level of Y.

5. **a.** $TC = 3000 + 2Q$, $TR = 5Q$
   $TC = TR$ at $Q_E = 1000$

   b. $TR = 9Q$, $TC = 540 + 6Q$
   $TR = TC$ at $Q = 180$

7. For price = \$8000, $Q_D = 400$; $Q_S = 67$.
   Excess demand = 400 – 67 = 333 (product shortage)

9. Graphs of supply and demand functions

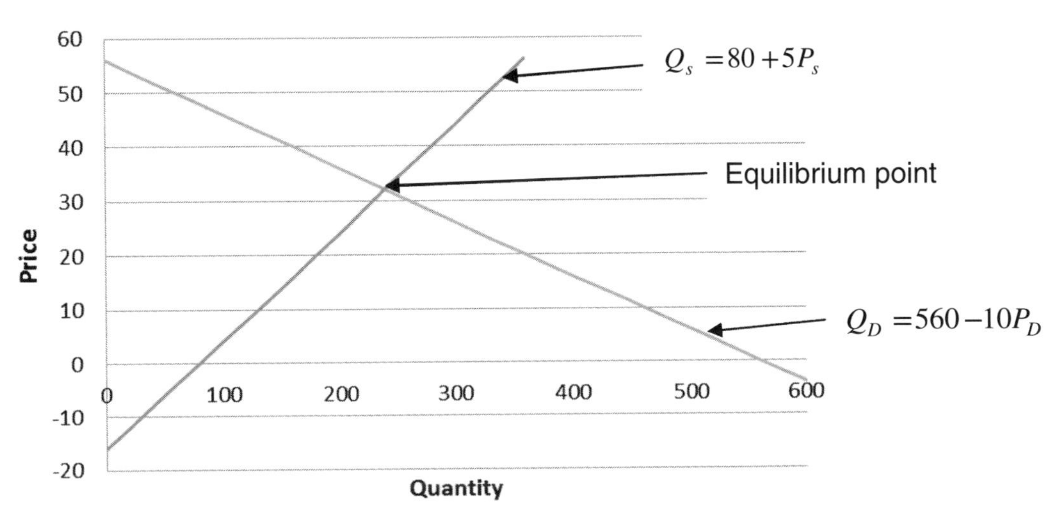

At equilibrium point $Q_s = Q_d = Q$ and $P_s = P_d = P$
Thus the demand and supply equations assume the form

$$Q = 560 - 10P \qquad \text{(demand)}$$
$$Q = 80 + 5P \qquad \text{(supply)}$$

Since each equals Q,
$80 + 5P = 560 - 10P$
$10P + 5P = 560 - 80$
$15P = 480$
$P = \$32$
At $P = 32$, $Q = 80 + 5(32) = 80 + 160 = 240$ units

11. $P_t = P_0 e^{0.1\times 10} = 1000e^1 = 1000 \times 2.71828 = 2718.28$

13. Present value of the pension plan $= 100,000 + \int_0^t 60,000e^{-0.05x}\,dx$
$= 100,000 + 60,000 \times \dfrac{1}{-0.05}[e^{-0.05x}]_0^t$
This is the case of perpetuity, and therefore we assume that t tends to ∞ and, because of this, the above integral value equals
$100,000 + \dfrac{60,000}{-0.05}\left[e^{-0.05t} - e^{-0.05(0)}\right] = 100,000 - 1,200,000[0 - 1]$ as $t \to \infty$
$= 100,000 + 1,200,000 = 1,300,000.$

Here it is assumed the interest rate is compounded continuously and the pension is paid to the retiring employee and the surviving heirs.

15. The total savings function, denoted by S(Y), is given by $\int \frac{1}{10}dY$, where Y stands for a particular value of income and dY represents change in income. The value of the above integral is $\frac{1}{10}Y + k$ where k is constant of integration. We know that $Y = S + C$ from the result of Exercise 3(d) above. Substituting, $Y = \frac{1}{10}Y + k + C$. When $Y = 0$ we obtain

$0 = \frac{1}{10}Y\ (0) + k + 15,000$, where 15,000 is assumed minimum consumption level (C). Thus the constant of integration $k = -15,000$, which can be called dissaving. Therefore, the savings function is $S(Y) = \frac{1}{10}Y - 15,000$.

17. The total revenue, denoted by TR (Q), is given by $\int(100000 - 2Q)dQ$ $- 100000Q - \frac{2Q^2}{2} + c = 100000Q - Q^2 + c$, where $c = 0$ since TR $(0) = 0$. Thus TR $(Q) = 100000Q - Q^2$.

The demand function $P = \dfrac{TR}{Q} = (100000Q - Q^2) \div Q = 100000 - Q$.

19. **a.** $AC = (540/Q) + 6$. At $Q = 100$, $AC = 11.4$
b. $AR = 9$
c. $AR = P(Q) = 40 - 0.1Q$. At $Q = 100$, $AR = 30$

21. Given the demand function $Q = 20 - 3P, \frac{dQ}{dP} = -3$, and thus

$E_p = \frac{P}{Q} \times \frac{dQ}{dP}$, where E_P is the price elasticity.

$E_p = \frac{2}{20 - 3(2)} \times (-3)$ when $P = 2$

$= \frac{-6}{14} = -0.429$

We can conclude that, as $|E_P| < 1$, the demand is inelastic at $P=2$, that is, with 1% change in price, there will be less than 1% change in the quantity demanded

At $P=\$3.33$,

$$E_p = \frac{P}{Q} \times \frac{dQ}{dP}$$

$$= \frac{3.33}{20 - 3(3.33)} \times (-3)$$

$$= \frac{-9.99}{10.01} = -0.998 \approx -1$$

We can conclude that, as $|E_P| = 1$, the demand is unit elastic at $P=3.33$, that is, with 1% change in price, there will be 1% change in the quantity demanded

At $P=\$4$

$$E_p = \frac{P}{Q} \times \frac{dQ}{dP}$$

$$= \frac{4}{20 - 3(4)} \times (-3) = \frac{-12}{8} = -1.5$$

We can conclude that, as $|E_P| > 1$, the demand is elastic at $P=4$, that is, with 1% change in price, there will be more than 1% change in the quantity demanded.

23. $E_p = 0.86. |E_p| < 1$ so that supply is inelastic. A 1% change in price leads to a less than 1% change in quantity supplied.

25. After obtaining the graph, one can see that there are two breakeven points. The first point is where $Q=1.13$ and the second point is where $Q=8.88$.

As you increase Q, at the first point, the loss will be zero and, at the second point, the profit will be zero. This is the difference between these two breakeven levels of output.

27.

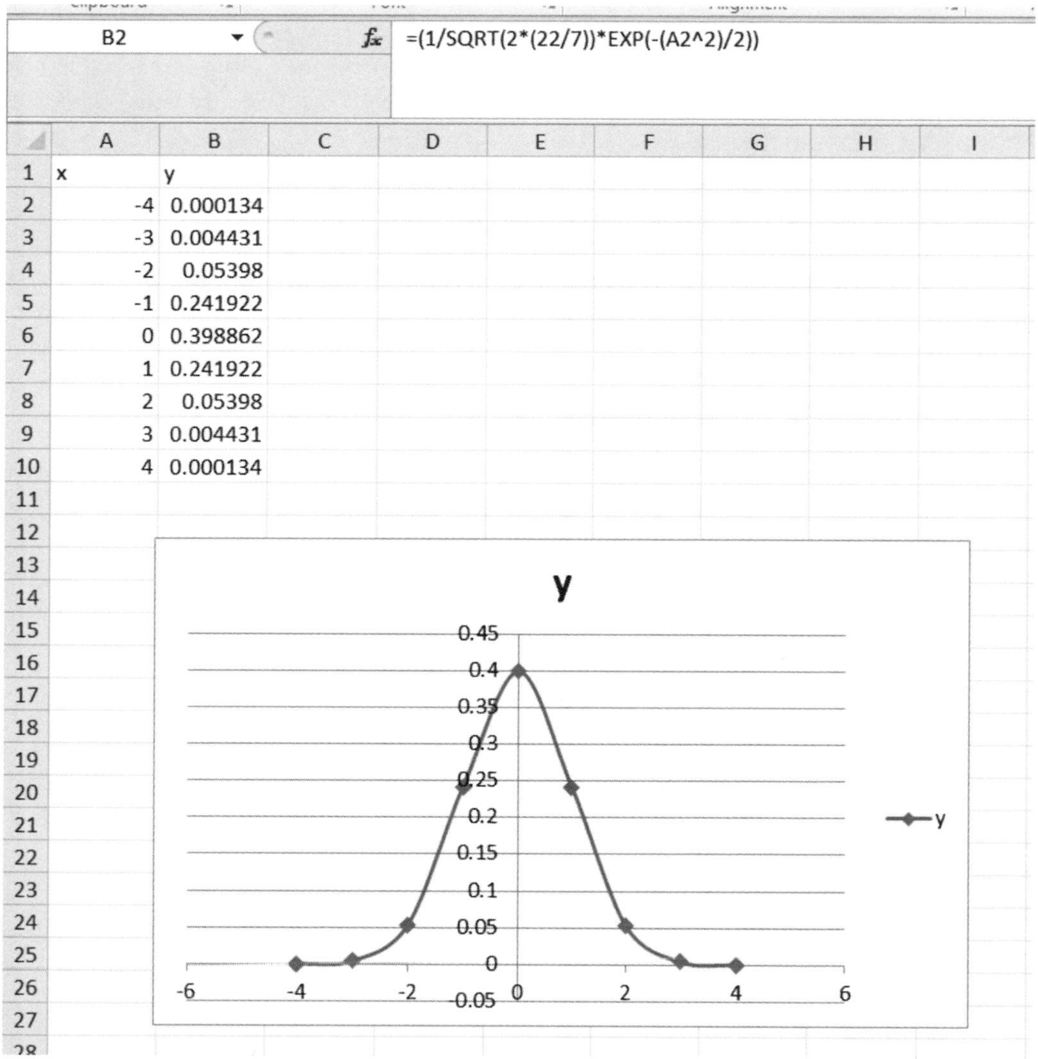

B2 =(1/SQRT(2*(22/7))*EXP(-(A2^2)/2))

	A	B	C	D	E	F	G	H	I
1	x	y							
2	-4	0.000134							
3	-3	0.004431							
4	-2	0.05398							
5	-1	0.241922							
6	0	0.398862							
7	1	0.241922							
8	2	0.05398							
9	3	0.004431							
10	4	0.000134							
11									
12									
13									
14									
15									
16									
17									
18									
19									
20									
21									
22									
23									
24									
25									
26									
27									
28									

Chapter 3

1. **a.** Set MR = MC to obtain Q = 20
b. P = 1200
c. TR – TC = 13000
d. Set MR = 0 to obtain Q = 25

3. **a.** $x = \dfrac{-5 \pm 10.44}{6}$, $x = -15.44$ or $x = 5.44$

b. $x = \dfrac{-3 \pm 7}{-4}$, $x = 2.5$ or $x = -1$

c. $x = \dfrac{1 \pm 5}{6}$, $x = -0.67$ or $x = 1$

5. a. $MC = C'(Q) = 0.03Q^2 - 6Q + 4$

And to find a minimum of MC set $MC' = 0$.

$MC'(Q) = 0.06Q - 6 = 0$, solving $Q = 6/0.06 = 100$. So at $Q = 100$ MC achieves an extreme value. To see whether this value of Q corresponds to a maximum or a minimum, consider $MC''(Q) = 0.06 \geq 0$ at any value of Q. So MC is a minimum at $Q = 100$.

Since $C''(100) = MC'(100) = 0.06 \times 100 - 6 = 0$ and $C'''(100) = MC''(100) = 0.06 \neq 0$, the output level of $Q = 100$ is a point of inflection for the function C(Q).

b. The AC(Q), average cost function $= C(Q)/Q = (0.01Q^3 - 3Q^2 + 4Q + 10)/Q = 0.01Q^2 - 3Q + 4 + 10/Q$.

For AC to have a minimum, set $AC'(Q) = 0$ and solve for Q.

Now $AC'(Q) = 0.02Q - 3 - 10Q^{-2} = 0$. Multiplying this equation on both sides by Q^2, we get $0.02Q^3 - 3Q^2 - 10 = 0$. This is a cubic equation. One of the methods to solve this equation is by graphing it using Excel and finding the Q value at which $AC'(Q)$ crosses Q axis where $AC' = 0$.

You will see that the curve AC' within the given range crosses the Q-axis around $Q = 150$, meaning that AC attains an extreme (minimum) value at $Q = 150$. That is, $AC'(150) = 0$. To see whether the sufficiency condition is satisfied, consider

$AC''(Q) = 0.06Q^2 - 6Q$ and $AC''(150) = 0.06 \times 150^2 - 6 \times 150 = 450 > 0$, confirming that AC attains a minimum around $Q = 150$.

This is shown below.

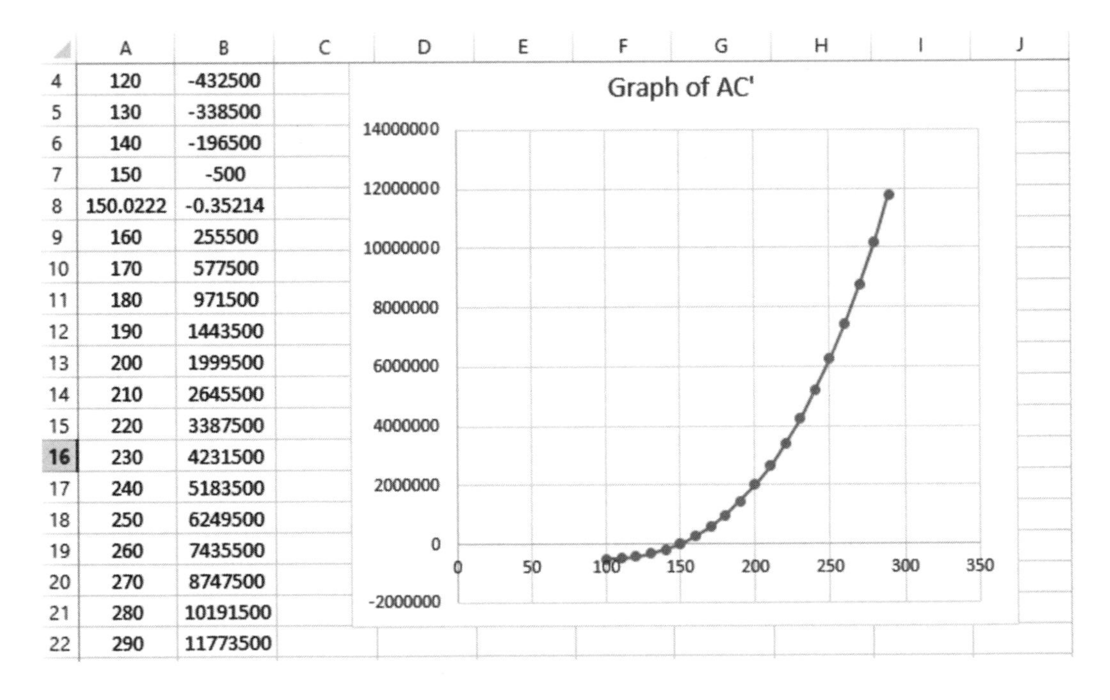

	A	B
4	120	-432500
5	130	-338500
6	140	-196500
7	150	-500
8	150.0222	-0.35214
9	160	255500
10	170	577500
11	180	971500
12	190	1443500
13	200	1999500
14	210	2645500
15	220	3387500
16	230	4231500
17	240	5183500
18	250	6249500
19	260	7435500
20	270	8747500
21	280	10191500
22	290	11773500

Instructions for drawing the above graph.

1. In column A enter Q values starting from 120 to say 290 with increments of 10.
2. In cell B4 enter formula = 0.02*highlightA4^3 – 3*highlightA4^2 – 10 and hit enter key to get the value – 432500 in B4 cell.
3. Go back to B4 cell and click and drag the fill handle (the right hand bottom corner when the cursor becomes black cross) to other cells in B column.
4. Notice the values in B column change from negative to positive attaining zero value in between.
5. You can draw a graph by highlighting the data in both the columns A and B and click on 'Insert' in the menu bar and select line chart to get the graph as shown above.

c. If the firm faces a constant price P = \$10, the profit function,

$$\prod(Q) = \text{Total Revenue} - \text{Total Cost}$$
$$= P \times Q - C(Q) = 10Q - (0.01Q^3 - 3Q^2 + 4Q + 10)$$
$$= -0.01Q^3 + 3Q^2 + 6Q - 10 \text{ is to be maximized.}$$

The first-order condition is $\Pi'(Q) = -0.03Q^2 + 6Q + 6 = 0$. Solving this quadratic equation, we get

$$Q = \frac{-6 \pm \sqrt{(6^2) - 4(-0.03)(6)}}{2(-0.03)} = \frac{-6 \pm 6.06}{-0.06}$$

Thus Q = 201 or Q = −1. The negative value of Q is ruled out since quantity sold is never negative.

The second-order condition for maximum profit is that

$\Pi''(Q) = -0.06Q + 6$ should be negative at Q = 201. Thus $\Pi''(201) = -0.06 \times 201 + 6 = -12.06 + 6 = -6.06 < 0$, confirming that the profit, Π, is maximized when Q = 201.

If the firm faces a demand (inverse) curve of the form P = 10 – 3Q, then $\Pi(Q) = P \times Q - C(Q) = (10 - 3Q)Q - (0.01Q^3 - 3Q^2 + 4Q + 10) = 10Q - 3Q^2 - 0.01Q^3 + 3Q^2 - 4Q - 10 = -0.01Q^3 + 6Q - 10$.

For profit maximization, set $\Pi'(Q) = 0$.

Thus $\Pi'(Q) = -0.03Q^2 + 6 - 0$ or $Q^2 = 6/\ 0.03 = 200$ or $Q - \pm 14.14$. Ruling out any negative value,

Q = 14.14 maximizes profit. To check if this value of Q indeed maximizes profit, consider $\Pi''(Q) = -0.06Q < 0$ for any positive Q and so for Q = 14.14, thus confirming that this value maximizes profit.

At maximum TR(Total Revenue), $TR'(Q) = 0$.

Now $TR(Q) = PQ = (10 - 3Q)Q = 10Q - 3Q^2$ and

$TR'(Q) = 10 - 6Q = 0$ or $Q = 10/6 = 1.67$
Second-order condition for maximum TR is
$TR''(Q) = -6 < 0$ for any value of Q, thus confirming that $Q = 1.67$ maximizes TR.

7. a. Annual holding cost $= \frac{Q}{2} \times 1$, annual ordering costs $= \frac{D}{Q} \times 50 = \frac{500,000}{Q}$.
Thus total inventory cost $= \frac{Q}{2} + \frac{500,000}{Q}$.

b. $Q^* = 1000$. At Q^*, total costs is $1000.

c. Annual holding cost = $750, annual ordering cost = $333.33, and total inventory cost = $1083.33

d. Total inventory cost obviously rises in part c.

9. $Q^* = \sqrt{\frac{2(12000)(60)}{1}}$ Order less frequently to minimize cost. New order size must be 1200.

11. Maximize $Z = 11X_1 + 15X_2$ s.t.
$2X_1 + 3X_2 \leq 12$ (labor)
$4X_1 + 5X_2 \leq 22$ (cash)
$X_1, X_2 \geq 0$

Extreme point	$Z = 11X_1 + 15X_2$
(0,0)	$11(0) + 15(0) = 0$
(5.5,0)	$11(5.5) + 15(0) = 60.5$
(3,2)	$11(3) + 15(2) = 63$
(0,4)	$11(0) + 15(4) = 60$

The optimal extreme point is (3,2), and total profit there is $63.

Slack variables

Constraint	Left side of the constraint	Right side of the constraint	Slack value
Labor	$2(3) + 3(2) = 12$	12	$X_3 = 0$
Cash	$4(3) + 5(2) = 22$	22	$X_4 = 0$

13. Minimize $Z = 10X_1 + 12X_2$ s.t.
$0.05X_1 + 0.3X_2 \geq 2$ (protein)
$0.6X_1 + 0.4X_2 \geq 4.8$ (fat)
$X_1, X_2 \geq 0$

Extreme point	$Z = 10X_1 + 12X_2$
(0,0)	10(0) + 12(0) = 0
(40,0)	10(40) + 12(0) = 400
(4,6)	10(4) + 12(6) = 112

The optimal extreme point is (4,6), and total cost there is $112.

Surplus variables

Constraint	Left side of the constraint	Right side of the constraint	Slack value
Protien	0.05(4) + 0.3(6) = 2	2	$X_3 = 0$
Fat	0.6(4) + 0.4(6) = 4.8	4.8	$X_4 = 0$

15.

	A	B	C	D	E	F
1	Ans(a)	692.8203	→	=sqrt(2*40*4800/0.8)		
2	Ans(b)	6.928203	→		=4800/B1	
3	Ans(c)	51.96152	→	=360/B2		
4	Ans(d)	66.66667	—	=5*4800/360		
5	Ans(e)	277.1281	→	=(B1/2)*08		
6	Ans(f)	277.1281	→		=40*B2	
7	Ans(g)	554.2563	→	=B5+B6		
8						

Ans means Answer:

17. a. (i) Profit maximizing output is $Q = 20$ where, $MR = MC$
(ii) Profit maximizing price is $P = 1200$ when $Q = 20$
(iii) Maximum total profit is $TP = TR - TC = 13000$.
(iv) Revenue maximizing output $Q = 25$ (approximately), where $MR = 0$.

b. (i) Profit maximizing output $Q = 100$, where $MR = MC$
(ii) Profit maximizing price is $P = 200$ when $Q = 100$
(iii) Maximum total profit $TP = TR - TC = 6000$.
(iv) Revenue maximizing output is $Q = 105$ (approximately), where $MR = 0$

19 a. Optimal decision is to buy 4 lbs of food 1, 6 lbs of food 2 and 0 lbs of food 3 at a minimum cost of $112 for nutritional requirements to be satisfied.

b. Optimal solution is $X_1 = 4.4375$ lbs, $X_2 = 4.59375$ lbs, and $X_3 = 1$ lb. Minimum cost is $115.5.

Chapter 4

1. a., b.

Table 4.3b Relative Frequency and Cumulative Frequency Table

Serial #	Class	Frequency	Relative Frequency	Cumulative Frequency
1	19–40	18	.18	18
2	41–62	40	.40	18 + 40 = 58
3	63–84	19	.19	58 + 19 = 77
4	85–106	14	.14	77 + 14 = 91
5	107–128	2	.02	91 + 2 = 93
6	129–150	4	.04	93 + 4 = 97
7	151–173	3	.03	97 + 3 = 100
Total		100	1.00	

c. 18.5 – 40.5
40.5 – 62.5
62.5 – 84.5
84.5 – 106.5
106.5 – 128.5
128.5 – 150.5
150.5 – 173.5

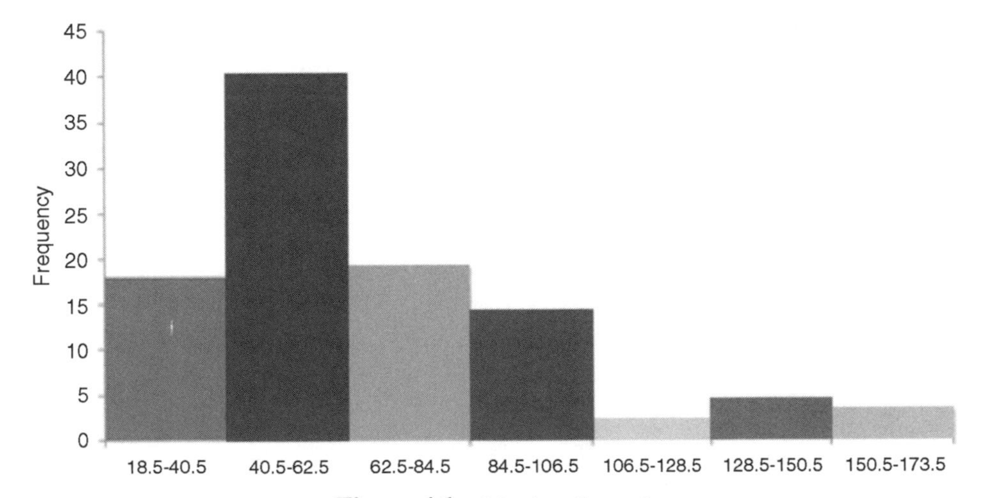

Figure 4.1 Merchandise trade

d. **Table 4.3c** Frequency Table with Midpoints

Serial #	Class	Midpoint	Frequency
1	19–40	29.5	18
2	41–62	51.5	40
3	63–84	73.5	19
4	85–106	95.5	14
5	107–128	117.5	2
6	129–150	139.5	4
7	151–173	162	3
Total			100

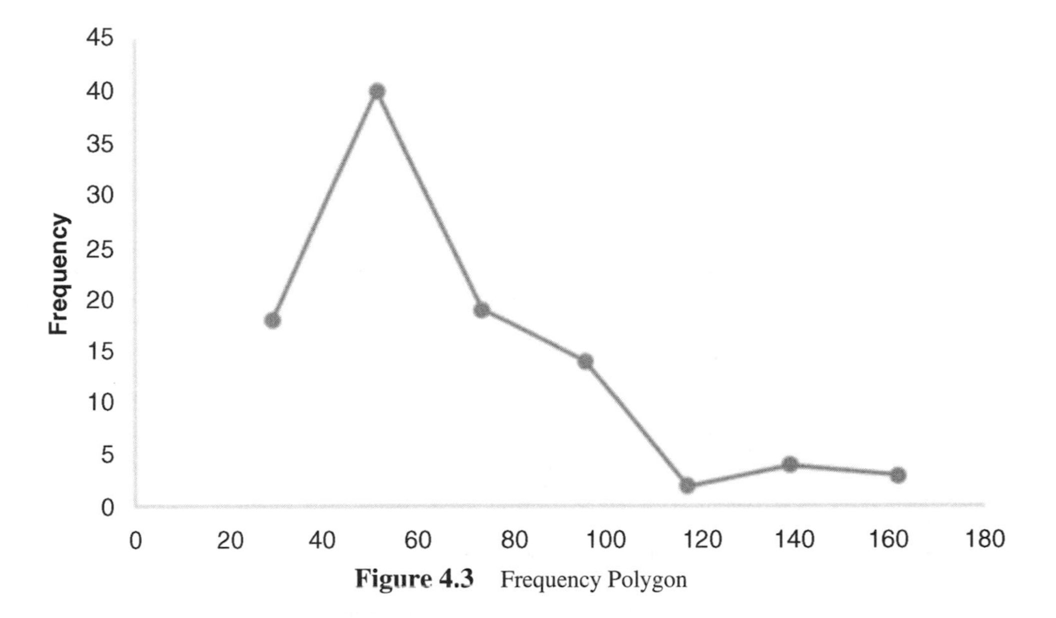

Figure 4.3 Frequency Polygon

e.

Table 4.4 Cumulative Frequency Values

Upper Class Boundaries (%)	Cumulative Frequencies (Less than or equal to)
40.5	18
62.5	58
84.5	77
106.5	91
128.5	93
150.5	97
173.5	100

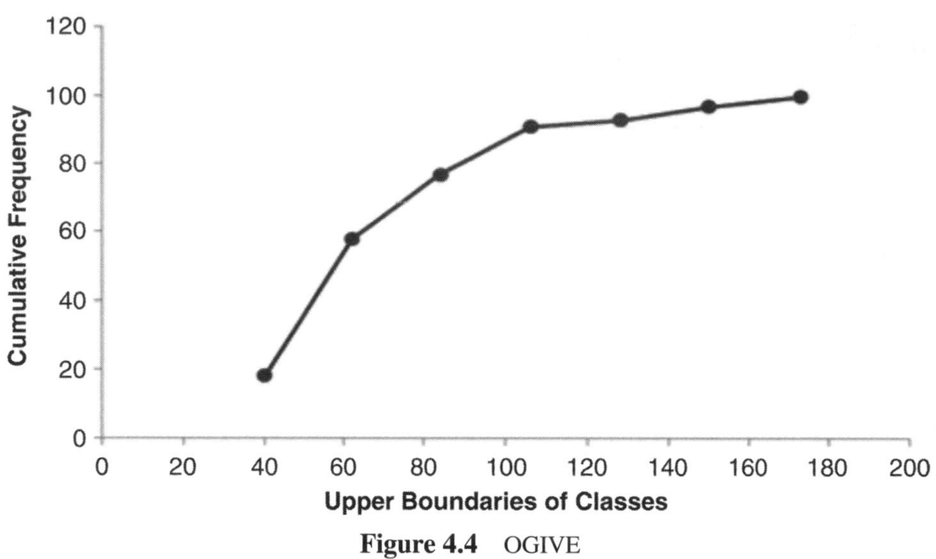

Figure 4.4 OGIVE

3. a.,b.

Table 4.8b Relative Frequency and Cumulative Frequency Table

Serial #	Mortgage Yields	Frequency	Relative Frequency	Cumulative Frequency
1	3.50–5.49	6	0.12	6
2	5.50 – 7.49	18	0.35	6 + 18 = 24
3	7.50 – 9.49	17	0.33	24 + 17= 41
4	9.50 – 11.49	5	0.01	41 + 5 = 46
5	11.50–13.49	4	0.08	46 + 4 = 50
6	13.50–15.49	2	0.04	50 + 2 = 52
Total		52	1.00	

c. 5.50 up to 7.50

7.50 up to 9.50

9.50 up to 11.50

11.50 up to 13.50

13.50 up to 15.50

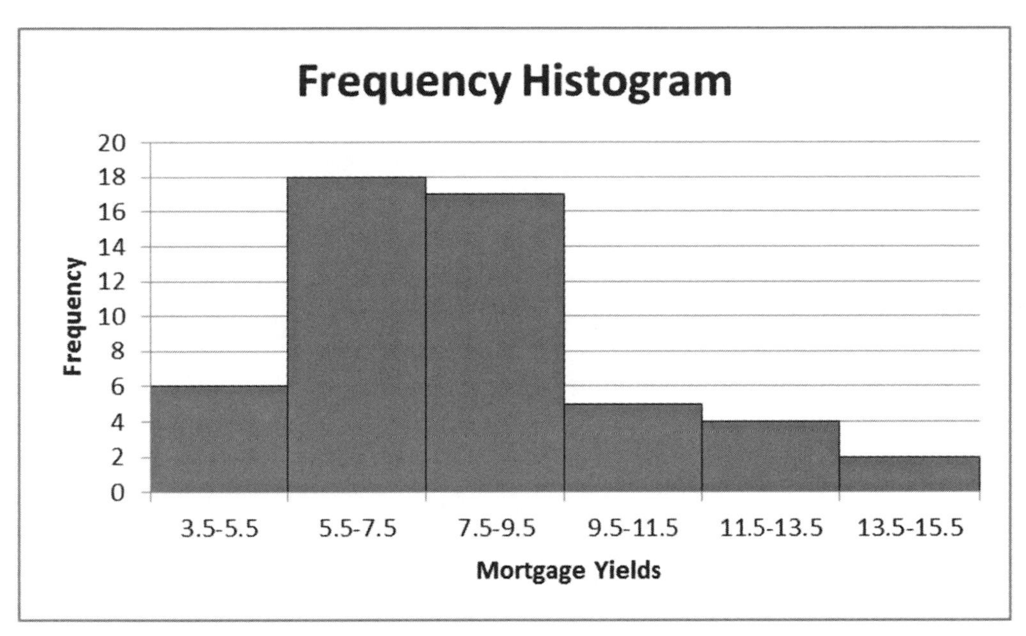

Figure 4.9 Frequency Histogram

d.

Table 4.8c Frequency Table with Midpoints

Serial #	Mortgage Yields	Midpoints	Frequency
1	3.50–5.50	4.50	6
2	5.50–7.50	6.50	18
3	7.50 –9.50	8.50	17
4	9.50 –11.50	10.50	5
5	11.50–13.50	12.50	4
6	13.50–15.50	14.50	2
Total			52

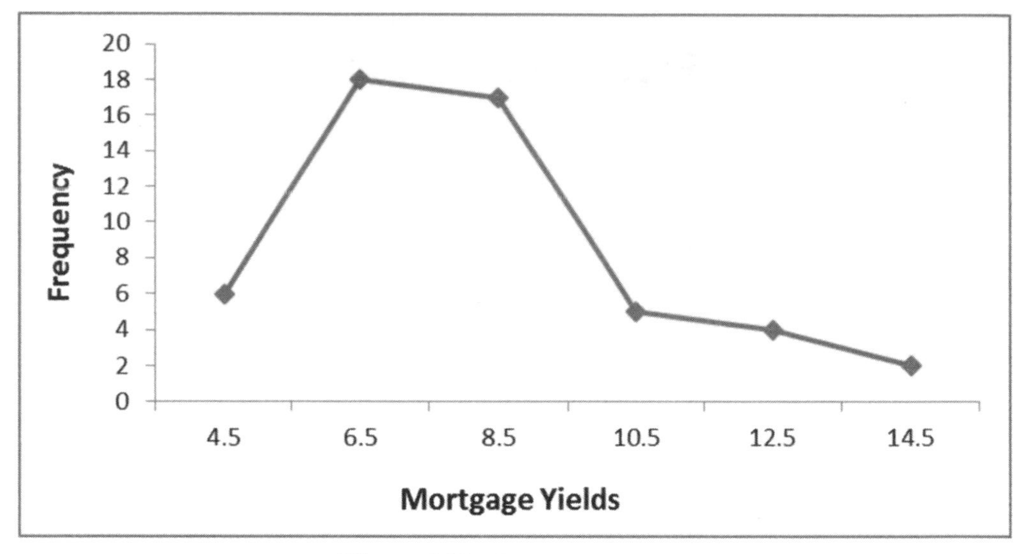

Figure 4.11 Frequency Polygon

e.

Table 4.8d Cumulative Frequency Values

Upper class boundaries	Cumulative Frequencies (Less than or equal to)
5.50	6
7.50	24
9.50	41
11.50	46
13.50	50
15.50	52

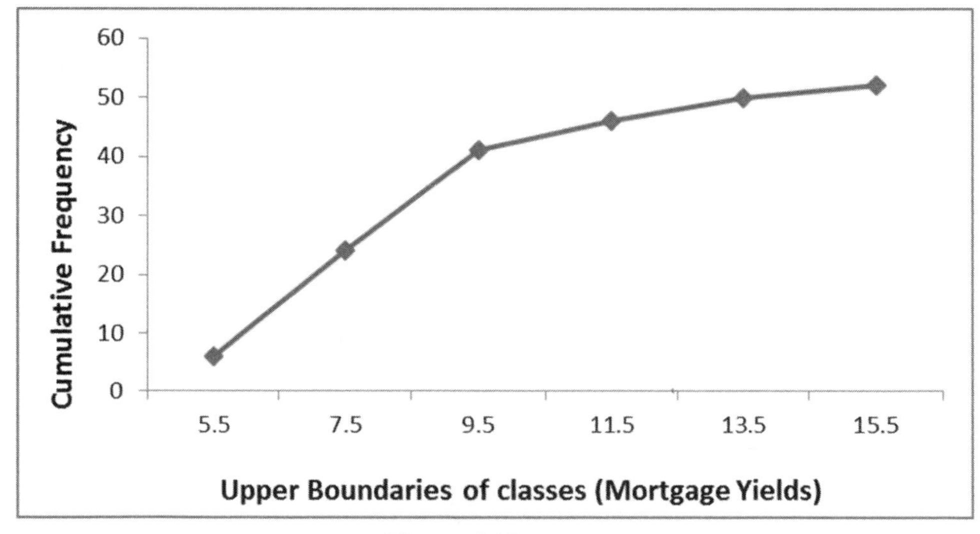

Figure 4.12 Ogive

5. $\overline{X} = 3.06/45 = 0.068$ or 6.8%

7. $X_g = \sqrt[3]{(1+0.04)(1+0.03)(1+0.05)} - 1 = 0.039968$ or 3.99%

9. Let the medical care cost in the year 1996 be X_1 and let R be the annual average growth rate. Then the medical care cost in the year 2005, $X_{10,}$ that is, after 10 years, will be

$X_{1(1+R)}{}^{10} = X_{10}$. Then the rate of growth over the period of 10 years is

$\frac{X_{10}}{X_1} = (1+R)^{10} = 1.42$, which is given in the problem. Solving for R, we get

$1 + R = 1.42^{\frac{1}{10}} = 1.42^{0.1} = 1.035688$ or $R = 1.035688 - 1 = 0.035688$ or 3.5688%, approximately.

11. Z for 3.30 is $(3.30 - 3.57)/0.23 = -1.17$;
Z for 3.84 is $(3.84 - 3.57)/0.23 = 1.17$;

At least $\left(1 - \frac{1}{K^2}\right) = 1 - \frac{1}{(1.17)^2} = 0.2695$. Thus 26.95% of the students are within a GPA of 3.30 to 3.84.

13. $\overline{X} - 2S = 3.53 - 2(0.25) = 3.03$.
$\overline{X} + 2S = 3.53 + 2(0.25) = 4.03$.
By the empirical rule, 95% of the gas prices fall between \$3.03 and \$4.03 per gallon.

15. $P_i = i(n+1)/100, i = 1, \ldots, 99$.

$P_{85} = \frac{85}{100}(11) = 9.35$th term. $P_{75} = \frac{75}{100}(11) = 8.25$th term.

Arranged data: 5, 7, 9, 11,11, 14, 15, 16, 26, 28. Then
85th percentile = 26 + 0.35 (2) = 26.7;
75th percentile = 16 + 0.25 (10) = 18.50.

17. High temperature modes: 85 (occurs twice); 80 (occurs twice).
Low temperature modes: 59 (occurs twice); 67 (occurs twice).

19. Enter the data in an Excel spreadsheet, and highlight the data. Click on the Icon

"Z below A" on the top right-hand side of the ribbon and select the option, "smallest to largest". This arranged data in the column can be transposed to a row by highlighting the data, right click, select 'copy' and choose a cell where you want to start the row. Then 'right click' and click on 'paste' and choose the option 'transpose'.

You see the following debt figures in ascending order of magnitude:

Country (2014) and external debt

Algeria	Nigeria	Bangladesh	Egypt	Pakistan	Vietnam	Venezuela
4,872	22,010	33,200	55,860	62,330	68,050	69,660
Malaysia	Indonesia	Turkey	India	Mexico	Russia	China
109,300	278,500	407,100	425,300	438,400	683,600	894,900

Since there is an even number of countries (14), the median is calculated as the mean of the middle two values of the external debt series. Thus, the mean of $\frac{n}{2} = \frac{14}{2} = 7$th value and $\frac{n}{2} + 1 = \frac{14}{2} + 1 = 7 + 1 = 8$th value must be calculated. That is, the median is (69,660 + 109,300)/2 = 178,960/2 = 89,480 millions of dollars.

There are 7 countries (Algeria, Nigeria, Bangladesh, Egypt, Pakistan, Vietnam, Venezuela) having an external debt equal to or less than 89,480 millions of dollars; and another 7 countries (Malaysia, Indonesia, Turkey, India, Mexico, Russia, and China) having an external debt above 89,480 millions of dollars.

(Note: The data can be arranged in ascending order in a column as follows: Open the data file **External debt.xlsx** that contains the data on external debt and sort the values in ascending order using Excel. To accomplish this, first highlight the data on external debt; press **control**, **shift** and **down arrow** simultaneously. Click on the Icon "Z below A" on the top right and then click on the option, "smallest to largest".)

21. A relative measure (coefficient) of skewness is (Mean –Mode)/Standard deviation = (287.86 -149)/295.85 = 0.469

Comment: Since the measure of skewness is positive, we can say that the distribution is "positively skewed" or we can say that "the distribution has an elongated right tail."

Note: Excel software also gives a measure of skewness, but it is based on a different formula.

23. The location of the 40th percentile is between the 19th and 20th values in the sorted data.

Then, it is interpolated as 1246 + 0.2(1260 – 1246) = 1246 + 0.2(14) = 1246 + 2.8 = 1248.8.

Interpretation: 40% of the scores are at or below a score of approximately 1248.8

Median is the 50th percentile and its location is given by $P_{50} = \frac{50}{100}(47 + 1) = \frac{1}{2}(48) = 24$

Therefore, the 24th value in the sorted data is the 50th percentile or median which is 1283.

Interpretation: 50% of the scores are at or below a score of 1283.

Third quartile is the same as the 75th percentile and its location is given by

$$P_{75} = \frac{75}{100}(47+1) = \frac{3}{4}(48) = 36$$

Therefore, the third quartile is the 36th value in the sorted data which is 1333.

Interpretation: 75% of the scores are at or below 1333.

25. Coefficient of Variation of Exports (X) is (482.95/1003.73)*100% = 48.12%
Coefficient of Variation of Imports (Y) is (709.09/1334.43)*100% = 53.14%
Comment: Imports show slightly higher variability (C.V. = 53.14%) than exports (C.V. = 48.12%) during the period 1986–2011.

27. **a.** The mean, median and the mode for male and female unemployment rates can be seen from the
Excel printout below:

Unemployment Rate Males		Unemployment Rate Females	
Mean	6.490322581	Mean	6.209677
Standard Error	0.336054106	Standard Error	0.268783
Median	6.1	Median	5.6
Mode	7.4	Mode	5.4
Standard Deviation	1.871070075	Standard Deviation	1.496519
Sample Variance	3.500903226	Sample Variance	2.23957
Kurtosis	−0.128460489	Kurtosis	−0.58231
Skewness	0.859740404	Skewness	0.664836
Range	6.6	Range	5.3
Minimum	3.9	Minimum	4.1
Maximum	10.5	Maximum	9.4
Sum	201.2	Sum	192.5
Count	31	Count	31

b. Based on the Skewness measures from the above table, unemployment rates for males and females are slightly positively skewed, meaning that they have an elongated right tail.

c. Variance of the unemployment rate for males is 3.5; and for females it is 2.24.

Standard deviation of the unemployment rate for males is 1.87; and for females it is 1.50.

Finally, the Coefficient of variation of the unemployment rate for males, computed as the standard deviation divided by the mean, equals 1.87/6.49 = 0.29 or 29%; and for females it is computed as 1.50/6.2 = 0.24 or 24%.

Chapter 5

1. 3 coins: There are 8 possible outcomes: HHH, HHT, HTH, HTT, THT, TTH, THH, and TTT. P (3 heads) $=\frac{1}{8}$. 4 coins: 16 possible outcomes. P (4 heads) $=\frac{1}{16}$.

3. 36/52

5. There are 29 even numbered balls from 1 to 59. Probability is $29/59 \times 28/58 \times 27/57 = 0.1124$

7. A – MBA graduate reads WSJ; B – MBA graduate reads NYT. $P(A \cup B) = P(A) + P(B) - P(A \cap B) = 0.80$. P (reads neither) $= 1 - P(P \cup B) = 0.20$

9. With replacement: probability is $45/100 \times 45/100 = 0.2025$.
Without replacement: probability is $45/100 \times 44/99 = 0.20$.

11. Probability is $\frac{1}{5} \times \frac{1}{5} \times \frac{1}{5} = 0.008$.

13. **a.** $P(R_2) = 0.1143$
b. $P(R_1 \cup (C_1 \cup C_2)) = P(R_1) + P(C_1 \cup C_2) - P(R_1 \cap (C_1 \cup C_2))$
$= 0.8393 + 0.7433 - 0.6202 = 0.9634$.
c. $P(C_4|R_2) = 0.0058/0.1143 = 0.0507$.
d. Let S_1, S_2 represent two people having a high school degree or less. $P(S_1 \cap S_2) = P(S_1) \cdot P(S_2|S_1) = 0.24066$ (without replacement). If sampling is done with replacement, $P(S_1 \cap S_2) = P(S_1) \cdot P(S_2) = 0.2406611$.
e. Pick a category from the rows and from the columns, e.g., $P(C_1|R_1) = P(C_1)$ under independence. Since $P(C_1|R_1) = 0.4849 \neq P(C_1) = 0.49057$, independence does not hold.

15. **a.** $P(R_1) = 0.113$
b. $P(R_2 \cup C_1) = P(R_2) + P(C_1) - P(R_2 \cap C_1) = 0.2565 + 0.4612 - 0.1353 = 0.5824$
c. $P(C_3|R_3) = 238/1667 = 0.1428$
d. Let S_1, S_2 represent to stores in the West selected at random.

$$P(S_1 \cap S_2) = P(S_1) \cdot P(S_2|S_1) = 0.1397 \times 0.1395 = 0.0195$$

17. The answers (Ans) for the questions can be seen in the lower part of the Excel printout.

	A	B	C	D	E	F	G	H
1		Discount Stores (C1)	Supercenters (C2)	SAM's Club (C3)	Neighborhood(C4)	Total		
2	Northeast (R1)	243	83	60	0	386		
3	Midwest (R2)	460	266	146	0	872		
4	South (R3)	591	790	238	48	1667		
5	West (R4)	274	119	81	1	475		
6	Total	1568	1258	525	49	3400		
7								
8	Probability table							
9								
10	Northeast (R1)	0.071470588	0.024411765	0.017647059	0	0.113529		
11	Midwest (R2)	0.135294118	0.078235294	0.042941176	0	0.256471		
12	South (R3)	0.173823529	0.232352941	0.07	0.014117647	0.490294		
13	West (R4)	0.080588235	0.035	0.023823529	0.000294118	0.139706		
14	Total	0.461176471	0.37	0.154411765	0.014411765	1		
15								
16	Ans a.	P(R1) = 0.113529	——————→		CellF10 Marginal probability			
17								
18	Ans b.	P(R2) + P(C1) - P(R2∩C1) = 0.256471 + 0.461176471 - 0.135294118 = 0.582353						
19								
20	Ans c.	P(C3 | R3) = P(C3∩R3)/P(R3) = 0.07/0.490294 = 0.142771						
21								
22	Ans d.	P(selecting one store in R4) * P(second store in R4 | the first is not replaced) = CellF13*(474/3399).						
23		0.139706*0.139453 = 0.0194824. Note that the second probability is from the contingency table.						
24								

Chapter 6

1. **a.** No. Negative probability
   b. No. Sum of probabilities exceeds 1
   c. Yes.

3. $E(X) = 1.5, \sigma^2 = 0.75, \sigma = 0.866$

5. $E(X) = 2.5$ or \$2,500.
   $E(Y) = \$80,000 + \$100,000 = \$180,000$
   $V(Y) = 1680, \sigma = 40.99$

7. Mean = \$706.50
   Variance = 128,257.75 squared dollars
   Standard Deviation = \$358.13

9. Mean = 5.45 rooms
   Variance = 4.06 squared rooms
   Standard Deviation = 2.015 rooms

11. a. **Table 6.12** Payoff Table

		State of the Economy			
		Down (0.2)	same as now (0.3)	Boom (0.5)	
Action	Demand (X)	10	20	30	Exp. Payoff
A_1	10	\$4 m	\$4 m	\$4 m	\$4 m
A_2	20	−7 m	8 m	8 m	5 m
A_3	30	−18 m	−3 m	12 m	1.5 m

Here X represents the demand in different states of the economy, with their probabilities in parentheses.

b. The optimal decision, corresponding to maximum expected payoff, is to construct 20 houses, and the builder's expected payoff is \$5 million.

c. Before constructing the opportunity loss table, we have to construct the maximum potential profit (corresponding to each possible state of the economy) table, which is as follows:

X	10	20	30
A(10)	4	8	12
A(20)	4	8	12
A(30)	4	8	12

Now the opportunity loss (the difference between the maximum potential profit for each possible state of the economy and the net profit received for each chosen action) table is as follows:

A(10)	4 - 4	8 - 4	(12 - 4)
A(20)	4 - (−7)	8 - 8	(12 - 8)
A(30)	4 - (−18)	8 - (− 3)	(12 - 12)

P(X)	0.2	0.3	0.5	Exp. loss
A (10)	0	4	8	5.2
A (20)	11	0	4	4.2
A (30)	22	11	0	7.7

d. Based on the opportunity loss table, the optimal decision corresponding to minimum expected opportunity loss is A(20) for which minimum expected opportunity loss is \$4.2 million.

Note that this is the same optimal decision based on the payoff table. Also note that none of the opportunity losses is less than zero.

13. a. $P(X = 2; n = 4, p = 0.2) = {}_4C_2(0.2)^2(0.8)^2 = 6(0.04)(0.64) = 0.1536.$

b. $P(X \geq 2; n = 4, p = 0.2) = {}_4C_2(0.2)^2(0.8)^2 + {}_4C_36(0.2)^3(0.8)^1$
$+ {}_4C_4(0.2)^4(0.8)^0 = 0.1536 + 0.0256 + 0.0016 = 0.1808.$

c. c. $P(X \leq 2; n = 4, p = 0.2) = {}_4C_2\,(0.2)^2\,(0.8)^2 + {}_4C_16\,(0.2)^1\,(0.8)^3$
$+ {}_4C_0(0.2)^0(0.8)^4 = 0.1536 + 0.4096 + 0.4096 = 0.9728.$

d. $P(X = 0; n = 4, p = 0.2) = {}_4C_0(0.2)^0(0.8)^4 = 0.4096.$

15. a. Two possible outcomes: either she is a CEO (success) or not a CEO (failure). P (success) = 0.10; P (failure) = 0.90. P (success) is a constant. Random selection is undertaken.

b. $n = 15, p = 0.1$

c. $P(X = 3) = 0.1285$

d. $P\,(X \leq 3) = P(X = 0) + P(X = 1) + P(X = 2) + P(X = 3) = 0.2059 + 0.3432$
$+ 0.2665 + 0.1285 = 0.9441$

e. $P(X \geq 4) = 1 - P(X \leq 3) = 1 - 0.9441 + = 0.0559$

f. $\mu = np = 15, \sigma = \sqrt{np(1 - p)} = 1.162$

Index

Introduction to Quantitative Methods in Business: With Applications Using Microsoft® Office Excel®, First Edition. Bharat Kolluri, Michael J. Panik, and Rao Singamsetti.

Companion website: www.wiley.com/go/Kolluri/QuantitativeMethods